Rapid prototyping of biomaterials

© Woodhead Publishing Limited, 2014

Related titles:

Nanomedicine
(ISBN 978-0-85709-233-5)

Diamond-based materials for biomedical applications
(ISBN 978-0-85709-340-0)

Biomedical hydrogels
(ISBN 978-1-84569-590-3)

Details of these books and a complete list of titles from Woodhead Publishing can be obtained by:

- visiting our web site at www.woodheadpublishing.com
- contacting Customer Services (e-mail: sales@woodheadpublishing.com; fax: +44 (0) 1223 832819; tel.: +44 (0) 1223 499140 ext. 130; address: Woodhead Publishing Limited, 80 High Street, Sawston, Cambridge CB22 3HJ, UK)
- in North America, contacting our US office (e-mail: usmarketing@woodheadpublishing.com; tel.: (215) 928 9112; address: Woodhead Publishing, 1518 Walnut Street, Suite 1100, Philadelphia, PA 19102–3406, USA)

If you would like e-versions of our content, please visit our online platform: www.woodheadpublishingonline.com. Please recommend it to your librarian so that everyone in your institution can benefit from the wealth of content on the site.

We are always happy to receive suggestions for new books from potential editors. To enquire about contributing to our Biomaterials series, please send your name, contact address and details of the topic/s you are interested in to laura.overend@woodheadpublishing.com. We look forward to hearing from you.

The team responsible for publishing this book:

Commissioning Editor: Laura Overend
Publications Coordinator: Ginny Mills
Project Editor: Elizabeth Moss
Editorial and Production Manager: Mary Campbell
Production Editor: Adam Hooper
Project Manager: Newgen Knowledge Works Pvt Ltd
Copyeditor: Newgen Knowledge Works Pvt Ltd
Proofreader: Newgen Knowledge Works Pvt Ltd
Cover Designer: Terry Callanan

© Woodhead Publishing Limited, 2014

Woodhead Publishing Series in Biomaterials: Number 70

Rapid prototyping of biomaterials

Principles and applications

Edited by
Roger Narayan

WP
WOODHEAD
PUBLISHING

Oxford   Cambridge   Philadelphia   New Delhi

© Woodhead Publishing Limited, 2014

Published by Woodhead Publishing Limited,
80 High Street, Sawston, Cambridge CB22 3HJ, UK
www.woodheadpublishing.com
www.woodheadpublishingonline.com

Woodhead Publishing, 1518 Walnut Street, Suite 1100, Philadelphia,
PA 19102-3406, USA

Woodhead Publishing India Private Limited, 303 Vardaan House, 7/28 Ansari Road,
Daryaganj, New Delhi – 110002, India
www.woodheadpublishingindia.com

First published 2014, Woodhead Publishing Limited
© Woodhead Publishing Limited, 2014. The publisher has made every effort to ensure that permission for copyright material has been obtained by authors wishing to use such material. The authors and the publisher will be glad to hear from any copyright holder it has not been possible to contact.
The authors have asserted their moral rights.

This book contains information obtained from authentic and highly regarded sources. Reprinted material is quoted with permission, and sources are indicated. Reasonable efforts have been made to publish reliable data and information, but the authors and the publishers cannot assume responsibility for the validity of all materials. Neither the authors nor the publishers, nor anyone else associated with this publication, shall be liable for any loss, damage or liability directly or indirectly caused or alleged to be caused by this book.
  Neither this book nor any part may be reproduced or transmitted in any form or by any means, electronic or mechanical, including photocopying, microfilming and recording, or by any information storage or retrieval system, without permission in writing from Woodhead Publishing Limited.
  The consent of Woodhead Publishing Limited does not extend to copying for general distribution, for promotion, for creating new works, or for resale. Specific permission must be obtained in writing from Woodhead Publishing Limited for such copying.

Trademark notice: Product or corporate names may be trademarks or registered trademarks, and are used only for identification and explanation, without intent to infringe.

British Library Cataloguing in Publication Data
A catalogue record for this book is available from the British Library.

Library of Congress Control Number: 2013951185

ISBN 978-0-85709-599-2 (print)
ISBN 978-0-85709-721-7 (online)
ISSN 2049-9485 Woodhead Publishing Series in Biomaterials (print)
ISSN 2049-9493 Woodhead Publishing Series in Biomaterials (online)

The publisher's policy is to use permanent paper from mills that operate a sustainable forestry policy, and which has been manufactured from pulp which is processed using acid-free and elemental chlorine-free practices. Furthermore, the publisher ensures that the text paper and cover board used have met acceptable environmental accreditation standards.

Typeset by Newgen Knowledge Works Pvt Ltd, India
Printed by Lightning Source

Cover image from 'Two-photon polymerization/micromolding of microscale barbs for medical application', R.D. Boehm, B. Chen, S.D. Gittard et al., *Journal of Adhesion Science and Technology*, Aug 10, 2012, Taylor & Francis Ltd. Reprinted by permission of the publisher (Taylor & Francis Ltd, http://www.tandf.co.uk/journals).

Contents

	Contributor contact details	*ix*
	Woodhead Publishing Series in Biomaterials	*xiii*
	Introduction	*xix*
1	**Introduction to rapid prototyping of biomaterials**	1
	C. K. Chua, K. F. Leong and J. An, Nanyang Technological University, Singapore	
1.1	Introduction	1
1.2	Definition of rapid prototyping (RP) systems	2
1.3	Basic process	3
1.4	Conventional RP systems and classification	5
1.5	RP of biomaterials	8
1.6	Conclusion and future trends	9
1.7	Sources of further information and advice	11
1.8	References	12
2	**Freeform fabrication of nanobiomaterials using 3D printing**	16
	M. Vaezi and S. Yang, University of Southampton, UK	
2.1	Introduction	16
2.2	Laser-based solid freeform fabrication (SFF) techniques	18
2.3	Droplet-based SFF techniques	25
2.4	Nozzle-based SFF techniques	30
2.5	Extrusion freeforming of biomaterials scaffold	35
2.6	Dry powder printing	57
2.7	Conclusion	62
2.8	References	63

3	Rapid prototyping techniques for the fabrication of biosensors	75
	K. Pataky and J. Brugger, Ecole Polytechnique Fédérale de Lausanne, Switzerland	
3.1	Introduction	75
3.2	Rapid prototyping (RP) of microfluidic systems	77
3.3	Functionalization	82
3.4	Biomaterials compatibility	89
3.5	Conclusion and future trends	90
3.6	Sources of further information and advice	91
3.7	References	91
4	Rapid prototyping technologies for tissue regeneration	97
	V. Tran and X. Wen, Virginia Commonwealth University, USA	
4.1	Introduction	97
4.2	Rapid prototyping (RP) technologies in tissue regeneration	100
4.3	Laser-assisted techniques	108
4.4	Extrusion-based techniques	119
4.5	Inkjet printing (IP)	129
4.6	Conclusion	137
4.7	References	144
5	Rapid prototyping of complex tissues with laser assisted bioprinting (LAB)	156
	B. Guillotin, S. Catros, V. Keriquel, A. Souquet, A. Fontaine, M. Remy, J.-C. Fricain and F. Guillemot, INSERM U1026, France and Université Bordeaux Segalen, France	
5.1	Introduction	156
5.2	Rationale for using laser assisted bioprinting (LAB) in tissue engineering	158
5.3	Terms of reference for LAB	160
5.4	LAB parameters for cell printing	164
5.5	High resolution and high throughput needs and limits	165
5.6	Applications of LAB	169
5.7	Conclusion	171
5.8	Acknowledgements	172
5.9	References	172

6	Scaffolding hydrogels for rapid prototyping based tissue engineering	176
	R. A. SHIRWAIKER, M. F. PURSER and R. A. WYSK, North Carolina State University, USA	
6.1	Introduction	176
6.2	Biomaterials in tissue engineering	178
6.3	Review of commonly used hydrogel-forming scaffolding biomaterials	182
6.4	Applications of scaffolding hydrogels	191
6.5	Conclusion	194
6.6	References	195
7	Bioprinting for constructing microvascular systems for organs	201
	T. XU, J. I. RODRIGUEZ-DEVORA, D. REYNA-SORIANO and B. MOHAMMOD, University of Texas at El Paso, USA, L. ZHU and K. WANG, Sun Yat-sen University, China and Y. YUAN, Medprin Regenerative Technologies, China	
7.1	Introduction	201
7.2	Biomimetic model for microvasculature printing	202
7.3	The bio-blueprint for microvasculature printing	203
7.4	Microvasculature printing strategies	208
7.5	Microvasculature post-printing stage	215
7.6	Future trends	217
7.7	Acknowledgements	218
7.8	References	218
8	Feasibility of 3D scaffolds for organs	221
	T. BURG and K. BURG, Clemson University, USA	
8.1	Introduction	221
8.2	Overview of organ fabrication	222
8.3	The right place: physical properties of the scaffold	225
8.4	The right time: temporal expectations on the scaffold	227
8.5	The right biomaterials: scaffold fabrication effects on non-scaffold components	228
8.6	The right characteristics: material types	230
8.7	The right process: biofabrication	232
8.8	Conclusion	233
8.9	Sources of further information and advice	234
8.10	References	234

9	3-D organ printing technologies for tissue engineering applications	236
	H.-W. Kang, C. Kengla, S. J. Lee, J. J. Yoo and A. Atala, Wake Forest School of Medicine, USA	
9.1	Introduction	236
9.2	Three-dimensional printing methods for organ printing	238
9.3	From medical imaging to organ printing	242
9.4	Applications in tissue engineering and regenerative medicine	243
9.5	Future trends	249
9.6	Conclusion	250
9.7	References	251
10	Rapid prototyping technology for bone regeneration	254
	J. Kundu, F. Pati, J.-H. Shim and D.-W. Cho, Pohang University of Science and Technology, Korea	
10.1	Introduction	254
10.2	Bone: properties, structure, and modeling	255
10.3	Engineering of bone tissue	259
10.4	Conventional scaffolds for bone regeneration	264
10.5	Cell printing technology for bone regeneration	271
10.6	Future trends	275
10.7	Conclusion	277
10.8	Acknowledgement	277
10.9	References	277
11	Additive manufacturing of a prosthetic limb	285
	S. Summit, Bespoke Products/3D Systems, USA	
11.1	Introduction	285
11.2	The aim in designing a prosthetic limb	287
11.3	A biomimetic approach to design	290
11.4	Integrating functionality	291
11.5	A 'greener' approach to design	292
11.6	Tactile dividends of additively manufactured parts	293
11.7	Vast design flexibility	294
11.8	Conclusion	295
	Index	*297*

Contributor contact details

(* = main contact)

Editor

Roger Narayan
UNC/NCSU
Joint Department of Biomedical
  Engineering
Box 7115
Raleigh, NC 27695–7115, USA

E-mail: roger_narayan@unc.edu;
  roger_naraya@ncsu.edu

Chapter 1

Chee Kai Chua*, Kah Fai Leong
  and Jia An
School of Mechanical & Aerospace
  Engineering
Nanyang Technological University
N3–2a-02, 50 Nanyang Avenue,
  Singapore 639798

E-mail: mckchua@ntu.edu.sg;
  mkfleong@ntu.edu.sg; anjia@ntu.edu.sg

Chapter 2

Mohammad Vaezi and Shoufeng
  Yang*
Faculty of Engineering & the
  Environment
University of Southampton
Southampton, SO17 1BJ, UK

E-mail: mv1y11@soton.ac.uk;
  s.yang@soton.ac.uk

Chapter 3

Kristopher Pataky* and
  J. Brugger
Ecole Polytechnique Federal de
  Lausanne (EPFL) STI IMT
  LMIS1
Station 17
1015 LAUSANNE, Switzerland

E-mail: kris.pataky@gmail.com

Chapter 4

Van Tran and Xuejun Wen*
Institute for Engineering and
  Medicine
Department of Chemical and Life
  Science Engineering
Virginia Commonwealth University
601 West Main Street
Richmond, Virginia 23284–3028,
  USA

E-mail: xwen@vcu.edu

Chapter 5

Bertrand Guillotin, Sylvain Catros,
  Virginie Keriquel, Agnès
  Souquet, Aurélien Fontaine,
  Murielle Remy, Jean-Christophe
  Fricain and Fabien Guillemot*

INSERM U1026
Bordeaux, F-33076, France

and

Université Bordeaux Segalen
Bordeaux, F-33076, France

E-mail: fabien.guillemot@inserm.fr

Chapter 6

Rohan A. Shirwaiker*, Molly F. Purser and Richard A. Wysk
Edward P. Fitts Department of Industrial & Systems Engineering
North Carolina State University
400 Daniels Hall
Raleigh
North Carolina 27695–7906, USA

E-mail: rashirwaiker@ncsu.edu

Chapter 7

Tao Xu*, Jorge I. Rodriguez-Devora, Daniel Reyna-Soriano, and Bhuyan Mohammod
Department of Mechanical Engineering and Biomedical Engineering Program
Engineering Bldg. A226
University of Texas at El Paso
TX 79968, USA

E-mail: txu2@utep.edu; dreyna2@miners.utep.edu; jirodriguez6@miners.utep.edu

Lei Zhu
Department of Plastic Surgery
Third Affiliated Hospital
Sun Yat-sen University
Guangzhou
Guangdong Province 510630, China

Kun Wang
Department of Orthopedic Surgery
Third Affiliated Hospital
Sun Yat-sen University
Guangzhou
Guangdong Province 510630, China

Yuyu Yuan
Medprin Regenerative Technologies Co. Ltd
Guangzhou
Guangdong Province 510663, China

Chapter 8

Timothy Burg*
Department of Electrical and Computer Engineering

and

Institute for Biological Interfaces of Engineering
Clemson University, USA

E-mail: tburg@clemson.edu

Karen Burg
Department of Bioengineering

and

Institute for Biological Interfaces of Engineering
Clemson University, USA

Chapter 9

Hyun-Wook Kang, Carlos Kengla, Sang Jin Lee, James J. Yoo and Anthony Atala*
Wake Forest Institute for Regenerative Medicine
Wake Forest School of Medicine

Medical Center Boulevard
Winston-Salem
NC 27157, USA

E-mail: aatala@wakehealth.edu; sjlee@wakehealth.edu

Chapter 10

Joydip Kundu, Falguni Pati, Jin-Hyung Shim and Dong-Woo Cho*
Department of Mechanical Engineering

and

Division of Integrative Biosciences and Biotechnology
Pohang University of Science and Technology (POSTECH)

San 31, Hyoja-dong, Nam-gu, Pohang
Kyungbuk 790–749, Korea

E-mail: dwcho@postech.ac.kr

Chapter 11

Scott Summit
Bespoke Products/3D Systems
431 Tehama Street
San Francisco
CA 94103, USA

E-mail: scott.summit@3dsystems.com

Woodhead Publishing Series in Biomaterials

1 **Sterilisation of tissues using ionising radiations**
   Edited by J. F. Kennedy, G. O. Phillips and P. A. Williams
2 **Surfaces and interfaces for biomaterials**
   Edited by P. Vadgama
3 **Molecular interfacial phenomena of polymers and biopolymers**
   Edited by C. Chen
4 **Biomaterials, artificial organs and tissue engineering**
   Edited by L. Hench and J. Jones
5 **Medical modelling**
   R. Bibb
6 **Artificial cells, cell engineering and therapy**
   Edited by S. Prakash
7 **Biomedical polymers**
   Edited by M. Jenkins
8 **Tissue engineering using ceramics and polymers**
   Edited by A. R. Boccaccini and J. Gough
9 **Bioceramics and their clinical applications**
   Edited by T. Kokubo
10 **Dental biomaterials**
   Edited by R. V. Curtis and T. F. Watson
11 **Joint replacement technology**
   Edited by P. A. Revell
12 **Natural-based polymers for biomedical applications**
   Edited by R. L. Reiss et al
13 **Degradation rate of bioresorbable materials**
   Edited by F. J. Buchanan
14 **Orthopaedic bone cements**
   Edited by S. Deb
15 **Shape memory alloys for biomedical applications**
   Edited by T. Yoneyama and S. Miyazaki
16 **Cellular response to biomaterials**
   Edited by L. Di Silvio

17 **Biomaterials for treating skin loss**
   Edited by D. P. Orgill and C. Blanco
18 **Biomaterials and tissue engineering in urology**
   Edited by J. Denstedt and A. Atala
19 **Materials science for dentistry**
   B. W. Darvell
20 **Bone repair biomaterials**
   Edited by J. A. Planell, S. M. Best, D. Lacroix and A. Merolli
21 **Biomedical composites**
   Edited by L. Ambrosio
22 **Drug–device combination products**
   Edited by A. Lewis
23 **Biomaterials and regenerative medicine in ophthalmology**
   Edited by T. V. Chirila
24 **Regenerative medicine and biomaterials for the repair of connective tissues**
   Edited by C. Archer and J. Ralphs
25 **Metals for biomedical devices**
   Edited by M. Ninomi
26 **Biointegration of medical implant materials: Science and design**
   Edited by C. P. Sharma
27 **Biomaterials and devices for the circulatory system**
   Edited by T. Gourlay and R. Black
28 **Surface modification of biomaterials: Methods analysis and applications**
   Edited by R. Williams
29 **Biomaterials for artificial organs**
   Edited by M. Lysaght and T. Webster
30 **Injectable biomaterials: Science and applications**
   Edited by B. Vernon
31 **Biomedical hydrogels: Biochemistry, manufacture and medical applications**
   Edited by S. Rimmer
32 **Preprosthetic and maxillofacial surgery: Biomaterials, bone grafting and tissue engineering**
   Edited by J. Ferri and E. Hunziker
33 **Bioactive materials in medicine: Design and applications**
   Edited by X. Zhao, J. M. Courtney and H. Qian
34 **Advanced wound repair therapies**
   Edited by D. Farrar
35 **Electrospinning for tissue regeneration**
   Edited by L. Bosworth and S. Downes

36 **Bioactive glasses: Materials, properties and applications**
    Edited by H. O. Ylänen
37 **Coatings for biomedical applications**
    Edited by M. Driver
38 **Progenitor and stem cell technologies and therapies**
    Edited by A. Atala
39 **Biomaterials for spinal surgery**
    Edited by L. Ambrosio and E. Tanner
40 **Minimized cardiopulmonary bypass techniques and technologies**
    Edited by T. Gourlay and S. Gunaydin
41 **Wear of orthopaedic implants and artificial joints**
    Edited by S. Affatato
42 **Biomaterials in plastic surgery: Breast implants**
    Edited by W. Peters, H. Brandon, K. L. Jerina, C. Wolf and V. L. Young
43 **MEMS for biomedical applications**
    Edited by S. Bhansali and A. Vasudev
44 **Durability and reliability of medical polymers**
    Edited by M. Jenkins and A. Stamboulis
45 **Biosensors for medical applications**
    Edited by S. Higson
46 **Sterilisation of biomaterials and medical devices**
    Edited by S. Lerouge and A. Simmons
47 **The hip resurfacing handbook: A practical guide to the use and management of modern hip resurfacings**
    Edited by K. De Smet, P. Campbell and C. Van Der Straeten
48 **Developments in tissue engineered and regenerative medicine products**
    J. Basu and J. W. Ludlow
49 **Nanomedicine: Technologies and applications**
    Edited by T. J. Webster
50 **Biocompatibility and performance of medical devices**
    Edited by J-P. Boutrand
51 **Medical robotics: Minimally invasive surgery**
    Edited by P. Gomes
52 **Implantable sensor systems for medical applications**
    Edited by A. Inmann and D. Hodgins
53 **Non-metallic biomaterials for tooth repair and replacement**
    Edited by P. Vallittu
54 **Joining and assembly of medical materials and devices**
    Edited by Y. (Norman) Zhou and M. D. Breyen
55 **Diamond-based materials for biomedical applications**
    Edited by R. Narayan

56 **Nanomaterials in tissue engineering: Fabrication and applications**
Edited by A. K. Gaharwar, S. Sant, M. J. Hancock and S. A. Hacking

57 **Biomimetic biomaterials: Structure and applications**
Edited by A. J. Ruys

58 **Standardisation in cell and tissue engineering: Methods and protocols**
Edited by V. Salih

59 **Inhaler devices: Fundamentals, design and drug delivery**
Edited by P. Prokopovich

60 **Bio-tribocorrosion in biomaterials and medical implants**
Edited by Y. Yan

61 **Microfluidic devices for biomedical applications**
Edited by X-J. James Li and Y. Zhou

62 **Decontamination in hospitals and healthcare**
Edited by J. T. Walker

63 **Biomedical imaging: Applications and advances**
Edited by P. Morris

64 **Characterization of biomaterials**
Edited by M. Jaffe, W. Hammond, P. Tolias and T. Arinzeh

65 **Biomaterials and medical tribology**
Edited by J. Paolo Davim

66 **Biomaterials for cancer therapeutics: Diagnosis, prevention and therapy**
Edited by K. Park

67 **New functional biomaterials for medicine and healthcare**
E.P. Ivanova, K. Bazaka and R. J. Crawford

68 **Porous silicon for biomedical applications**
Edited by H. A. Santos

69 **A practical approach to spinal trauma**
Edited by H. N. Bajaj and S. Katoch

70 **Rapid prototyping of biomaterials: Principles and applications**
Edited by R. Narayan

71 **Cardiac regeneration and repair Volume 1: Pathology and therapies**
Edited by R-K. Li and R. D. Weisel

72 **Cardiac regeneration and repair Volume 2: Biomaterials and tissue engineering**
Edited by R-K. Li and R. D. Weisel

73 **Semiconducting silicon nanowires for biomedical applications**
Edited by J. L. Coffer

74 **Silk biomaterials for tissue engineering and regenerative medicine**
Edited by S. Kundu

75 **Novel biomaterials for bone regeneration: Novel techniques and applications**
Edited by P. Dubruel and S. Van Vlierberghe

76 **Biomedical foams for tissue engineering applications**
  Edited by P. Netti
77 **Precious metals for biomedical applications**
  Edited by N. Baltzer and T. Copponnex
78 **Bone substitute biomaterials**
  Edited by K. Mallick
79 **Regulatory affairs for biomaterials and medical devices**
  Edited by S. Amato and R. Ezzell
80 **Joint replacement technology Second edition**
  Edited by P. A. Revell
81 **Computational modelling of biomechanics and biotribology in the musculoskeletal system: Biomaterials and tissues**
  Edited by Z. Jin
82 **Biophotonics for medical applications**
  Edited by I. Meglinski
83 **Modelling degradation of bioresorbable polymeric medical devices**
  Edited by J. Pan
84 **Perspectives in total hip arthroplasty: Advances in biomaterials and their tribological interactions**
  S. Affatato
85 **Tissue engineering using ceramics and polymers Second edition**
  Edited by A.R. Boccaccini and P.X. Ma

Introduction

R. J. NARAYAN,
UNC/NCSU Joint Department of Biomedical Engineering, USA

Rapid prototyping, also referred to as additive manufacturing, layer manufacturing, or solid freeform fabrication, is a materials processing approach that involves additive building of a three-dimensional object in a layer-by-layer manner. Over the past three decades, several rapid prototyping techniques that involve processing of material in solid, powder, or liquid form have been developed.[1] In 1986, one of the earliest patents involving rapid prototyping was awarded to Charles Hull, who is now considered by many to be the father of rapid prototyping.[2] The patent, entitled 'Apparatus for production of three-dimensional objects by stereolithography', describes the use of energy in the form of chemical reaction, particle bombardment, or radiation to build layers of material on a surface in order to create an object.[3] Charles Hull founded a company in 1986 called 3D Systems which produced the first commercial instrument for rapid prototyping, the SLA®-1.[3,4] Around that time, other rapid methods were described, including fused deposition modeling, laminated object manufacturing, selective laser sintering, solid ground curing, and 3D printing.[5] In recent years, improvements to existing techniques have been introduced; for example, microstereolithography methods that enable fabrication of small-scale three-dimensional structures have been developed.[6]

Use of rapid prototyping to create objects for medical and dental applications, including implants and external prostheses, has been explored by numerous researchers over the past three decades.[4] In this book, several innovations involving the use of rapid prototyping in biomedical applications are considered. Chua et al. introduce the topic of rapid prototyping and consider rapid prototyping of biomaterials. Vaezi and Yang discuss the use of rapid prototyping techniques such as extrusion-based freeform and dry powder printing techniques for processing of nanostructured biomaterials. Pataky and Brugger describe use of rapid prototyping techniques such as stereolithography and 3D printing for processing of biosensors. Tran and Wen consider the use of rapid prototyping techniques, including inkjet-based, laser-based, and dispensing/extrusion-based technologies, for creating artificial tissues and organs. Guillotin et al. discuss the use of a rapid prototyping technique known as laser assisted bioprinting for patterning of liquids and viable cells; utilization of this technique to create heterogeneous artificial tissues is described. Shirwaiker *et al.* consider the use of rapid prototyping techniques to create three-dimensional hydrogel structures for use as scaffolds in tissue engineering. Xu *et al.* describe the use of a rapid prototyping technique known as bioprinting for creating microscale vascular structures; the challenges associated with creating microscale vascular structures for use in artificial tissues and organs are described. Burg and Burg consider the physical and biological requirements for tissue engineering scaffolds besides discussing the materials and processes used in rapid prototyping of scaffolds. Kang *et al.* describe several rapid prototyping techniques, including inkjet-based, laser-based, and extrusion-based techniques, used for creating three-dimensional organs; recent developments in rapid prototyping of artificial tissues and organs are discussed. Kundu *et al.* discuss the use of rapid prototyping techniques such as fused deposition modeling and precise extrusion manufacturing to create scaffolds for artificial bone tissue. Finally, Summit describes the use of rapid prototyping techniques, including selective laser sintering, to create prostheses for amputees.

Despite the advances in rapid prototyping technology in recent years, many technical challenges and commercialization-related challenges associated with the translation of rapid prototyping techniques to clinical medicine and dentistry remain. The successful commercial translation of rapid prototyping-produced orthodontic appliances suggests that there is tremendous demand among patients and health care providers for patient-specific devices created via rapid prototyping techniques. Improving the mechanical, chemical, and biological properties of rapid prototyping-produced products, as well as reducing the cost of rapid prototyping instrumentation and feedstock materials, are challenges that the rapid prototyping community is working to overcome. It is hoped that this book will help students and researchers to understand the limitations of current rapid prototyping techniques and to develop solutions that will enable commercialization and clinical translation of rapid prototyping-produced structures.

References

1. Narayan RJ, Doraiswamy A, Chrisey DB and Chichkov BN. (2010) Medical prototyping using two photon polymerization. *Materials Today*. **13**: 44–50.
2. Grimm T. (2004) *User's Guide to Rapid Prototyping. Society of Manufacturing Engineers*. Dearborn, MI, p. 15 .
3. Hull CW. (1986) Apparatus for production of three-dimensional objects by stereolithography. Patent US4575330 A. Published 11 March 1986.

4. The Journey of a Lifetime. http://www.3dsystems.com/30-years-innovation. Retrieved on 12 July 2013.
5. Levy GN, Schindel R and Kruth JP. (2003) Rapid manufacturing and rapid tooling with layer manufacturing (LM) technologies, state of the art and future perspectives. *CIRP Annals – Manufacturing Technology.* **52**: 589–609.
6. Skoog SA and Narayan RJ. (2013) Stereolithography in medical device fabrication. *Advanced Materials & Processes.* **171**: 32–36.

1
Introduction to rapid prototyping of biomaterials

C. K. CHUA, K. F. LEONG and J. AN,
Nanyang Technological University, Singapore

DOI: 10.1533/9780857097217.1

Abstract: Rapid prototyping (RP) or additive manufacturing (AM) is a group of technologies that create three dimensional objects additively in a layer-by-layer manner. This chapter first discusses the definition of RP. It goes on to describe the basic process, the classification of RP and typical RP systems. Finally, the chapter discusses how RP can be explored for processing biomaterials and suggests possible research directions. It is the intention of the authors that this chapter will give readers a general view on rapid prototyping of biomaterials.

Key words: rapid prototyping, additive manufacturing, 3D printing, biofabrication, tissue engineering.

1.1 Introduction

Rapid prototyping (RP) is a group of technologies that create three dimensional objects additively in a layer-by-layer manner. The origin of the idea of RP can be traced back to 1890s, but RP was commercially launched only in the 1980s (Chua *et al.*, 2010a). The development of RP is closely tied with the development of the computer and software industry. In particular, the existence of computer-aided design (CAD) plays a critical role in the emergence of most, if not all, of current RP systems. The primary function of RP systems is to fabricate prototypes within a short period of time (usually within hours or days) to accelerate product development. However, after three decades of development, the applications of RP extend far beyond building prototypes. In fact it has been adapted to meet the needs of a variety of industries, including design, manufacturing, automotive, aerospace, biomedical, jewelry, coin, etc. Currently, RP is rejuvenating itself in the defense and manufacturing industries, especially with the establishment of National Additive Manufacturing Innovation Institute (NAMII) in the United States (The White House, 2012) and concurrently, the Nanyang Additive Manufacturing Centre (NAMC) in Singapore. But what is important in the twenty-first century is that RP, being explored to process biomaterials and biological materials, is establishing a significant role in the emerging biofabrication industry to address the huge clinical demand of human tissues and organs for transplantation.

1.2 Definition of rapid prototyping (RP) systems

For historical reasons, there are many terms used today to describe layer-by-layer fabrication techniques. These terms include layered manufacturing, rapid prototyping, rapid tooling, solid freeform fabrication, direct writing, additive manufacturing, 3D printing, biofabrication, and so on. In the recent past, RP was used most favorably over other names because the application of prototyping dominated the use of these technologies. Because of the rapid expansion of the RP industry over the years, many new processes have been created, and each was uniquely named for easy distinction, though all of them are based on a similar layer-by-layer principle. This unprecedented proliferation of names had led to terminological complexity, ambiguity and even confusion, causing a problem of classification and standardization in the field. To standardize the use of terminology, 'additive manufacturing' was finally adopted as the recognized name for all processes that are additive, as opposed to those which are subtractive in nature. This has been clearly spelled out in the Standard Terminology for Additive Manufacturing Technologies (ASTM 2972). Unlike application-based names such as RP, rapid tooling, 3D printing, or others, AM reflects the fundamental principle of RP and is able to represent and include all kinds of layer-by-layer fabrication technologies.

However, what has attracted media attention and hence spurred public awareness of AM is the term '3D printing', which is often portrayed as an office or home printer with the capability of printing in a third dimension. Actually, 3D printers represent only a small group of RP systems that have low cost or compromised overall capability for low end applications like building a form study model in product development. According to Wohlers Report 2010, an annual report that reviews the AM industry worldwide, 3D printing is defined as 'fabrication of objects through the deposition of a material using a print head, nozzle, or another printer technology; terms often used synonymously with additive manufacturing, in particular associated with machines that are low end in price and/or overall capability.'

Technically, AM is defined as the 'process of joining materials to make objects from 3D model data, usually layer upon layer, as opposed to subtractive manufacturing methodologies; synonyms include additive fabrication, additive processes, additive techniques, additive layer manufacturing, layer manufacturing, and freeform fabrication', and RP is defined as 'additive manufacturing of a design, often iterative, for form, fit, or functional testing, or combination thereof.' These two names are equivalent in terms of intended representation and will co-exist during the transition stage that moves towards the use of AM.

Although RP has long been applied in biomedical applications such as surgical models and prostheses, there was no specific term to name such applications in the past. About a decade ago, biomedical applications of RP underwent a fast and extensive expansion by including drug delivery devices, tissue engineering scaffolds, and bioprinting of living tissues. The involvement of biomaterials, especially biopolymers and biological cells, created a paradigm shift in the biomedical branch of RP, making it more interdisciplinary, much wider in its application, and more specialized than ever. The term 'biofabrication', which refers to the production of complex living and non-living biological products from raw materials such as living cells, molecules, extracellular matrices (ECMs), and biomaterials (Mironov *et al.*, 2009), gradually emerged in the literature to flag this rising field, especially when a new journal called *Biofabrication* was launched in 2009. Synonyms of biofabrication include 3D bioprinting, organ printing, organ biofabrication, and so forth. Although the principle of biofabrication is still very much based on the layer-by-layer technique, the group of biofabrication technologies has a distinct characteristic in comparison to other RP technologies, that is, the use of biomaterials or biological materials for prototyping. This extends far beyond the scope of conventional engineering materials used by most RP systems, such as metals, ceramics, polymers, and composites. Therefore, biofabrication represents one of the industries of the future.

1.3 Basic process

A complete RP process includes the creation and slicing of a virtual computer model usually followed by the layer-by-layer fabrication process. In general, a typical RP process consists of five steps.

First, a CAD model is created or captured from a physical object by digital means, for example, a 3D laser scanner. The CAD model is usually generated by using standard software packages such as AutoCAD, Pro/Engineer, or Solid Works.

Secondly, the CAD model is converted to a format that allows virtual slicing of the computer model, usually in a stereolithography (STL) file format (Leong *et al.*, 1996a, b). The STL file contains an unordered list of triangular facets and the outward normal of each triangle in Cartesian coordinates. These triangular facets approximate the outside surface of a precise CAD model of interest. However, an STL file is not the only type that can represent a CAD model. There are several other available file formats, such as initial graphics exchange specification (IGES), Hewlett-Packard graphics language (HP/GL), computerized tomography (CT) scan data, and stereolithography contour (SLC) (Chua *et al.*, 1997a, b). The preferred use of STL file format is mainly due to its simplicity and its ability to provide data transfer for certain shapes and it has become the *de facto* standard in many RP systems. Nevertheless, the size of STL files is usually much larger than the original CAD data, causing a longer slicing time, and sometimes repair software is necessary in order to correct geometric flaws that occur during the conversion.

Thirdly, the STL file is sliced digitally into cross-sectional layers. Because the CAD model is numerically represented in an STL file, it is possible to slice the model into layers, with each representing a cross-sectional contour of the CAD model. However, the slicing process usually requires proprietary software developed by the respective RP vendor. By using this software, users can set different parameters for a CAD model, such as build direction, part size, number of duplicates, layer thickness, and so on. Layer thickness is an important parameter in the slicing process because it affects both model accuracy and build time. Decreasing layer thickness will increase model accuracy but prolong build time. The proprietary software also identifies features such as overhangs and voids in the model and generates necessary support structures. Further, the material volume and build time can be estimated and displayed by the software before the actual building. This third step in a typical RP process is sometimes referred to as 'pre-processing'.

Fourthly, the prototype item is built a layer at a time on top of another previously laid layer. After pre-processing, the digital data of the sliced layers is sequentially sent to the RP machine to be processed by its computer. The machine then acts as a 'printer' to print an actual physical layer of the model, usually one at a time. Once a layer is finished, the workstation, where the physical model is built, lowers down (or raises up, depending on the design of the RP machine) by a distance of one layer thickness. The process repeats itself until the entire model is complete. For most RP machines, once the building processing is initiated, it rarely stops during the process.

Finally the model is cleaned and finished. This step is also known as 'post-processing'. Depending on the design of a CAD model as well as the nature of the RP process, support material may be required during the fabrication, and removal of these support materials must be appropriate. Also, depending on the intended use of the prototype (for functional tests or visualization purposes), post-curing, hardening, or surface treatment such as sanding, sealing, or painting may be required. The last step is equally important to the previous steps, because a perfectly fabricated physical prototype can be irreversibly damaged by inappropriate finishing and cleaning. Usually, instructions are given in the manual of each RP system.

For RP of biomaterials, there is little change in the basic process described above. Whether it is a tissue engineering scaffold or a vascular segment, a virtual model must be available and sliceable before the actual fabrication. For example, in order to model tissue engineering scaffolds with customized structures, a scaffold library can be developed (Chua et al., 2003a, b). Similarly, to make a prototype of an organ, a 'blueprint' of the organ must first be obtained. Existing methods include 3D surface scan data (Cheah et al., 2003a, b), CT scan data (Simpson et al., 2008; Ciocca et al., 2009) or magnetic resonance imaging (MRI) scan data (Sodian et al., 2005). Alternatively, virtual models of human organs can be purchased from 3D design software companies such as Uformia. However, most of existing RP systems can process only proprietary materials supplied by the vendors, and most of these materials are not biocompatible. This limitation of materials presents a huge research opportunity to explore various biomaterials for direct or indirect fabrication. Progress has been made in some commercial systems, including selective laser sintering (SLS) (Tan et al., 2003; Chua et al., 2004; Tan et al., 2005; Wiria et al., 2007; Simpson et al., 2008; Chua et al., 2011), fused deposition modeling (FDM) (Hutmacher et al., 2001; Zein et al., 2002), and some ink-jet printing systems, by using either an easily processable biopolymer such as polycaprolactone or biocompatible materials developed by the vendor such as MED610™ for the Objet system. Besides direct fabrication using biomaterials, another approach is employing indirect fabrication. In this approach, the fabricated non-biomaterial prototype is used as a mold, usually a sacrificial one, to cast a biomaterial prototype (Tan et al., 2010). This approach always requires two or more steps for fabrication. Therefore, it is named indirect as compared to the one-step direct method. Despite the progress in RP of synthetic biomaterials, fabricating a high-resolution prototype of natural or synthetic biomaterial and maintaining its shape and structure during and after the layer-by-layer process still remains a challenge. In addition, cleaning and finishing may also present a challenge to some RP systems. For example, powder particles are frequently trapped inside the pores of scaffolds and are difficult to remove.

1.4 Conventional RP systems and classification

Currently, there are more than thirty commercial RP systems. One convenient way to classify RP systems is to base it on the form of model material used in each system, namely liquid, solid and powder. Liquid-based RP systems generally use a photosensitive liquid polymer as a model material. The liquid polymer solidifies upon exposure to laser (usually with wavelength in the ultraviolet (UV) range) or to UV light. This process is called photopolymerization. Polymerization is a process of linking random monomers into long chains, and photopolymerization is a polymerization process initiated by photon energy. In photopolymerization, photoinitiators in the liquid polymer absorb photon energy, reaching a higher energy or an excited state and becoming free radicals. Some of these free radicals, in turn, act as polymerization-initiating molecules by binding to other monomers in the liquid polymer. These polymerization-initiating molecules react with other monomers to form molecules with longer chains, and the monomer-adding process propagates until it is terminated by an inhibition process. The length of each polymer chain depends on how long the polymerization reaction is sustained, and if the monomers contain more than three reactive groups, they can react with other polymer chains and result in a network of cross-linked polymer. When all monomers are linked into long chains and the long chains are crosslinked, the molecules cannot easily be separated from each other, and therefore the liquid polymer becomes solidified. This photopolymerization process occurs instantaneously upon exposure to photon energy, and repetition of this process in a layer-by-layer manner can result in stacked layers of solid. When the contour of each solid layer is controlled by either the path of the laser or a photomask, a 3D solid with defined surface geometry can then be fabricated.

Stereolithography apparatus (SLA) is the first commercial RP system and operates based on a liquid polymer. It can process a wide range of materials such as epoxy-based systems and acrylates to produce a part with good accuracy and good surface finish. The system consists of a computer, a control panel, a laser, an optical system, and a process chamber. Briefly, the SLA process is described as follows:

1. The operator loads a STL file of a CAD model into the computer, where the software slices the CAD model into a series of cross-sections.
2. The computer-controlled optical system directs the laser beam to scan across the surface of a liquid polymer corresponding to the contour of each cross-section of the CAD model, generating a layer of solid polymer according to the defined geometry.
3. The elevator then drops down equivalent to one layer thickness so that liquid polymer covers the solid layer again before the laser beam traces the next cross-section.
4. The process repeats itself until an entire part is complete.

Fundamentally, liquid-based systems rely on photopolymerization. Depending on the type of light, method of exposure, type of liquid polymer, and type of optical system, there are many variations to this fundamental process. Some other liquid-based RP systems include Objet Geometries Polyjet, Solid Creation System (SCS), Solid Object Ultraviolet-laser Printer (SOUP) and so on.

Solid-based RP systems use a solid as the primary medium for prototyping, even though some systems do also use laser for processing. Since all materials are already in a solid form, the solid-based systems differ from each other mainly in the method of fabricating cross-sections. One typical example of such systems is FDM. In this process, the cross-section is fabricated from filaments. The underlying principle is that a two dimensional cross-section can be drawn by many parallel lines, much like hatching in technical drawings. The material used in the FDM system is in spools of filaments. The system has an extrusion head which heats the fed filament into semi-liquid state. Then the semi-liquid filament is deposited onto the workstation line by line, like hatching. Because the temperature of the surrounding air is much lower than the melting point of the filament, after deposition, the semi-liquid extrusion gradually solidifies in air. Once a single layer cross-section is finished, the workstation lowers the equivalent of one layer thickness and the extrusion head draws the next cross-section. A dual extrusion head mechanism is used in FDM systems, one for dispensing model material and the other for support material. This allows minimal wastage of material and easy change of material. However, a part built by FDM usually has compromised accuracy due to the constraint imposed by the diameter of the filament and the distance between laid down filaments. Moreover, because the extrusion head moves more slowly than laser scanning, build time of FDM is relatively longer. Furthermore, filament shrinkage may occur due to the heating and cooling introduced in the process.

Besides the filament method to fabricate a cross-section, carbon dioxide laser can be used to cut a cross-section out of a layer of solid. The system based on this principle is called laminated object manufacturing (LOM). In this system, the medium for prototyping is sheets of paper. The computer guides a carbon dioxide laser to precisely cut the outline of the cross-section. Once complete, a new layer of paper is laminated on top of the cross-section outlined and rolled over by a heated roller to allow it to bond to the previous layer. The process repeats itself until the entire part is complete. LOM has a fast build time and a high precision in X-Y plane, but is not well suited to making thin-walled structures in the Z-direction. Other solid-based RP systems include Solidscape Inc.'s benchtop 3D printer, Multi-jet Modeling System (MMS), and plastic sheet lamination (PSL).

A special class of solid-based RP systems utilizes powder to create prototypes and hence they are classified as powder-based systems. Typical examples include SLS selective laser melting (SLM) (Chua et al., 2010b; Liu et al., 2010, 2012), and 3D printer (3DP). The SLS process uses a carbon dioxide laser to heat powder particles to their glass transition temperature, which is usually near their melting point, fusing the particles to form a solid. This process is called sintering, in which the powder material coalesces into a solid mass directly, without entering into the melt phase. The surrounding powder does not melt and therefore serves as a support to the part being built. After one layer is sintered, the workstation is lowered by one layer thickness and a new layer of powder is spread over the as-sintered layer *via* a roller mechanism, preparing for the next round of laser scan for the following layer. The process repeats until a part is complete. SLS is one of the very few technologies that can directly process a wide range of engineering materials, such as polyamide, thermoplastic elastomer, polycarbonate, nylon, metal, and ceramics. However, since powder serves as both model and support material, the powder consumption is relatively high.

3DP is a printer-like RP system. The machine prints binder solution ('ink') onto a layer of loose powder ('paper') to form a solid cross-section. The powder adheres firmly where the binder solution is deposited. The surrounding powder does not bind and similarly serves as support. The process alternates between spreading a layer of powder and depositing a layer of binder solution until the entire part is complete. 3DP consumes a large amount of powder, but compared to SLS, the availability of different types of materials is limited (starch and plaster-based materials only) and the built part is weak and has a poor finish. Post-processing such as wax infiltration and application of hardeners is usually required. However, the simplicity of operation and the ability to print color material make 3DP very popular across a range of industries such as automotive, packaging, footwear, and telecommunications. Some other powder-based RP systems include the EOSINT system, laser engineered net shaping (LENS), LaserCUSING, SLM, selective mask sintering (SMS), etc.

1.5 RP of biomaterials

Rapid prototyping has long been applied along with biomedical imaging technologies for surgery planning owing to its ability to provide detailed visualizations of the anatomy of patient injuries or trauma (Chua *et al.*, 1998). However, these tissue or organ models are mainly for visualization and have no real effect on the tissues and organs of the patient. It is only when a biomaterial is used as the model material, which is at least biocompatible and incurs minimal or no tissue response, that the prototype can be used clinically, for example, a customized tracheobronchial stent (Lim *et al.*, 2002). Since engineering materials are usually classified as metals, ceramics, polymers, and composites, bioengineering materials that can be utilized in RP systems include biometals, bioceramics, biopolymers, and biocomposites. Among these bioengineering materials, biopolymers and biocomposites are the most intensively investigated, especially for tissue engineering and regenerative medicine applications. This is because some of these materials can degrade into metabolic intermediates in the human body, potentially eliminating many problems associated with permanent implants. However, the capability of RP technologies is not just limited to fabricating biopolymers and biocomposites; it can also fabricate biocompatible metals and ceramics; titanium implants and prostheses are two examples (Winder *et al.*, 1999; Singare *et al.*, 2004; Ovsianikov *et al.*, 2007).

Biopolymers can be in a natural or synthetic form. Direct processing of natural biopolymers using conventional RP systems is challenging as these systems were not originally designed for such a purpose. Instead, indirect processing may be considered, for example, using a RP-made mold to cast out a collagen scaffold with an internal network of microchannels (Tan et al., 2010). In order to directly process natural biopolymers, new RP systems capable of dispensing liquid or pastes at an un-elevated temperature (room temperature or body temperature) have to be designed and developed. Landers et al. (2002) first reported using such a system to fabricate hydrogel scaffolds with a well-defined external shape and internal structure, which partially expanded the choice of material from the synthetic to the natural range and laid down a principle for later studies that incorporated biological cells into the construction of 3D tissues. RP of natural biopolymers is also known as organ printing, 3D bioprinting or biofabrication (Mironov et al., 2003, 2009).

On the other hand, synthetic biopolymers, especially biodegradable ones, such as poly(glycolic acid), poly(lactic acid), polycaprolactone, and their various hybrids or copolymers, can be readily processed by conventional RP techniques when prepared in an appropriate form, that is, photocurable polymer for SLA, powder for SLS, and filament for FDM. For example, poly(lactic acid) oligomers with different arms could be synthesized and end-functionalized with methacryloyl chloride for photocrosslinking, and using this resin, porous constructs were accurately fabricated by STL (Melchels et al., 2009). Also, three-armed polycaprolactone oligomers could be synthesized and functionalized with methacrylic anhydride for photocrosslinking. The fabricated constructs precisely matched the CAD model with no observable material shrinkage (Elomaa et al., 2011). However, primarily for cost reasons, only polycaprolactone has been commonly used in SLS and FDM. This material preference could be a factor for most of the SLS and FDM-made 3D scaffolds being directed for hard tissue applications such as bones. Nevertheless, there are RP-made scaffolds for soft tissue application, for example, cardiac patches (Yeong et al., 2010). Recently, these synthetic scaffolds have evolved to comprise a functional gradient to better mimic the anatomical structure and mechanical function of bones (Leong et al., 2008; Chua et al., 2011; Sudarmadji et al., 2011).

1.6 Conclusion and future trends

RP technologies have a wide range of applications, in fields such as engineering, manufacturing, aerospace, automotive, jewelry, coin, tableware, arts, and architecture. They also play an increasingly important role in the biomedical industry, such as surgery planning, customized implants and prostheses, drug delivery devices, and tissue engineering. The versatile applications of

RP is largely due to its merit of being able to fabricate parts with complex structures and intricate details. However, the basis for these versatile applications is not that RP is a single technology, powerful enough to process all types of materials. It is because there is a wide range of RP systems catering to a wide range of industrial material requirements. Fundamentally, RP is a material-dependent process. Each RP system is designed to process selective materials with optimized performance. Frequently, optimizing material performance requires modification of material composition and often leads to the development of many proprietary materials supplied only by vendors. This limitation of materials may not be apparent to existing industries as long as material performance satisfies their intrinsic requirements. However, this challenge becomes very obvious when applying a specific RP system to emerging industries such as biofabrication. To date there is no general agreement on which RP system is the best suited for such applications. Researchers usually work independently on an individual system such as SLA, SLS, or FDM and attempt to process different synthetic biomaterials into constructs with desirable functionalities. Therefore, one trend in RP of biomaterials in future is to design and develop new biomaterials that suit the capability of current RP systems, such as photocurable synthetic biomaterials for SLA or cryogenic prototyping (Lim et al., 2010).

Besides new biomaterials, another critical issue regarding RP systems for biofabrication is their resolution. Tissue engineering scaffolds are expected to provide cells with a microenvironment similar to ECMs at the micron or submicron scale. Most of the current RP systems can fabricate porous structures with only macroscale struts, and a direct fabrication of biomaterial ECM analogs at the nanoscale has yet to be realized. The highest resolution achieved so far is via a technique called two-photon polymerization, a liquid-based RP system but able to create objects with submicron features. In this process, a femtosecond laser is used, and photopolymerization occurs only at the focal point where the light is absorbed the most. Using this technique, highly organized fibrin scaffolds with submicron features could be fabricated via an indirect approach, that is, the two-photon polymerization process merely produced a submicron mold (Koroleva et al., 2012b). Recently, photocurable poly(lactic acid) was prepared and used for direct fabrication (Koroleva et al., 2012a). In future, more studies in line with this trend can be expected. An ideal RP system for scaffold fabrication should be able to 'print' ECM consistently. In fact, consistency at the submicron scale is one of the challenges faced by another technique called electrospinning, in which ECM-like nanofibers are obtained in a random fashion.

Developing new RP systems specifically for processing biological materials is another major research direction for RP in biomedical applications (Bártolo et al., 2009; An et al., 2013). These systems are called 'bioprinters', because they dispense hydrogels or a mixture of hydrogel and cells in a line-by-line and layer-by-layer manner. For example, cells could be prepared in the form of tissue spheroids by an automated robotic system (Mehesz *et al.*, 2011). Once printed one by one in a defined layout such as a ring or a branched structure, these tissue spheroids can fuse and integrate to form a tissue accordingly (Mironov *et al.*, 2009). However, one limitation of the bioprinting approach is the weak mechanical strength of hydrogels (Billiet *et al.*, 2012). To fabricate and maintain the shape and structure of the bioprinted 3D tissue, the hydrogel struts must hold their own weight and form without mixing or breaking. Moreover, the number of studies on bioprinting of multiple cell types is currently limited due to biological challenges in cell culturing. However, two approaches could be considered for fabricating constructs with multiple cell types: (1) deposition of multiple types of cells through multiple nozzles; or (2) deposition of tissue spheroids that already contain a mixture of multiple cell types. Nevertheless, these approaches may only apparently address the issue of how to aggregate multiple types of cells. At a deeper level, culture and growth of these cells after the aggregation will be a huge challenge ahead. Solving such a challenge will require time and a multidisciplinary effort.

1.7 Sources of further information and advice

Interested readers may refer to the book *Rapid Prototyping: Principles and Applications* (3rd edition), authored by Professor Chee Kai Chua, Associate Professor Kah Fai Leong and Associate Professor Chu Sing Lim at Nanyang Technological University (Singapore), for a further understanding of RP fundamentals and a comprehensive understanding of various RP systems and their applications.

To understand the current status of the AM industries worldwide, readers may consider reading the most recent *Wohlers Report*, which is published annually by Wohlers Associates Inc:
- Website: http://www. wohlersassociates.com

Alternatively, the reader may wish to look up two internationally peer-reviewed journals, *Rapid Prototyping Journal* and *Virtual & Physical Prototyping*, published by Emerald and Taylor & Francis, respectively, for the most recent research news on RP and AM. Readers who intend to pursue more on organ printing or biofabrication may wish to look up the journal *Biofabrication* for the latest research publications:

- *Rapid Prototyping Journal*
  Website: http://www.tandfonline.com/toc/nvpp20/current
- *Virtual & Physical Prototyping*
  Website: http://www.emeraldinsight.com/journals.htm?issn=1355-2546
- *Biofabrication*
  Website: http://iopscience.iop.org/1758-5090

1.8 References

An, J., Chua, C. K., Yu, T., Li, H. and Tan, L. P. (2013). Advanced nanobiomaterial strategies for the development of organized tissue engineering constructs. *Nanomedicine*, **8**, 591–602.

Bártolo, P. J., Chua, C. K., Almeida, H. A., Chou, S. M. and Lim, A. S. C. (2009). Biomanufacturing for tissue engineering: Present and future trends. *Virtual and Physical Prototyping*, **4**, 203–216.

Billiet, T., Vandenhaute, M., Schelfhout, J., VAN Vlierberghe, S. and Dubruel, P. (2012). A review of trends and limitations in hydrogel-rapid prototyping for tissue engineering. *Biomaterials*, **33**, 6020–6041.

Cheah, C. M., Chua, C. K. and Tan, K. H. (2003a). Integration of laser surface digitizing with CAD/CAM techniques for developing facial prostheses. Part 2: Development of molding techniques for casting prosthetic parts. *International Journal of Prosthodontics*, **16**, 543–548.

Cheah, C. M., Chua, C. K., Tan, K. H. and Teo, C. K. (2003b). Integration of laser surface digitizing with CAD/CAM techniques for developing facial prostheses. Part 1: Design and fabrication of prosthesis replicas. *International Journal of Prosthodontics*, **16**, 435–441.

Chua, C. K., Chou, S. M., Lin, S. C., Eu, K. H. and Lew, K. F. (1998). Biomedical applications of rapid prototyping systems. *Automedica*, **17**, 29–40.

Chua, C. K., Jacob, G. G. K. and Mei, T. (1997a). Interface between CAD and rapid prototyping systems. Part 1: A study of existing interfaces. *International Journal of Advanced Manufacturing Technology*, **13**, 566–570.

Chua, C. K., Jacob, G. G. K. and Mei, T. (1997b). Interface between CAD and rapid prototyping systems. Part 2: LMI – An improved interface. *International Journal of Advanced Manufacturing Technology*, **13**, 571–576.

Chua, C. K., Leong, K. F., Cheah, C. M. and Chua, S. W. (2003a). Development of a tissue engineering scaffold structure library for rapid prototyping. Part 1: Investigation and classification. *International Journal of Advanced Manufacturing Technology*, **21**, 291–301.

Chua, C. K., Leong, K. F., Cheah, C. M. and Chua, S. W. (2003b). Development of a tissue engineering scaffold structure library for rapid prototyping. Part 2: Parametric library and assembly program. *International Journal of Advanced Manufacturing Technology*, **21**, 302–312.

Chua, C. K., Leong, K. F. and Lim, C. S. (2010a). *Rapid Prototyping: Principles and Applications*, 3rd edition, World Scientific, Singapore.

Chua, C. K., Leong, K. F., Sudarmadji, N., Liu, M. J. J. and Chou, S. M. (2011). Selective laser sintering of functionally graded tissue scaffolds. *MRS Bulletin*, **36**, 1006–1014.

Chua, C. K., Leong, K. F., Tan, K. H., Wiria, F. E. and Cheah, C. M. (2004). Development of tissue scaffolds using selective laser sintering of polyvinyl alcohol/hydroxyapatite biocomposite for craniofacial and joint defects. *Journal of Materials Science: Materials in Medicine*, **15**, 1113–1121.

Chua, C. K., Liu, A. and Leong, K. F. (2010b). A practical approach on temperature variation in Selective Laser Melting with a novel heat transfer model. Proceedings of 4th International Conference on Advanced Research in Virtual and Rapid Prototyping, 6–10 October, 2009, Leiria, Portugal, 363–367.

Ciocca, L., DE Crescenzio, F., Fantini, M. and Scotti, R. (2009). CAD/CAM and rapid prototyped scaffold construction for bone regenerative medicine and surgical transfer of virtual planning: A pilot study. *Computerized Medical Imaging and Graphics*, **33**, 58–62.

Elomaa, L., Teixeira, S., Hakala, R., Korhonen, H., Grijpma, D. W. and Seppälä, J. V. (2011). Preparation of poly(ε-caprolactone)-based tissue engineering scaffolds by stereolithography. *Acta Biomaterialia*, **7**, 3850–3856.

Hutmacher, D. W., Schantz, T., Zein, I., Ng, K. W., Teoh, S. H. and Tan, K. C. (2001). Mechanical properties and cell cultural response of polycaprolactone scaffolds designed and fabricated via fused deposition modeling. *Journal of Biomedical Materials Research*, **55**, 203–216.

Koroleva, A., Gill, A. A., Ortega, I., Haycock, J. W., Schlie, S., Gittard, S. D., Chichkov, B. N. and Claeyssens, F. (2012a). Two-photon polymerization-generated and micromolding-replicated 3D scaffolds for peripheral neural tissue engineering applications. *Biofabrication*, **4**, 025005.

Koroleva, A., Gittard, S., Schlie, S., Deiwick, A., Jockenhoevel, S. and Chichkov, B. (2012b). Fabrication of fibrin scaffolds with controlled microscale architecture by a two-photon polymerization-micromolding technique. *Biofabrication*, **4**, 015001.

Landers, R., Hübner, U., Schmelzeisen, R. and Mülhaupt, R. (2002). Rapid prototyping of scaffolds derived from thermoreversible hydrogels and tailored for applications in tissue engineering. *Biomaterials*, **23**, 4437–4447.

Leong, K. F., Chua, C. K. and Ng, Y. M. (1996a). A study of stereolithography file errors and repair. Part 1. Generic solution. *International Journal of Advanced Manufacturing Technology*, **12**, 407–414.

Leong, K. F., Chua, C. K. and Ng, Y. M. (1996b). A study of stereolithography file errors and repair. Part 2. Special cases. *International Journal of Advanced Manufacturing Technology*, **12**, 415–422.

Leong, K. F., Chua, C. K., Sudarmadji, N. and Yeong, W. Y. (2008). Engineering functionally graded tissue engineering scaffolds. *Journal of the Mechanical Behavior of Biomedical Materials*, **1**, 140–152.

Lim, C. S., Eng, P., Lin, S. C., Chua, C. K. and Lee, Y. T. (2002). Rapid prototyping and tooling of custom-made tracheobronchial stents. *International Journal of Advanced Manufacturing Technology*, **20**, 44–49.

Lim, T. C., Chian, K. S. and Leong, K. F. (2010). Cryogenic prototyping of chitosan scaffolds with controlled micro and macro architecture and their effect on in vivo neo-vascularization and cellular infiltration. *Journal of Biomedical Materials Research – Part A*, **94**, 1303–1311.

Liu, A., Chua, C. K. and Leong, K. F. (2010). Properties of test coupons fabricated by selective laser melting. *Key Engineering Materials*, **447–448**, 780–784.

Liu, Z. H., Chua, C. K., Leong, K. F., Kempen, K., Thijs, L., Yasa, E., VAN-Humbeeck, J. and Kruth, J. P. (2012). A preliminary investigation on selective laser melting of M2 high speed steel. Proceedings of 5th International Conference on Advanced Research in Virtual and Rapid Prototyping, 28 September–1 October 2011, Leiria, Portugal, 339–346.

Mehesz, A. N., Brown, J., Hajdu, Z., Beaver, W., DA Silva, J. V. L., Visconti, R. P., Markwald, R. R. and Mironov, V. (2011). Scalable robotic biofabrication of tissue spheroids. *Biofabrication*, **3**, 025002.

Melchels, F. P. W., Feijen, J. and Grijpma, D. W. (2009). A poly (D,L-lactide) resin for the preparation of tissue engineering scaffolds by stereolithography. *Biomaterials*, **30**, 3801–3809.

Mironov, V., Boland, T., Trusk, T., Forgacs, G. and Markwald, R. R. (2003). Organ printing: Computer-aided jet-based 3D tissue engineering. *Trends in Biotechnology*, **21**, 157–161.

Mironov, V., Trusk, T., Kasyanov, V., Little, S., Swaja, R. and Markwald, R. (2009). Biofabrication: A 21st century manufacturing paradigm. *Biofabrication*, **1**, 022001.

Ovsianikov, A., Chichkov, B., Adunka, O., Pillsbury, H., Doraiswamy, A. and Narayan, R. J. (2007). Rapid prototyping of ossicular replacement prostheses. *Applied Surface Science*, **253**, 6603–6607.

Simpson, R. L., Wiria, F. E., Amis, A. A., Chua, C. K., Leong, K. F., Hansen, U. N., Chandrasekaran, M. and Lee, M. W. (2008). Development of a 95/5 poly (L-lactide-co-glycolide)/hydroxylapatite and β-tricalcium phosphate scaffold as bone replacement material via selective laser sintering. *Journal of Biomedical Materials Research – Part B Applied Biomaterials*, **84**, 17–25.

Singare, S., Dichen, L., Bingheng, L., Yanpu, L., Zhenyu, G. and Yaxiong, L. (2004). Design and fabrication of custom mandible titanium tray based on rapid prototyping. *Medical Engineering and Physics*, **26**, 671–676.

Sodian, R., Fu, P., Lueders, C., Szymanski, D., Pritsche, C., Gutberlet, M., Hoerstrup, S. P., Hausmann, H., Lueth, T. and Hetzer, R. (2005). Tissue engineering of vascular conduits: Fabrication of custom-made scaffolds using rapid prototyping techniques. *Thoracic and Cardiovascular Surgeon*, **53**, 144–149.

Sudarmadji, N., Tan, J. Y., Leong, K. F., Chua, C. K. and Loh, Y. T. (2011). Investigation of the mechanical properties and porosity relationships in selective laser-sintered polyhedral for functionally graded scaffolds. *Acta Biomaterialia*, **7**, 530–537.

Tan, J. Y., Chua, C. K. and Leong, K. F. (2010). Indirect fabrication of gelatin scaffolds using rapid prototyping technology. *Virtual and Physical Prototyping*, **5**, 45–53.

Tan, K. H., Chua, C. K., Leong, K. F., Cheah, C. M., Cheang, P., ABU Bakar, M. S. and Cha, S. W. (2003). Scaffold development using selective laser sintering of polyetheretherketone-hydroxyapatite biocomposite blends. *Biomaterials*, **24**, 3115–3123.

Tan, K. H., Chua, C. K., Leong, K. F., Cheah, C. M., Gui, W. S., Tan, W. S. and Wiria, F. E. (2005). Selective laser sintering of biocompatible polymers for applications in tissue engineering. *Bio-Medical Materials and Engineering*, **15**, 113–124.

The White House. (2012). *We Can't Wait: Obama Administration Announces New Public-Private Partnership to Support* [Online]. Available: http://www.whitehouse.gov/the-press-office/2012/08/16/we-can-t-wait-obama-administration-announces-new-public-private-partners (Accessed 16 August 2012).

Winder, J., Cooke, R. S., Gray, J., Fannin, T. and Fegan, T. (1999). Medical rapid prototyping and 3D CT in the manufacture of custom made cranial titanium plates. *Journal of Medical Engineering and Technology*, **23**, 26–28.

Wiria, F. E., Leong, K. F., Chua, C. K. and Liu, Y. (2007). Poly-ε-caprolactone/hydroxyapatite for tissue engineering scaffold fabrication via selective laser sintering. *Acta Biomaterialia*, **3**, 1–12.

Yeong, W. Y., Sudarmadji, N., Yu, H. Y., Chua, C. K., Leong, K. F., Venkatraman, S. S., Boey, Y. C. F. and Tan, L. P. (2010). Porous polycaprolactone scaffold for cardiac tissue engineering fabricated by selective laser sintering. *Acta Biomaterialia*, **6**, 2028–2034.

Zein, I., Hutmacher, D. W., Tan, K. C. and Teoh, S. H. (2002). Fused deposition modeling of novel scaffold architectures for tissue engineering applications. *Biomaterials*, **23**, 1169–1185.

2
Freeform fabrication of nanobiomaterials using 3D printing

M. VAEZI and S. YANG, University of Southampton, UK

DOI: 10.1533/9780857097217.16

Abstract: Nanobiomaterials play an important role in nanobiotechnology and have made a great contribution to biomedical research and healthcare. Recent progress in nanobiomaterials has increased demand for multidisciplinary approaches from physical, biological and engineering sciences. Solid freeform fabrication (SFF) technologies are based on layer-by-layer deposition of materials which bring about new application possibilities for processing nanobiomaterials. The aim of this chapter is to provide a comprehensive overview of SFF techniques suitable for processing nanobiomaterials; current state and limitations regarding the techniques are discussed. Overviews of subjects including biofabrication of tissue engineering (TE) scaffolds using extrusion-based freeforming and dry powder printing of nanobiomaterials are also provided.

Key words: solid freeform fabrication, nanobiomaterials, extrusion freeforming, 3D printing, dry powder printing.

2.1 Introduction

Recent advances on nanotechnology have created new frontiers, terminologies, and possibilities which have led to breakthroughs in several distinct and multidisciplinary sciences. In particular, recent developments in the field of nanobiotechnology have significantly improved the area of nanomedicine, biomedical and healthcare sciences. Conventional biomedical applications have taken advantage of nanotechnology science in different areas such as intelligent systems, controlled release systems, tissue engineering (TE), biosensors and nanocomposites used in orthopaedic implants (Bartolo and Bidanda, 2008). Unique and useful characteristics over conventional materials have been observed from nanobiomaterials due to the size and surface effects that can be employed in various medical applications. Improved reactivity, bioactivity, electrical and optical properties, strength and magnetic characteristics are of particular interest to the biomedical field (Ferrari, 2005; Qin *et al.*, 1999; Vasir *et al.*, 2005). Moreover, there is the possibility of controlling some other properties such as melting point and solubility of nanobiomaterials by altering their particle size.

Nanofillers such as nanofibres and nanoparticles (NPs) have been widely presented in different biomedical fields with new applications in mind. Polymeric nanofibres with unique and intrinsic properties, resulting from their high surface area to volume ratio, are attractive for many practical applications, and intensive studies have been conducted on this class of nanobiomaterials (Hasirci et al., 2006; Huang et al., 2003; Zhang et al., 2005). It has been observed that human cells tend to attach, grow and proliferate on fibres with diameters smaller than those of cells (Teixeira et al., 2003). In this way, polymeric nanofibres have great potential to be exploited in TE scaffolds, and their intrinsic properties and dimensions make them attractive carriers. The use of polymeric nanofibres in cartilage, nerve, bone, skin, skeletal muscle and blood vessel TE have being extensively studied (Mengyan et al., 2005; Xu et al., 2004; Yoshimoto et al., 2003; Zong et al., 2005). Nanofibres and NPs can also be exploited to reinforce the composite structures of various biomedical applications such as dental restorations (Chen et al., 2006) and production of orthodontic composites with enhanced elastic modulus and flexural and tensile strengths (Fong, 2004; Price et al., 2003). Reinforcement of composites using either nanofibres or carbon nanotubes (CNTs) has been of great interest in recent years. Exploitation of different NPs in hard tissue implants for the purpose of reinforcement is an interesting subject that is under development. Carbon nanofibres (CNFs) have been employed to reinforce poly (ether ether ketone) (PEEK) structures, and nanocomposites with superior properties were obtained (Sandler et al., 2002).

Advances in nanobiomaterials and the advent of new possibilities and applications have led to demands for the development of new fabrication and processing techniques, as conventional techniques are inefficient or unable to meet new requirements. Broad ranges of processing technologies have been developed with different applications and capabilities as their fundamentals are very diverse. Among different attainable methods, SFF processes based on layer-by-layer manufacturing are identified as effective approaches worth further investigation. SFF is a fabrication technique used for building three dimensional (3D) parts layer-by-layer directly from computer-aided design (CAD) data in a short time. The combined potential of nanobiomaterials and SFF technologies has been an exciting route in nanobiotechnology and health sciences over the past decade.

More than 30 different SFF methods are being exploited worldwide in various industries; around 20 are able to process biomaterials and have found biomedical applications (Chua et al., 2004). In addition to the processing of biomaterials, there has been another significant trend in processing a variety of nanobiomaterials using SFF methods. The additive nature of SFF technologies ensures minimal waste of scarce and expensive nanobiomaterials. The use of CAD data enables fabrication of customized parts from nanobiomaterials, offers a high level of control over the architecture, and guarantees reproducibility. In addition to the complex shape of parts, the composition of the nanobiomaterials can be controlled in the parts, resulting in the potential to produce a variety of bio-nanocomposites efficiently. However, a specific class of SFF methods is currently suited for direct processing of nanobiomaterials (Fig. 2.1). Processes such as stereolithography (SL), nanocomposite deposition system (NCDS), selective laser sintering (SLS), inkjet printing, aerosol jet printing and extrusion-based systems are among the most widely used SFF approaches for nanobiomaterials processing. Moreover, SFF techniques can be used indirectly (building negative parts to be used as a mould) to produce final parts from nanobiomaterials (Dong-Woo *et al.*, 2007).

In this chapter the key SFF methods suited for the direct processing of nanobiomaterials are discussed comprehensively, and emphasis will be placed on recent advancements in the respective techniques as a working principle of each process described in Chapter 1. In addition, the use of extrusion-based freeforming in TE scaffold fabrication, and nanobiomaterials dry powder printing are reviewed extensively.

2.2 Laser-based solid freeform fabrication (SFF) techniques

The key SFF systems suitable for direct processing of nano-biomaterials can be classified into three main groups (Fig. 2.1). The first group is laser-based SFF techniques including SL and SLS processes. This section provides a comprehensive overview on these processes and their current applications and possibilities.

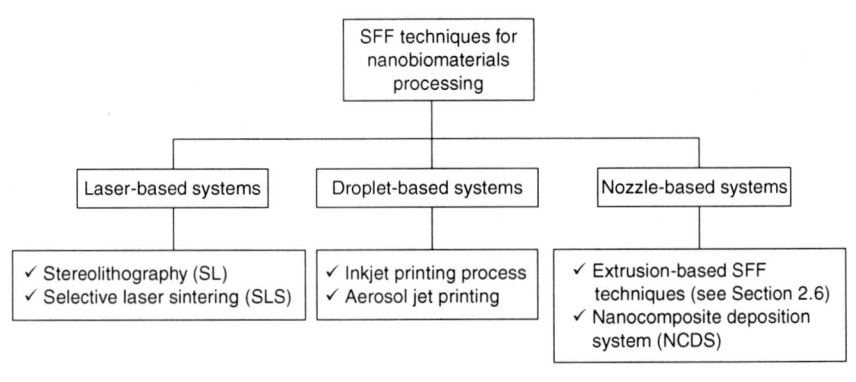

2.1 Classification of different SFF systems suitable for processing of nanobiomaterials.

2.2.1 Stereolithography (SL) process

Stereolithography was the first commercial SFF process developed by 3D Systems Inc. and is based on layer-by-layer polymerization of photosensitive resin using ultraviolet (UV) light. Two main SL techniques, namely scanning SL and projection SL, have been developed depending on the beam delivery system. Scanning SL solidifies the photopolymer (including UV photo-initiator, monomer and other additives) in a point-by-point and line-by-line style in each layer. In projection SL, build time is saved significantly as whole layer of the photopolymer is cured at once via exposure through the provided mask. A digital micro-mirror device (DMD), embedded in digital light processing (DLP) projectors, is normally applied as the dynamic mask in projection SL systems.

The SL process can be used for fabrication of 3D nanocomposite parts from resins, based on the insertion of a high load of bio-nanofillers in a photosensitive polymer matrix which acts as the binder material. Use of resins containing nanofillers in the SL process is mostly for the purpose of reinforcing the final nanocomposite part. The use of NPs with lower density and smaller particle size is associated with an increase in resin viscosity (Bartolo and Gaspar, 2008; Gaspar *et al.*, 2008). The resulting nanocomposite objects produced in this way can also be subjected to debinding (via an appropriate thermal treatment) and sintering steps to be converted into pure 3D part. During the debinding and sintering steps, the shape of the part remains unaltered but the part is subjected to shrinkage. The load of nanoparticle in the resin should be controlled accurately and should be sufficiently high (up to 80 wt.% for alumina NPs) to avoid part deformation and crack generation. Nanoparticle content should be high enough (solid loading more than 50 wt.%) to obtain satisfying characteristics in the final, dense part. However, increasing nanoparticle content results in increasing viscosity of the suspension, and subsequently it becomes difficult to recoat the suspension layers during the SL process. Therefore, innovative approaches either using special recoating or through developing a low viscosity suspension are required to build parts from high NPs-loaded resin in the SL process. Instead of trying to develop low viscosity suspensions, Doreau and co-workers (2000) used a special scraper, patented by Optoform Inc., to spread a paste containing a photo-curable resin, high loads of ceramic particles (50–60 vol.%), dispersants and a thickener. In a similar way, Bertsch *et al.* (2004) gained success in processing a special paste containing very high loads of alumina NPs (up to 80 wt.%), a UV photoinitiator, a low viscosity monomer (Polyethylene glycol 400 diacrylate) and a silane (3-glycidoxypropyltrimethoxysilane) in the SL process. Silane served to prevent alumina particle agglomeration and to stabilize the formulation. As mentioned before, parts produced in this way keep their shape and show no

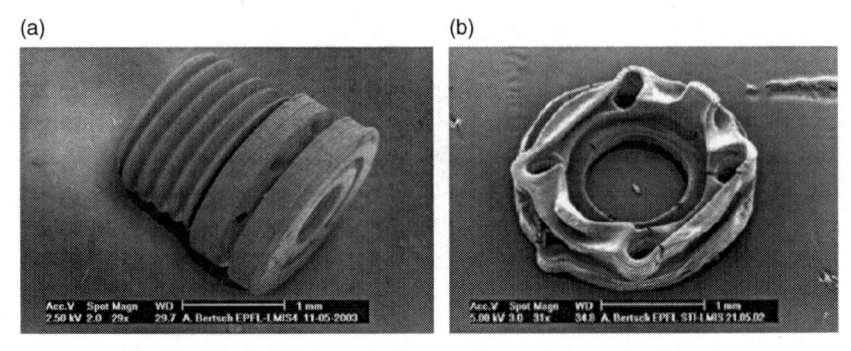

2.2 SEM images of two identical micro parts produced via micro SL from composites of different nanoalumina loadings: (a) 75 wt.% alumina loading (~20% part shrinkage and 8% porosity after sintering but original shape kept) (Bertsch *et al.*, 2004); (b) 50 wt.% alumina loading (~27% part shrinkage and 30% porosity after sintering, part deformation and cracks were observed due to low solid particle loading) (Bartolo, 2011).

deformations or cracks once sintered if high ceramic NP loading is used, but some shrinkage and a residual porosity have been observed. Figure 2.2 depicts two different micro parts built by Bertsch and co-workers from alumina NPs with 50 wt.% and 75 wt.% loading.

Different research groups have fabricated parts by inserting nano-/micron-sized ceramic particles such as silica and silicon nitride (Griffith and Halloran, 1996), hydroxyapatite (HA) (Griffith *et al.*, 1995), alumina (Greco *et al.*, 2001; Hinczewski *et al.*, 1998), etc., into water-based or acrylate-based, photocurable resins. Acrylate prepolymers such as 1,6-hexanediol diacrylate (HDDA, a low viscosity acrylate monomer) is normally used as matrix for alumina and HA NPs. An appropriate dispersant needs to be included to prevent nanoparticle agglomeration and to decrease the viscosity of the prepolymer.

In addition to NPs, CNTs can be exploited as nanosized fillers with the aim of improving mechanical properties of nanocomposites made by SL. Sandoval *et al.* (2007) dispersed controlled amounts of multiwalled carbon nanotubes (MWCNTs) in epoxy-based resins and made complex 3D nanocomposite parts with enhanced mechanical properties. Their electron microscopy results showed affinity between the constituents of the nanocomposite. In the meantime, buckled and collapsed MWCNTs in several micrographs of samples that were previously pulled in tensile tests were observed (Fig. 2.3). It was thought that the buckling and collapsing phenomena of the MWCNTs were a result of the photopolymerization (in the SL machine and in the UV oven) and thermal effects introduced by the SL system laser.

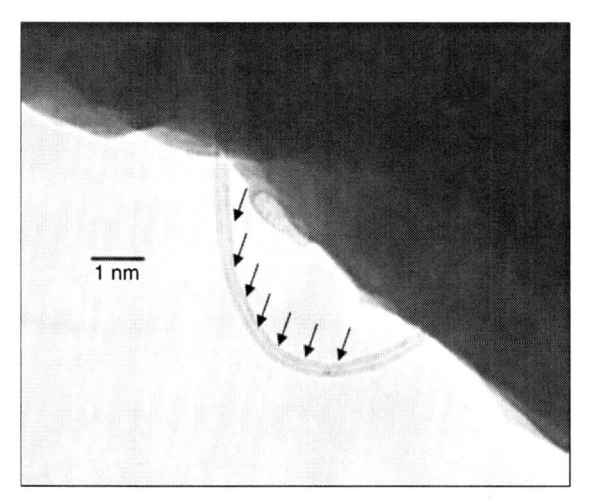

2.3 Collapsed MWCNTs at the fracture surface of the nanocomposite produced by SL (Sandoval *et al.*, 2007).

Stereolithography has been used directly and indirectly to make biodegradable TE nanocomposite scaffolds. Jin Woo *et al.* (2009) used a suspension containing poly (propylene fumarate) (PPF), diethyl fumarate (DEF) (to reduce the viscosity) and HA NPs to fabricate a nano-/micro-scale PPF/DEF-HA composite scaffold directly (Fig. 2.4c and 2.4d). In addition, they produced a negative scaffold model (Fig. 2.4a) as a mould with an internal pore size of 250 μm and a line width of about 350 μm from SL5180 resin (Huntsman). Then the mould was filled with HA nanopowders of 500 nm particle size, and the final scaffold was produced through a sintering process (Fig. 2.4b) (Dong-Woo *et al.*, 2007).

2.2.2 Selective laser sintering (SLS) process

Selective laser sintering utilizes a CO_2 laser beam to sinter thin layers of polymers or their composite powders selectively to build 3D parts. Different biomaterials ranging from biopolymers and bioceramics to various biocomposites have been processed by SLS for possible medical applications. Powders outside the part boundary fuse during processing because of 'growth' effect phenomena that result in inaccuracy and rough parts with micropores on the surface which may promote cell attachment and growth (Yang *et al.*, 2002). However, the powdery surface of SLS parts induces some difficulties in terms of sterilization and cell culture.

SLS has been recognized as a useful tool initially for fabrication of bone implants from poly (methyl methacrylate) coated calcium phosphate (Lee and Barlow, 1993), and further it was used to process some new high

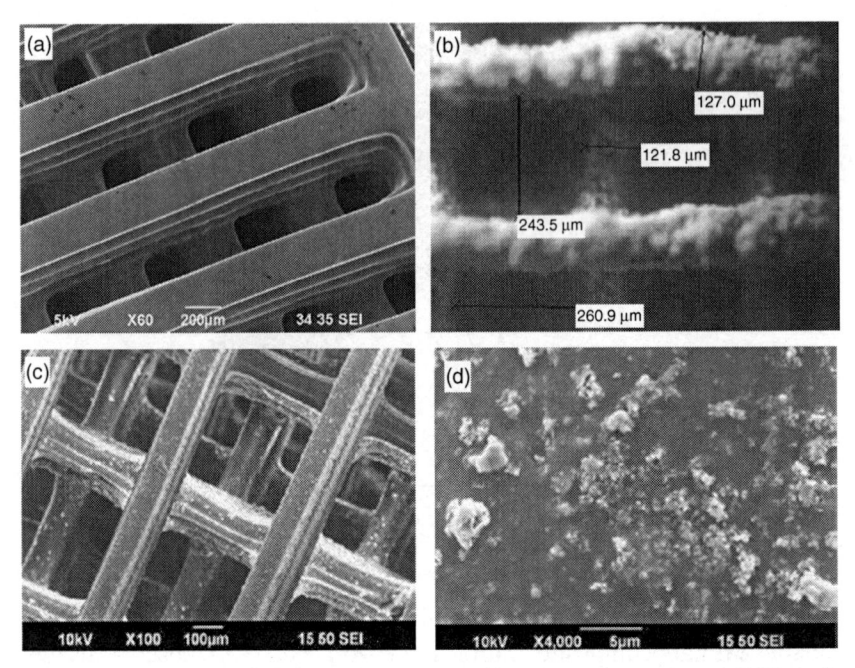

2.4 Direct and indirect fabrication of bio-nanocomposite scaffolds using the SL process. (a) SEM image of fabricated negative scaffold model as a mould (Dong-Woo *et al.*, 2007). (b) Internal shape of final bone scaffold after casting of HA nanoparticles (Dong-Woo *et al.*, 2007). (c) SEM image of a PPF/DEF-HA scaffold fabricated directly using SL (Jin Woo *et al.*, 2009). (d) Surface morphology of a PPF/DEF-HA scaffold in which HA nanoparticles have generated a nano-/micro-scale topology (Jin Woo *et al.*, 2009).

performance biomaterials such as HA-reinforced polyethylene composites for bone implants (Hao *et al.*, 2006). In particular, SLS, along with fused deposition modelling (FDM), have been recognized to be advantageous for fabrication of TE scaffolds among various SFF technologies because of their ability to process different kinds of biocompatible and biodegradable materials. Non-biodegradable polymers including ultrahigh molecular weight polyethylene (UHMWPE) (Rimell and Marquis, 2000) and PEEK (Schmidt *et al.*, 2007) have been employed to build TE scaffolds. As for biodegradable polymers, scaffolds from poly (e-caprolactone) (PCL) (Williams *et al.*, 2005) and poly (L-lactic acid) (PLLA) (Tan *et al.*, 2005) have been produced using SLS. Moreover, different composite scaffolds (biodegradable polymers and bioactive ceramics) including HA/PCL, HA/poly (L-lactide–co-glycolide) (PLGA) and β-tricalcium phosphate (β-TCP)/PLGA have been produced using SLS (Simpson *et al.*, 2008; Wiria *et al.*, 2007).

Bio-nanocomposites comprising biopolymer and different types of nanofillers are of particular interest with the SLS process as significant changes in biological or mechanical properties can be obtained with the use of only a small amount of nanofillers. Nanofillers in the form of NPs or nanofibres are used to control biodegradability and bioactivity due to the high surface to volume ratio. By using nanofillers some other properties such as mechanical properties, optical properties, thermal conductivity and heat resistance can be enhanced. NPs such as nanosilica (Chung and Das, 2008) and nanoalumina (Haizhong *et al.*, 2006) are commonly used within a biopolymer matrix to improve the mechanical properties of nanocomposites. Nanofillers can offer improved strength in the x–y direction of a part, but typically offer little or no additional strength in the z-direction as they do not span the divide between build layers. Positive effects such as decreasing the required laser energy have been observed using nanofillers specifically due to the fact that they can absorb laser power more efficiently (Tolochko *et al.*, 2000). Ho and co-workers (2002) showed that graphite fillers have the most significant effect on improving the absorptance of the laser sintering polycarbonate (PC) powder among other examined fillers, including quartz, silica and talc, since it was proposed that graphite powder could minimize thermally related problems since less laser energy is required for sintering and less energy is transmitted through the graphite powders.

Uniform base powder (as matrix), nanofiller distribution and good interfacial adhesion between them are two very important factors in SLS achieving a high performance nanocomposite part (Jain *et al.*, 2010). Mechanical mixing of the filler with the base powder is normally used for most biopolymer nanocomposites. However, mechanical mixing does not seem to be a sufficiently effective approach for uniform mixing of two powders with different sizes (especially when one is nanosized) and different densities (e.g., biopolymer and metallic nanofiller). Nanofillers can be coated with the base polymer for homogeneous dispersion and preventing nanofiller accumulation.

Calcium phosphate (Ca-P) nanofillers including HA and TCP NPs and nanofibres have been considered widely in the development of biomaterials in recent years due to their osteoconductivity, nano size effects and biomimetic resemblance to natural bone structure when mixed with biopolymers such as chitosan, collagen and PLLA. Biodegradable, osteoconductive nanocomposite scaffolds for bone tissue regeneration comprising a biodegradable polymer matrix such as PLLA and poly (hydroxybutyrate–co-hydroxyvalerate) (PHBV) with bioactive Ca-P nanofillers have been successfully built via a SLS process (Bin *et al.*, 2010). Cheung *et al.* (2008) used carbonated hydroxyapatite (CHAp) NPs within a PLLA matrix to produce nanocomposite TE scaffolds using a modified SLS machine (Fig. 2.5).

In an interesting study, Lin and co-workers (2009) used CNTs as filler and β-TCP NPs (average particle size 20.1 nm) as the main material to produce bone TE scaffolds with enhanced mechanical performance. β-TCP NPs,

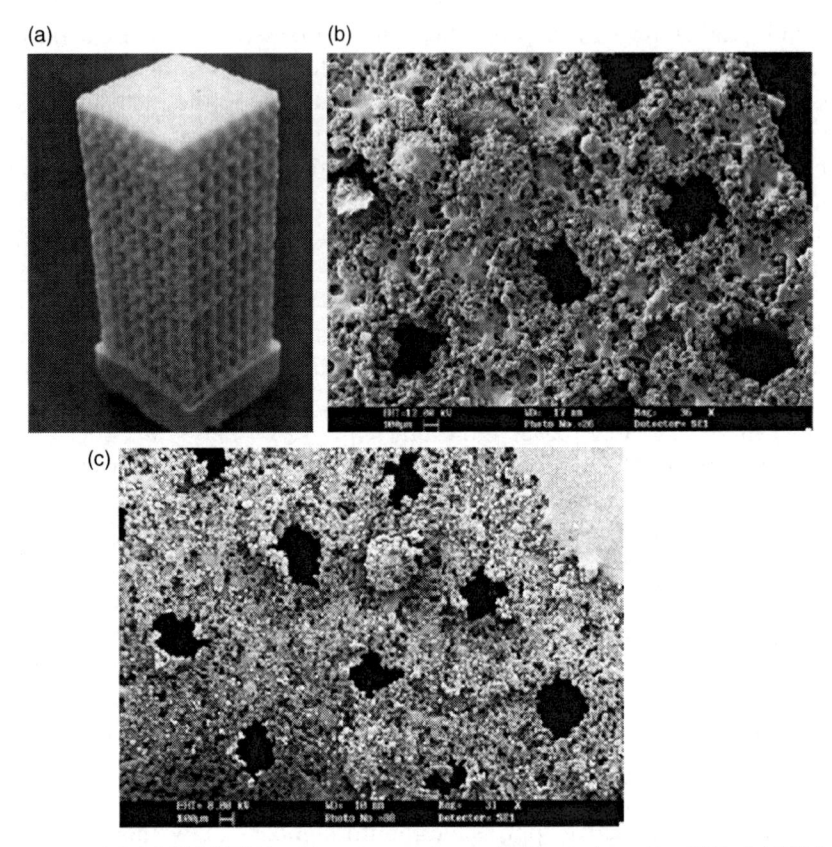

2.5 (a) PLLA/CHAp nanocomposite scaffold produced by SLS. (b) SEM image of the layer structures of a PLLA scaffold. (c) SEM image of the layer structures of a PLLA/CHAp nanocomposite scaffold. It can also be seen from Fig. 2.4b and 2.4c that the degree of fusion of the PLLA/CHAp nanocomposite is lower than that of the pure PLLA powder as the CHAp nanoparticles on the powder surface might act as a barrier against fusion (WenYou et al., 2008).

binder materials (particle size ~110 μm) and CNTs with the quality percentage of 0.1%, 0.2% and 0.3% individually were mixed using a four tank mixer. By increasing the amount of CNTs gradually, the scaffold strength first increased and then decreased. Results showed that the strength of scaffold mixed with 0.2% CNTs reached 0.819 MPa, which is an improvement of 85.7% compared with that without CNTs. The reduction in composite materials strength is thought to be caused by non-uniform dispersion of CNT aggregates. CNTs have a large aspect ratio, high surface energy and can easily form aggregates in the matrix material. Ko et al. (2007) proposed that the combination of SLS and inkjet printing processes would be an asset to increase resolution of the existing bio-nanoparticle inkjet printing. They set up a device for SLS of inkjet-printed Au nanoparticle solution on a polymer substrate by scanning with a focused continuous laser.

2.3 Droplet-based SFF techniques

The second group of SFF systems suitable for processing of nano-biomaterials is droplet-based SFF techniques including inkjet printing and aerosol jet processes. In this section, the principles and the recent progress of each process toward the processing of nano-biomaterials are described.

2.3.1 Inkjet printing process

In recent years there has been a propensity to mutate inkjet printing into a tool that can be applied in different manufacturing processes such as soldering microelectronics or fabricating micro-optical components using photocurable resins. Furthermore, inkjet printing technology has been used in a layer-by-layer process for direct freeforming of complex 3D structures pioneered by Evans and his group. In inkjet printing, liquid material (in droplet form) often turns into solid following the deposition process via cooling (e.g., by crystallization or vitrification), chemical changes (e.g., through the cross-linking of a polymer) or solvent evaporation (Hon *et al.*, 2008). Two different modes are prevalently utilized for droplet creation, including drop on demand (DOD) and continuous inkjet (CIJ). Generally, CIJ systems use fluids with lower viscosity at higher drop velocity than DOD and are mostly used where printing speed is important. In contrast, DOD is used where smaller drop size and higher accuracy are required, and it has fewer limitations on ink properties as compared with CIJ. In DOD, ink droplets are ejected from a reservoir through a nozzle using an acoustic pulse which can be induced either thermally or piezoelectrically. In thermal DOD, a vapour bubble which is generated by local heating of the ink causes droplet ejection. Thermal DOD is greatly restricted to using water as a solvent and thus compels strict limitations on the number of polymers that can be processed (de Gans and Schubert, 2003). In piezoelectric DOD, deformation of a piezoelectric membrane results in generation of acoustic pulses, and consequently ejection of the droplets. Piezoelectric DOD is an appropriate technique for a variety of solvents, and thus suited for different nanobiotechnology applications.

Inkjet printing of ceramics using both piezoelectric and thermal printers has been reported for various 3D micropatterning applications such as creating internal cavities (Mott *et al.*, 1999), functional gradients (Mott and Evans, 1999) and arrays of pillars (Evans *et al.*, 2001, Lejeune *et al.*, 2009).

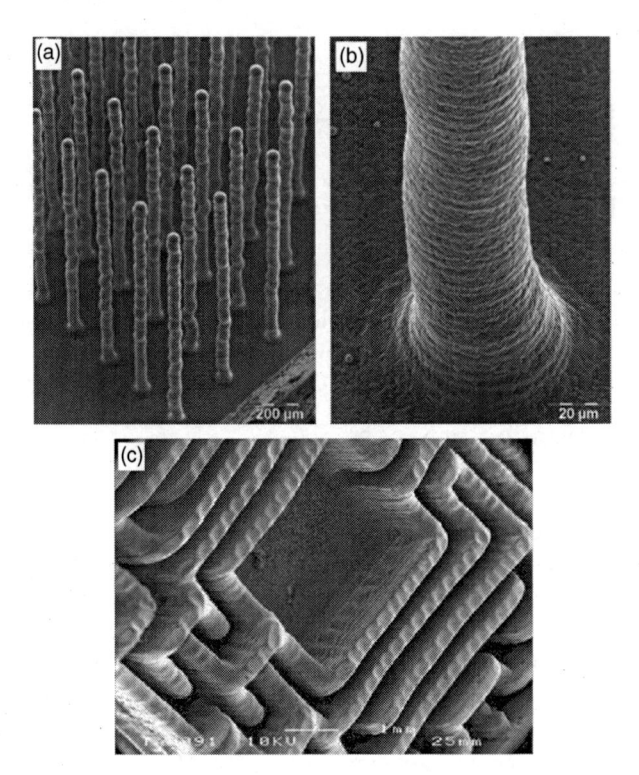

2.6 (a) and (b) SEM images of sintered TiO$_2$ micro-pillars using inkjet printer with 52 µm nozzle diameter (Lejeune *et al.*, 2009). (c) 3D ceramic micropattern made using inkjet printer from zirconia and sintered at 1450°C (Zhao *et al.*, 2002).

Piezoelectric DOD printers that print molten waxes at about 120°C have also been used to deposit suspensions with up to 40 vol.% ceramic powder (Seerden *et al.*, 2001). Figure 2.6 depicts some micropatterns produced by ceramic inkjet printing.

The physical properties of the chosen ink are probably the most vital aspects of inkjet printing. Viscosity, surface tension and inertia are the three main factors which affect behaviour of droplets and liquid jets. The viscosity of ink should be adequately low since the power produced by the piezoelectric diaphragm is limited. On the other hand, surface tension should be sufficiently high to avoid ink dripping from the nozzle. Some dimensionless parameters such as Reynolds number (*Re*), Weber number (*We*) and Ohnesorge number (*Oh*) are used for describing and analysing jetting and breakup phenomena in droplet generation. The Reynolds number is a characteristic which describes the ratio between inertial and viscous forces and is obtained by $Re = \rho d v/\eta$, where ρ is fluid density, d is specific length (droplet diameter), v is fluid velocity and η is dynamic viscosity. Weber number

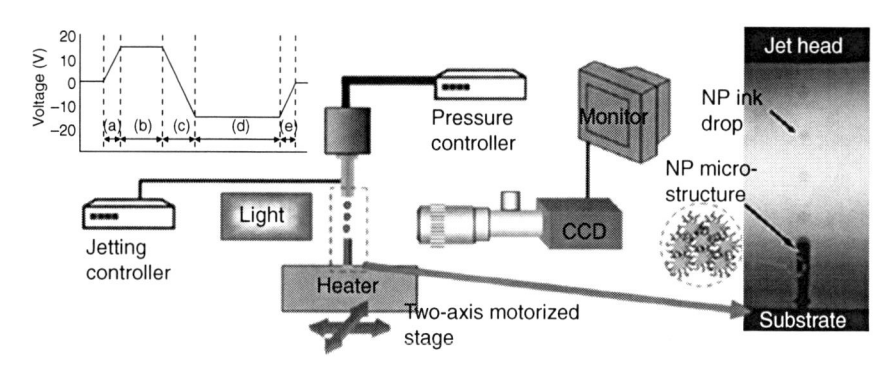

2.7 Typical piezoelectric DOD printing system for nanobioparticle printing (Ko et al., 2010).

is a characteristic which describes the ratio between kinetic energy and surface energy and is obtained by $We = \rho d v^2/\sigma$, where σ is surface tension. In addition, the Ohnesorge number is a characteristic which describes the relative importance of viscous and surface forces and is obtained by $Oh = We^{1/2}/Re = \eta/(\rho \sigma d)^{1/2}$ (Hon et al., 2008). According to research work by Wang and Derby (Tianming and Derby, 2005), for $Oh > 1$ fluid viscous dissipation results in nozzle clogging and impedes ejection of drops and also for $Oh < 0.1$ multiple drops are produced instead of a single, well-defined drop. So in practice, jettability criterion for precision DOD printing is $1 > Oh > 0.1$ and correspondingly droplet velocity should be 5–10 m/s. It should be noted that for non-Newtonian fluids other parameters such as the Weissenberg number (Wi) are used to consider the effects of viscoelasticity. The Wi value can be obtained from $Wi = tv/d$, where t is a characteristic relaxation time of droplet (Hon et al., 2008).

There has been a significant trend towards inkjet printing of inks containing bio-NPs in recent years. A large proportion of the atoms is in the surface of NPs which results in favourable properties such as a reduction in the melting point of metal biomaterials. The size of NPs should normally be 100 times less than the diameter of the jetting nozzle to avoid nozzle clogging (Kosmala et al., 2011). At the same time, ink containing bio-NPs should be non-viscous, and volatility of solvent should be adequately low to prevent nozzle clogging. Figure 2.7 shows a typical experimental set-up for printing inks containing NPs using piezoelectric DOD. The build plate can move in x and y directions and NP droplets can be observed via a CCD camera. In such a DOD experimental system, droplets are ejected via voltage waveform changes (Fig. 2.7, inset diagram). In short, the first rising voltage expands the glass capillary and a droplet is pushed through the nozzle due to the falling voltage. The final rising voltage cancels some of the residual acoustic oscillations that remain after droplet ejection and may cause satellite droplets.

The CCD camera captures images at the droplet generation frequency (Ko et al., 2010).

Droplet formation, its break-up and corresponding tail are related to the viscosity of the nanoparticle-based ink; the shape of the droplets (i.e., spherical) is influenced by surface tension. As for inks containing bio-NPs, viscosities in the range of 2–30 mPa s and surface tensions up to 60 mN/m are acceptable (Magdassi, 2010). In the meantime, proper substrate temperature allowing sufficient drying of the bio-nanoinks in each layer is essential for successful 3D printing. In short, the basic conditions required for successful 3D bio-nano inkjet printing are: ink properties (viscosity, surface tension); jetting parameters (signal width, voltage magnitude, jetting frequency); and environment (pressure, environmental and substrate temperature, humidity) (Ko et al., 2010).

Overcoming the strong agglomeration of the NPs in solution is the main challenge in making printable and stable bio-nanoink. Basically, nanoparticles tend to aggregate and cluster which results in fewer, larger particles in the solution, viscosity increment and fluctuation during storage. Viscosity measurement during storage time is normally performed to determine agglomeration rate and to investigate stability of nanoinks. A well-dispersed nanoink should be stable for at least one week at room temperature with no particle sedimentation. Some surface modifications on bio-NPs can be performed to avoid or delay aggregation of particles. For example, gold nanoparticles have been protected via coating with two different polymers, namely, poly(vinylpyrrolidone) (PVP) and acrylic resin on the surface of the particles to make the ink stable for a long time (1 year) even at gold concentration higher than 20% (Wenjuan et al., 2010).

To date, different bio-nanoinks have been successfully processed by inkjet printing processes. Nanobioceramic ink containing nanotitanium dioxide (TiO_2) has been inkjet printed on glass substrate (Hosseini and Soleimani-Gorgani). Gold (Fuller et al., 2002) and silver (Kosmala et al., 2011) NPs have been extensively investigated for inkjet printing. Hwan Ko et al. (2010) used ink containing gold NPs to produce true 3D parts including micro-pillar arrays, micro-helix, and micro-zigzag using linear and rotary tables. Inkjet printing of nanoinks containing single-wall carbon nanotubes (SWCNTs) (Chen et al., 2010; Nobusa et al., 2011, Song et al., 2008), MWCNTs (Kordas et al., 2006) and graphene (O'Connell et al., 2008) has also been reported. SWCNT and graphene are normally dispersed in dimethylfolmamide (DMF) (Song et al., 2008) and dichloroethane (DCE)/poly (mphenylenevinyleneco-2,5-dioctoxy-p-phyenylene) (PmPV) (O'Connell et al., 2008) suspensions, respectively, to make a stable ink for inkjet printing. In the meantime, sonication and centrifugation should be applied to remove heavy particles. Inkjet printing of antibiotic- and calcium-eluting 2D micropatterns was explored by Yexin et al. (2012) as a novel approach for

Freeform fabrication of nanobiomaterials using 3D printing 29

facilitating osteogenic cell development on orthopaedic titanium implant surfaces and preventing the formation of biofilm colonies. Using a commercial inkjet printer (Dimatix Materials Printer, DMP2800, FujiFilm Dimatix, Santa Clara, CA), circular dots with ~50 μm diameter were printed in arrays with ~150 μm distance from inks containing rifampicin (RFP) and PLGA dissolved in an organic solvent with ~100 nm biphasic calcium phosphate (BCP) NPs suspended in the solution.

2.3.2 Aerosol jet process

The aerosol jet process is a type of direct writing method which uses a focused aerosol stream instead of liquid ink droplets (as is used in inkjet printing) to deposit a wide range of materials. The process was developed and commercialized by OPTOMEC® under the trademark of M^3D which stands for Maskless Mesoscale Material Deposition. Figure 2.8 depicts a schematic of the aerosol jet printing process. First, composite suspension is aerosolized in an atomizer to make a dense aerosol of tiny droplets (normally 1–5 μm in diameter but droplets as fine as 20 nm have been obtained). Next, the aerosol is transported to the deposition head via a carrier gas flow (usually N_2 gas flow), and within the aerosol head, the aerosol is focused using a flow guidance deposition head, which creates an annular flow of sheath gas to collimate the aerosol. The high velocity co-axial aerosol stream is sprayed onto a substrate layer by layer (minimum layer thickness of 100 nm) to create 3D parts (Hon *et al.*, 2008). The high exit velocity of the aerosol stream enables a relatively large separation between the print head and the substrate, typically 2–5 mm. The aerosol stream remains tightly focused over this distance, resulting in the ability to print conformal patterns on 3D substrates. Writing speeds of up to 200 mm/s, line widths from 5 μm to 5 mm, inks with viscosity from 0.7 to 2500 mPa s and maximum

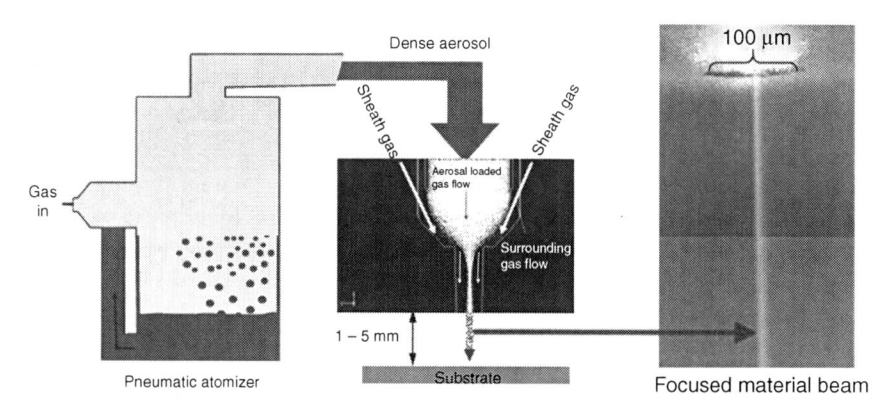

2.8 Schematic illustration of aerosol jet process. (*Source*: Courtesy of OPTOMEC Inc.)

volumetric deposition rate of 0.25 mm^3/s have been reported. Depending on the ink and substrate materials used, furnace, infrared laser and UV-curing (for polymers) can be used post-processing to achieve the desired mechanical and electrical properties.

Since aerosol jet printing is a low temperature process and the droplet size is of the order of a few femtolitres, it is a good candidate for biomanufacturing. The kinetic energy of droplets is so small that it will not demolish living cells due to their tiny mass. Aerosol jet inks can include polymers, ceramics, metals and biomaterials in the form of solutions, nanoparticle suspensions, etc. Materials including metals (bio-nanoinks containing Ag, Au and Pt NPs, Pd and Cu inks), resistors (carbon polymer thick film (PTF), ruthenium oxide), dielectrics (polyimide, polyester, polytetrafluoroethylene (PTFE), etc.) and biomaterials (such as protein and antibody solutions, DNA and biocompatible polymers like PLGA) have been employed successfully in the aerosol jet process (Hon et al., 2008).

Aerosol jet printing was first developed for 2 and 2.5D direct writing purposes, but with recent process developments there is possibility of using this process efficiently for true 3D nanobiomaterials manufacturing. Typical characteristics of nanoparticle based inks for aerosol jet systems are: solvent with low evaporation rate; NP size should be less than 500 nm (<200 nm preferred); solids content within the range 5–70 wt.% is possible; and viscosity of ink should be within the range of 1–1000 cP at ambient temperature. Aerosol jetting has been used successfully to produce bioceramic/polymer nanocomposite scaffolds for bone TE applications. Liu and Webster (2011) reported the use of an aerosol jet process for fabrication of 3D nanostructured titania/PLGA nanocomposite scaffolds for orthopaedic applications. *In vitro* cytocompatibility tests were conducted and the results demonstrated that the 3D nanocomposite scaffold they produced enhanced osteoblast infiltration into porous 3D structures in comparison to previous nanostructured surfaces. SWCNT inks have been formulated successfully as well for different biomedical applications.

2.4 Nozzle-based SFF techniques

The third group of SFF systems suitable for processing of nano-biomaterials is nozzle-based SFF techniques. This section will address these processes and provide detailed descriptions of their applications and their main advantages and limitations

2.4.1 Nanocomposite deposition system (NCDS)

The NCDS process was developed by Won-Shik and co-workers (2007) to overcome the recoating problem of nanofiller-loaded resin in SL. This process uses a nozzle to deposit and cure bio-nanoparticle filled biodegradable resin layer by layer. NCDS is a hybrid SFF technique which consists of two main operations in each layer, including nanoparticle-filled resin deposition and further material removal using micro-machining. Biocompatible UV-curable resins are used as matrix and various bio-nanofillers are used to form composite materials. Figure 2.9 shows a typical NCDS hardware and schematic NCDS process sequence.

The machine (1) deposits a thin layer of UV-curable resin containing bio-nanofillers onto the substrate using a micro-needle, (2) solidifies the deposited resin via UV (λ = 365 nm) and (3) removes the unnecessary deposited materials via micro-machining. The nanocomposite can be deposited into 10–100 μm layers to produce a near-net shape and further precise micro-machining is used to obtain the net shape. NCDS shows anisotropic compressive properties like some other SFF processes such as FDM and 3D printing (Ahn *et al.*, 2007b).

Different nanobiomaterials have been processed by NCDS to produce nanocomposites. Biocompatible acrylated polyurethane resin is normally used as a matrix of nanocomposite with different nanofillers such as HA

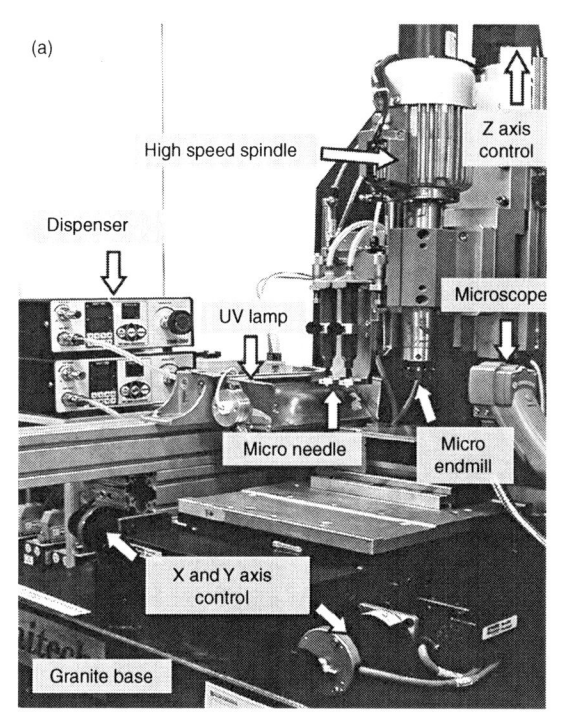

2.9 (a) NCDS hardware, (b) components of NCDS deposition head, (c) fabrication process sequence (Ahn *et al.*, 2007b).

(*Continued*)

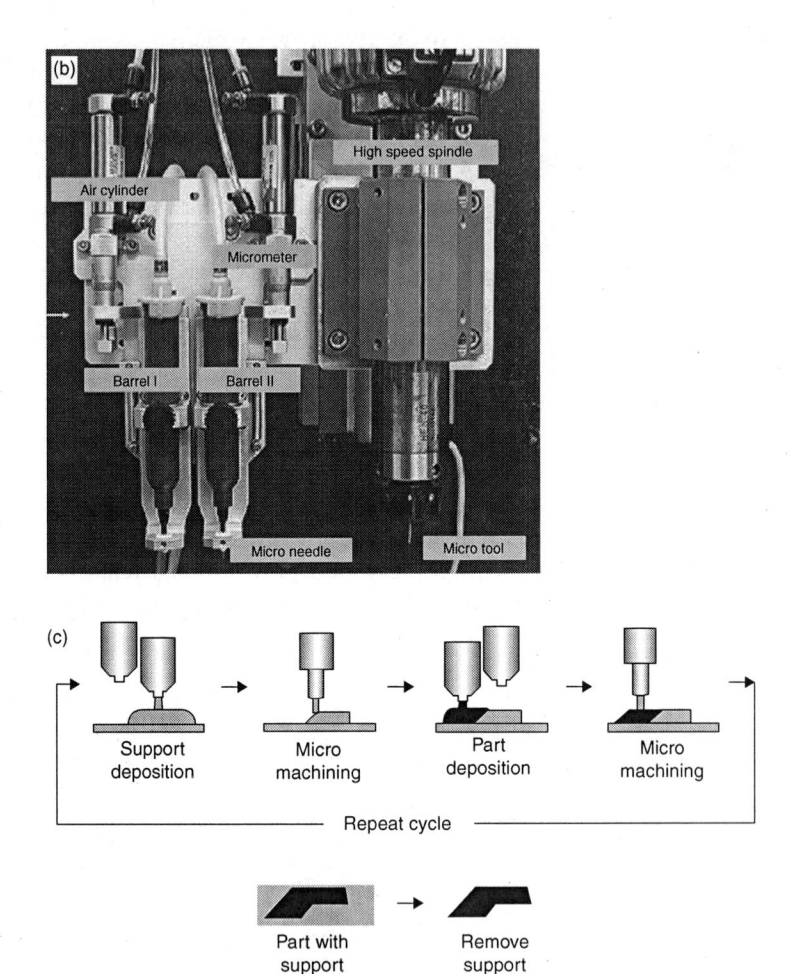

2.9 Continued

NPs and MWCNT, etc. Ahn and co-workers (2007) mixed acrylated polyurethane resin with HA NPs with an average diameter of ~100–300 nm and MWCNT with an average diameter of about 40 nm and an aspect ratio of over 1000. Both composite mixtures had a viscosity near 100 000 cPs (±20 000), which could flow through the micro-needle when pressurized by the dispensing system. Nanocomposite scaffold-type drug delivery devices have also been fabricated with the use of PLGA as a biodegradable base polymer, HA NPs as a release modifier and 5-fluorouracil (5-FU) as a model drug (Chu et al., 2007; Lee et al., 2008; Park et al., 2010).

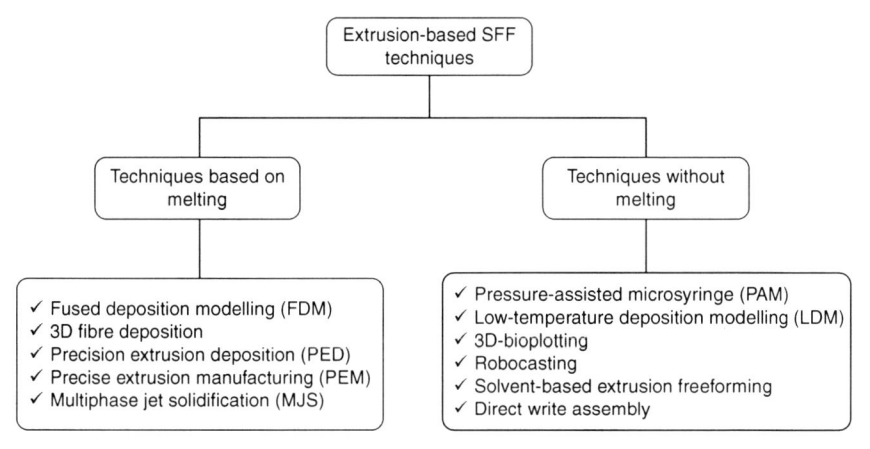

```
                    ┌─────────────────────┐
                    │ Extrusion-based SFF │
                    │     techniques      │
                    └──────────┬──────────┘
                 ┌─────────────┴─────────────┐
    ┌────────────┴─────────┐      ┌──────────┴───────────┐
    │  Techniques based on │      │  Techniques without  │
    │       melting        │      │       melting        │
    └────────────┬─────────┘      └──────────┬───────────┘
```

- Techniques based on melting:
  ✓ Fused deposition modelling (FDM)
  ✓ 3D fibre deposition
  ✓ Precision extrusion deposition (PED)
  ✓ Precise extrusion manufacturing (PEM)
  ✓ Multiphase jet solidification (MJS)

- Techniques without melting:
  ✓ Pressure-assisted microsyringe (PAM)
  ✓ Low-temperature deposition modelling (LDM)
  ✓ 3D-bioplotting
  ✓ Robocasting
  ✓ Solvent-based extrusion freeforming
  ✓ Direct write assembly

2.10 Different extrusion-based freeforming techniques.

2.4.2 Extrusion-based SFF systems

The term 'direct ink writing' was first proposed by Lewis (Lewis and Gratson, 2004) for a class of manufacturing techniques that utilize a computer-controlled ink deposition nozzle to create patterns and 3D objects with controlled composition and architecture. Direct ink writing techniques are classified into two main sub-groups: extrusion-based and droplet-based systems. Droplet-based systems deposit inks in the form of droplets to form final patterns and shapes. Inkjet printing and aerosol jet writing discussed earlier are the two main droplet-based direct ink writing systems. Extrusion-based systems have the same working principle barring that they deposit material in a continuous flow and are characterized by an extensive diversification.

Extrusion-based systems can basically be classified into two main subgroups as shown in Fig. 2.10: processes based on material melting and processes without material melting. Fused deposition modelling, precision extrusion deposition (PED) (Wang *et al.*, 2004), 3D fibre deposition (Woodfield *et al.*, 2004), precise extrusion manufacturing (PEM) (Xiong *et al.*, 2001) and multiphase jet solidification (MJS) (Greulich *et al.*, 1995) are SFF techniques based on material melting. Pressure-assisted microsyringe (PAM) (Vozzi *et al.*, 2002); low-temperature deposition manufacturing (LDM) (Zhuo *et al.*, 2002); 3D-bioplotting (Landers and Mulhaupt, 2000); robocasting (Cesarano, 1999); direct-write assembly (pH-controlled gelled ceramic colloid or polymers freeforming) (Smay *et al.*, 2002); and solvent-based extrusion freeforming (Grida and Evans, 2003) are the most commonly used SFF techniques without material melting. Four major nozzle designs have been exploited in non-heating processes: pressure-actuated, volume-driven injection nozzles (normally using a stepper motor), solenoid and piezoelectric-actuated, whereas the two main nozzle designs including filament driving wheels, and mini-screw extruder have been used in processes with material melting. More details on different extrusion-based freeforming techniques are provided in Section 2.5.2.

Much attention has been paid to extrusion-based systems in recent years as they are mechanically simple processes in comparison to other SFF techniques, and a wide range of biomaterials can be processed effectively. Different nanofillers have been employed to improve mechanical properties, bioactivity, etc. Nanofibre-reinforced biopolymers are fabricated by FDM so that nanofibres are well distributed in the matrix with no porosity in the bio-nanocomposite. Pressure-assisted microsyringing has been used to fabricate nanoHA/PCL bio-nanocomposite bone TE scaffolds (Heo et al., 2009). The use of HA NPs (particle size 20–90 nm) resulted in a higher compressive module and better attachment and proliferation of MG-63 cells in comparison to composite PCL scaffold fabricated using micron size HA particles and conventional PCL scaffold. Ye et al. (2010) reported the use of a PED technique to build bone bio-nanocomposite scaffold from nano-non-stoichiometric apatite (ns-AP) and PCL matrix. Jinku et al. (2012) used FDM to build PLGA/β-TCP and HA nanocomposite scaffolds. Mattioli-Belmonte et al. (2012) fabricated PCL/CNT composite scaffolds using a PAM process and reported the mechanical, thermal and biological characterization of the produced scaffold in which both the quantity of nanotubes in the matrix as well as the scaffold design were varied in order to tune the mechanical properties of the material. Their PCL/CNT nanocomposites were able to sustain osteoblast proliferation and modulate cell morphology. Their work shows the potential of custom designed CNT nanocomposites for bone TE. Dorj et al. (2012) used robocasting to produce a novel nanocomposite scaffold made of chitosan and nanobioactive glass (nBG) retaining a dual-pore structure. Robocasting was carried out under a cooled bath containing dry ice, to rapidly solidify the scaffold because the dispensed solution was hard to solidify at ambient conditions. The chitosan/nBG nanocomposite scaffolds were well-constructed, with an aligned, macro-channelled pore structure (Fig. 2.11a). As seen in Fig. 2.11b, the scaffold filament presented a highly microporous structure, with pore sizes of a few to 10 μm, demonstrating micro/macro dual-pore structure. Closer examination of the surface revealed the presence of the nanofibrous component of the nBG (arrowed in Fig. 2.11c) embedded within and enclosed by the chitosan dense matrix. The pure chitosan scaffold robocast under the same conditions as the chitosan/nBG scaffold also showed a highly microporous and macro-channelled structure (Fig. 2.11d). The chitosan micropores were shown to be more elongated and a little larger than those of the nanocomposite scaffold (Fig. 2.11e) (Dorj et al., 2012).

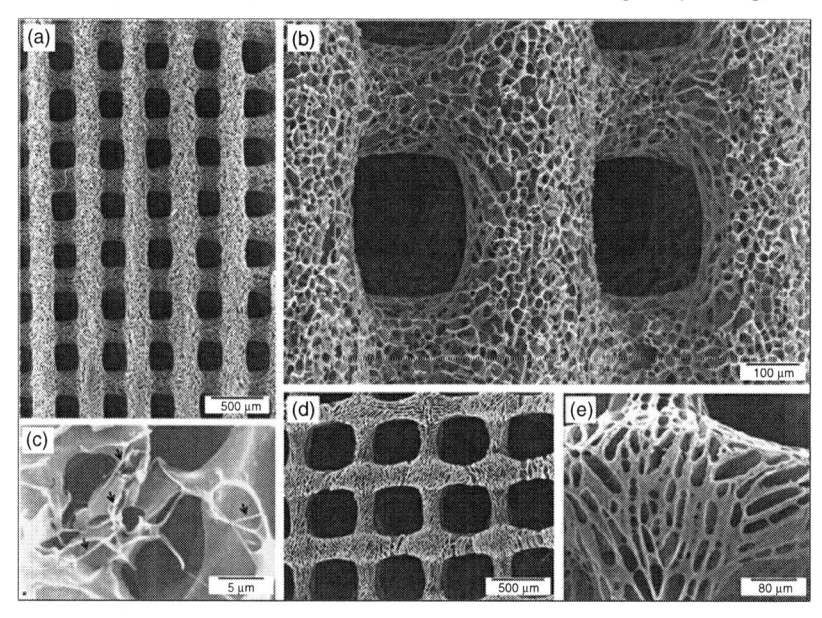

2.11 SEM morphologies of the scaffolds produced by a robocasting process in a cooled bath containing dry ice; (a–c) chitosan/nBG and (d, e) chitosan. Micropores were well-developed in both types of scaffolds, through the rapid-solidification of ice crystals and the subsequent freeze-drying process. The existence of nBG was noticed in the nanocomposite scaffold (c) (Dorj *et al.*, 2012)

2.5 Extrusion freeforming of biomaterials scaffold

One of the principle methods behind tissue engineering involves growing the relevant cells *in vitro* into the required 3D organ or tissue. However, cells are unable to grow in favoured 3D orientations and thus define the anatomical shape of the tissue. Instead, they randomly migrate to form a two-dimensional (2D) layer of cells. Scaffold is a porous structure that acts as a synthetic extracellular matrix (ECM), which permits cells to grow in favoured orientations and facilitates cell adhesion, proliferation and differentiation (Sachlos and Czernuszka, 2003). In the following sections the main procedures for scaffold design and fabrication using extrusion freeforming techniques will be described.

2.5.1 Scaffold materials and macro-microstructure design

One of the most important challenges in TE is to design an ideal scaffold. In addition to mimicking the structure and biological functions of ECM, TE scaffolds should provide a good environment for cell attachment, proliferation and differentiation. In this way, several aspects should be taken into consideration in the design of scaffolds for TE. In addition to being biocompatible (both in bulk and degraded form), TE scaffolds should provide a correct stress environment for the neotissues through their appropriate mechanical properties. Scaffolds also need to be sufficiently porous to allow the introduction of cells and nutrients and removal of waste materials, and should possess an appropriate surface chemistry for cell attachment. Two main aspects for scaffold design are scaffold material and scaffold macro-microstructure.

Selection of suitable biomaterials for fabrication of the TE scaffolds is a significant step as the scaffold characteristics are greatly determined by the inherent characteristics of the scaffold materials. Basically, two main requirements for scaffold material are: biocompatibility (i.e., not to stimulate any unwelcome tissue response to scaffold, and simultaneously to have an appropriate surface chemistry so as to assist cell attachment and proliferation) and biodegradability (i.e., capable of being broken down into non-toxic products after a specific period). TE scaffolds are normally produced from polymers, ceramics or polymer/ceramic composites.

To date, bioceramics such as the calcium phosphate family, including HA, β-TCP and bioactive glasses, have demonstrated good bioactivity and biocompatibility and are widely used as artificial bone matrix and filler material for bone injury repair. Bioceramics suitable for TE scaffolds can be classified into three main groups: bioinert ceramics such as alumina and zirconia; bioactive ceramics such as bioglass, sintered HA (s-HA) and alumina-wollastonite glass ceramic (AWGC); and bioresorbable ceramics such as sintered HA (u-HA), α or β-TCP, tetracalcium phosphate (TTCP) and octacalcium phosphate (OCP). Bioresorbable ceramics are an appropriate choice of bioceramics as reinforcing element where biodegradability is important. Their rates of biodegradation are in the following order: OCP > α-TCP > β-TCP > u-HA (Yang et al., 2001).

Natural polymers including chitosan, glycosaminoglycan, collagens, starch and chitin have been exploited for regeneration of different tissues such as cartilage, bone, nerves and skin. While naturally occurring biomaterials may most closely simulate the native cellular milieu, the main limitation for their wider application is large batch-to-batch variations upon isolation from biological tissues (Yang et al., 2001). Apart from that, scaffolds made from natural polymers such as collagen and chitin had poor and inadequate mechanical performance. To overcome these limitations of natural polymers, synthetic resorbable polymers such as polyphosphazens, polyanhydrides (PAs), poly (a-hydroxy esters), and polyorthoesters were utilized. Poly(a-hydroxy ester)s and copolyesters of lactic acid and glycolic acid form a significant group of synthetic biodegradable polymers. Polylactic acid (PLA), polyglycolide (PGA), PCL, PAs, polyorthoesters, polydioxanone and copolymers thereof are biodegradable, synthetic polymers which have been used for years in surgical sutures, and have a long and proven clinical record. Some biodegradable polymers such as PLLA can be used in composite systems with ceramic fillers due to their initial high strength. PLLA and PGA exhibit a high degree of crystallinity and degrade relatively slowly, while copolymers of PLLA and PGA (i.e., PLGA) are amorphous and degrade rapidly (Yang et al., 2001). Table 2.1 shows the properties of some biodegradable polymers.

Scaffolds made from some other biodegradable polymers such as PPF, poly (1,8)octanediol citrate (POC), and poly (glycerol-sebacate) (PGS) have exhibited good results in terms of mechanical properties. Hydrogel can also be used for tissue regeneration scaffolds; it is a biodegradable biomaterial in the form of a colloidal gel in which water is the dispersion medium and it is formed by the cross-linking network of hydrophilic polymer chains (Bartolo and Bidanda, 2008; Nguyen and West, 2002). The structure of hydrogels can be compared with elastin and collagen which form the natural tissue. Hydrogels are normally formulated from a wide range of materials including silicon, cellulose derivatives, poly (vinyl alcohol), poly (ethylene glycol), calcium alginate and the most widely used poly (hydroxyethyl methacrylate) (PHEMA) (Bartolo and Bidanda, 2008).

Apart from the advantages mentioned for polymers, they are ductile and not adequately rigid for some TE applications. Scaffolds with properties closer to natural load bearing tissues can be obtained by combining polymers with bioceramics which are too stiff and brittle by themselves.

Biocompatible metals such as stainless steels, and cobalt- and titanium-based alloys have been employed extensively for different biomedical applications like surgical implants. However, lack of biodegradability and processability are the two main obstacles for biocompatible metals in TE applications. Thus, polymers and polymer/ceramic composites have been considered more by researchers for TE scaffold applications.

Both macro- and micro-structure of scaffolds need to be designed based on the desirable performance and application. Scaffold may have simple or complicated macrostructure depending on the application. Macrostructure would be complex where reconstruction of a damaged organ/tissue of a patient is required on the basis of acquired medical images such as magnetic resonance imaging (MRI) or computed tomography (CT) scans. The main advantage of SFF techniques is that they are able to fabricate TE scaffolds with both predefined macro- and micro-structures.

It has been proved that porosity and pore size of the supporting 3D structures are two important factors which affect regeneration of specific tissues using synthetic substances. Cell attachment and ingrowth are promoted by increasing surface area. Highly porous scaffolds are favourable for improving nutrient diffusion and removing waste products and also for more efficient vascularization. In particular, mass transport control is an important

Table 2.1 Properties of biodegradable polymers suitable for TE scaffolds

Polymer	Melting point (°C)	Glass transition temperature (°C)	Degradation time (months)[a]	Density (g/cm^3)	Tensile strength (MPa)	Elongation (%)	Modulus (GPa)
PLGA	Amorphous	45–55	Adjustable	1.27–1.34	41.4–55.2	3–10	1.4–2.8
DL-PLA	Amorphous	55–60	12–16	1.25	27.6–41.4	3–10	1.4–2.8
L-PLA	173–178	60–65	>24	1.24	55.2–82.7	5–10	2.8–4.2
PGA	225–230	35–40	6–12	1.53	>68.9	15–20	>6.9
PCL	58–63	−65	>24	1.11	20.7–34.5	300–500	0.21–0.34

[a] Time to complete mass loss.
Source: Yang et al., 2001.

issue for bone TE scaffolds since the high rates of nutrient and oxygen transfer at the surface of the scaffold promote the mineralization of the scaffold surface, further limiting the mass transfer to the interior of the scaffold (Sachlos and Czernuszka, 2003). As a consequence, cells would be able to survive only close to the surface. Using larger pore size in the external areas of the scaffold would not be an effective measure since it may result in degradation of mechanical properties of the bone scaffolds. The minimum pore size of a scaffold is usually defined by the diameter of cells in suspension which differs from one cell type to another. Experiments show optimum pore size of 5 μm for neovascularization, 5–15 μm for fibroblast ingrowth, close to 20 μm for the ingrowth of hepatocytes, 20–125 μm for regeneration of adult mammalian skin, 40–100 μm for osteoid ingrowth and 100–350 μm for regeneration of bone (Yang et al., 2001). For rapid vascularization and survival of transplanted cells fibrovascular tissues need to possess pore sizes greater than 500 μm (Wake et al., 1994). Interconnectivity of pores within the scaffold is another important issue that should be taken into consideration. If the pores are not interconnected within the scaffold, mass transport (that is permeability and diffusion) and cell migration will not happen appropriately, even if the porosity of the scaffold is high.

Techniques in macro-microstructure design for patient-specific scaffold must be able to firstly offer hierarchical porous structures so that the required mechanical function and mass transport are satisfied and, secondly, these structures must be embedded in complicated and arbitrary 3D anatomical shapes. Computational topology design (CTD) is an effective design procedure which can be integrated with SFF for design and fabrication of TE scaffold. CTD-based scaffold design may start with the creation of unit cell libraries that can be assembled to form scaffold architectures. Unit cells libraries may be created by either using approaches based on CAD (Cheah et al., 2004; Fang et al., 2005; Van Cleynenbreugel et al., 2002) or using image-based design approaches (Hollister et al., 2000, 2002; Lin et al., 2004b). Homogenization theory (Hollister and Kikuchi, 1994) can then be exploited to calculate effective properties based on these unit-cell designs. Surpassing effective property calculation from defined microstructures, topology optimization approaches (Lin et al., 2004a; Sigmund, 1994) actually compute new microstructures to attain desired properties. There is the possibility of either optimizing functional elastic properties with a constraint on porosity or maximizing permeability with a constraint on required elastic properties (Hollister, 2006). Scaffold architecture creation within a complex 3D anatomic shape is the final stage of the scaffold design process. 3D anatomic shape is generated from CT or MRI images of the patient. Both CT and MRI produce structured voxel datasets where patient anatomy is defined by density distribution. These datasets can be used in the design process either by converting the voxel anatomic data into solid geometric models for use in CAD (Cheah *et al.*, 2004; Fang *et al.*, 2005; Van Cleynenbreugel *et al.*, 2002) or by directly using voxel database structures in image-based methods (Hollister *et al.*, 2000, 2002). Finally, Boolean techniques are used to intersect the defined anatomic defected shape with the microstructure design database, resulting in the final patient-specific scaffold design (Hollister, 2006).

2.5.2 Scaffold fabrication using extrusion freeforming

Various fabrication methods including traditional chemical engineering methods and advanced SFF techniques are currently used for construction of TE scaffolds. Traditional techniques to fabricate TE scaffolds include: solvent casting/salt leaching, phase separation, foaming and textile meshes. These techniques have several limitations as they cannot usually control pore size, pore geometry or spatial distribution of pores properly. In contrast, SFF advanced techniques can simply control the internal and external structure of scaffolds and overcome some intrinsic limitations of conventional methods such as shape restrictions, manual intervention and inconsistent and inflexible processing procedures. In order to take advantage of these breakthroughs, there has been a trend in recent years towards fabrication of TE scaffolds using SFF processes directly (building the final scaffold) or indirectly (building a negative scaffold for use as a mould). In particular, extrusion-based SFF systems have been widely investigated for making TE scaffolds due to their ability to process different biomaterials, and to manufacture scaffolds in a cell-friendly environment, their high reproducibility and flexibility, and their simple process control in comparison with other SFF techniques. As mentioned earlier, extrusion-based freeforming techniques are classified into two main groups: processes with and without material melting. Figure 2.12 illustrates the working principle of the key extrusion-based methods schematically. In the following section, the principles of each process are described and the main issues associated with each process will be highlighted.

Extrusion freeforming with material melting

FDM is the first SFF system based on extrusion of polymer melts. Thermoplastic materials in the form of filament are used as feedstock and a pinch roller feed mechanism is used to push the filament into the liquefier, with subsequent extrusion from a computer-controlled nozzle. By repeating the extrusion process the final part is fabricated layer by layer. There is demand for precise temperature control system to achieve desirable accuracy. Several modified FDM systems have been developed for fabrication of 3D scaffolds with micron size pores and filaments. Recent specialized FDM systems for 3D scaffold mostly use a screw feed instead of a pinch roller feed mechanism to enhance extrusion accuracy.

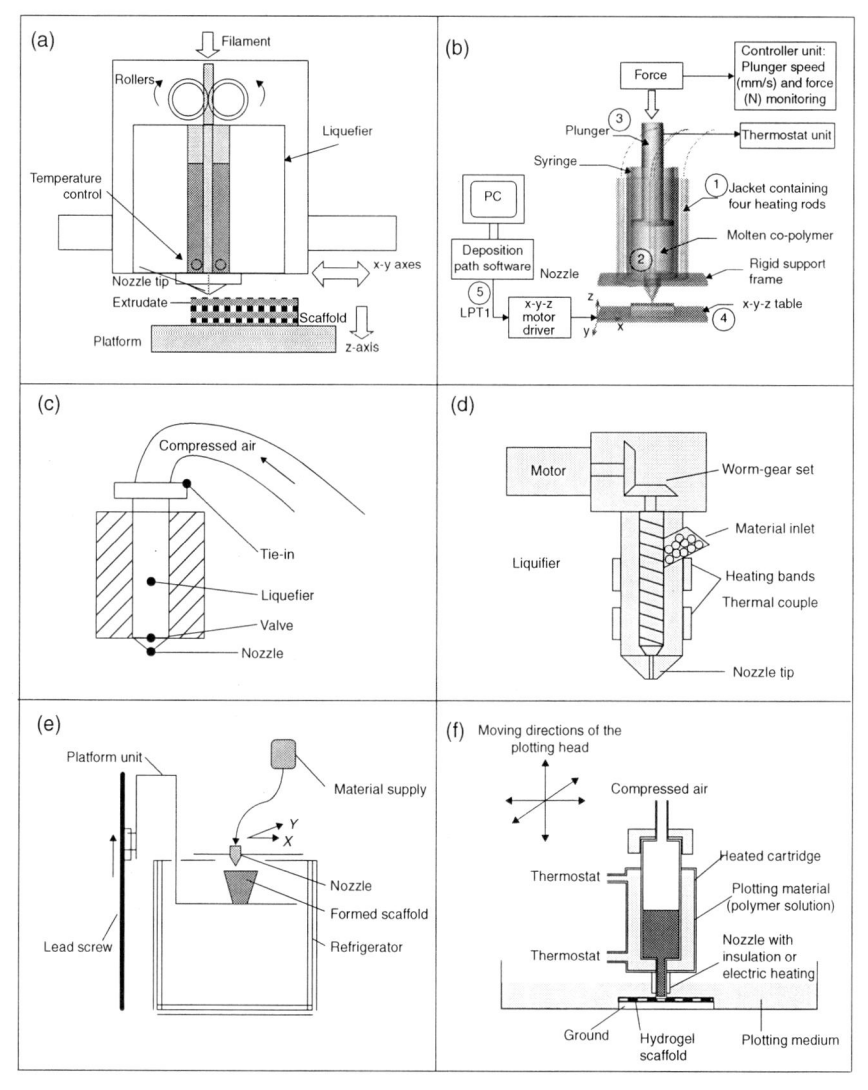

2.12 Schematic illustration of different extrusion-based systems, including processes with and without material melting: (a) FDM process (Zein et al., 2002); (b) 3D fibre deposition process (Woodfield et al., 2004); (c) PEM process (Xiong et al., 2001); (d) PED process (Wang et al., 2004); (e) LDM process (Zhuo et al., 2002); (f) 3D bioplotting process (Landers et al., 2002); (g) pressure-assisted writing processes such as PAM and direct-write assembly techniques (Vozzi et al., 2002); (h) paste extrusion techniques such as robocasting and solvent-based extrusion freeforming techniques (Miranda et al., 2008). It should be noted that robocasting process can be carried out under proper coagulation reservoirs such as oil bath or cooled bath containing dry ice, whereas solvent-based extrusion freeforming deposits biomaterials at room temperature.

(*Continued*)

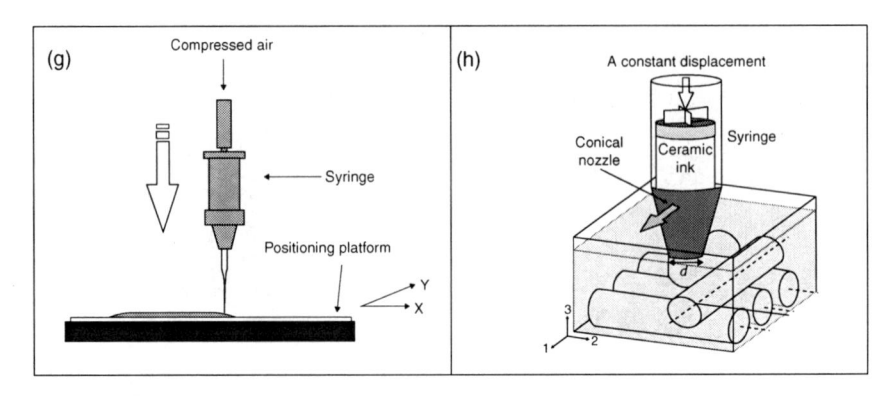

2.12 Continued

FDM has been used successfully to produce scaffolds in PCL, polypropylene (PP)/TCP, PCL/HA, PCL/TCP with a resolution of 250 μm. Hutmacher and his group (Zein *et al.*, 2002) have extensively investigated the process parameters for PCL and they fabricated several composites including PCL/HA and PCL/TCP scaffolds using FDM. Bone TE scaffolds produced from polymer and calcium phosphate (CaP) using FDM have exhibited good mechanical and degradation properties, improved cell seeding and enhanced incorporation and immobilization of growth factors. As for mechanical properties, the existence of the CaP phase brings about higher structural strength and the polymer phase provides plasticity and toughness to the scaffold. Kalita *et al.* (2003) produced controlled porosity polymer-ceramic composite PP/TCP scaffolds, with 3D interconnectivity designed to promote a richer supply of blood, oxygen and nutrients for healthy ingrowth of bone cells. Controlled porosity alumina and β-TCP ceramic scaffolds with pore sizes in the range 300–500 μm and pore volumes of 25%–45% have been produced using the indirect fused deposition process (Bose *et al.*, 2003). Safari and his group (Allahverdi *et al.*, 2001) produced a hybrid scaffold from alumina and wax (as the support structure) directly using multi-nozzle fused deposition of ceramics (FDC). Highly porous PLGA scaffolds for cartilage TE were fabricated by Hung-Jen *et al.* (2009) using FDM and were further modified by type II collagen. Tellis *et al.* (2008) used micro computed tomography (CT) to create biomimetic polybutylene terephthalate (PBT) trabecular scaffolds using an FDM process.

Two major limitations of FDM processes are the need to use filamentary materials as feedstock and the high heat effect on raw biomaterial. In particular, preparing filamentary feedstock makes it difficult/time consuming to process new biomaterial. To overcome this problem, some alternative processes with new configurations such as 3D fibre deposition, PED, PEM and MJS processes have been proposed.

In 2004, Woodfield et al. (2004) developed an extrusion-based system called 3D fibre deposition with the aim of extruding highly viscous polymer. The technique allowed them to make scaffolds by accurately controlling the deposition of molten co-polymer fibres from a pressure-driven syringe onto a computer controlled xyz table. As seen in Fig. 2.12b, the 3D fibre deposition consisted of five main components: (1) a thermostatically controlled heating jacket; (2) a molten copolymer dispensing unit consisting of a syringe and nozzle; (3) a force controlled plunger to regulate flow of molten co-polymer (4); a stepper motor-driven xyz table; and (5) a positional control unit consisting of stepper motor drivers linked to a personal computer containing software for generating fibre deposition paths. Woodfield and co-workers produced 3D poly(ethylene glycol)-terephthalate/poly(butylene terephthalate) (PEGT/PBT) block co-polymer scaffolds (Woodfield et al., 2004) and polyethyleneoxide terephthalate (PEOT) and PBT scaffolds (Moroni et al., 2006) with a 100% interconnecting pore network.

The PED process is another material melting extrusion-based process, developed by researchers at Drexel University (Wang et al., 2004). The working principle of PED is similar to FDM barring that material in the form of pellets or granules is liquefied in a chamber and a rotating screw (mini-extruder) forces the material through the nozzle. Shor et al. (2007) used this method to build PCL and PCL/HA 3D scaffolds with uniform pore size of 250 μm. The test results reported for PCL scaffolds produced by PED proved the structural integrity, controlled pore size, pore interconnectivity, favourable mechanical properties and basic biocompatibility (Shor et al., 2009). Hoque et al. (2009) developed a desktop robot-based rapid prototyping (DRBRP) system with the ability to process a wide range of synthetic polymers in the form of pellet, lump or powder to build 3D scaffolds. Their biocompatibility tests using rabbit smooth muscle cells proved excellent performance of fabricated scaffolds in terms of cell adhesion and tissue formation.

Xiong et al. (2001) have developed a PEM process in which compressed air is used instead of a piston or rotating screw to push the melted biomaterial through the deposition nozzle. With the computer-controlled digital valve upon the deposition nozzle, the switch response speed is high and filaments can be deposited with sufficient accuracy onto the substrate. Xiong and his group fabricated different porous PLLA and PLLA/TCP bone tissue engineering scaffolds with different properties and with controlled architecture and geometry through this PEM process.

The MJS process was developed at the Fraunhofer-Gessellschaft research centre to produce high density metallic or ceramic parts using low melting point alloys or a powder-binder mixture (Greulich et al., 1995). Heated paste is pushed out through a nozzle and deposited onto a computer-controlled build table. The feedstock is normally supplied as powder, pellet or bar and the extrusion temperature of the molten material can reach up to 200°C. Powder-binder feedstock is heated in a process chamber above the melting point of the binder, and thus only the binder is liquefied during the process. A piston is used to push out the low viscous flow through the nozzle and the material is deposited layer by layer. MJS was used to build 3D scaffolds made of poly (D, L-lactide) (PDLLA) for bone and cartilage tissue engineering. The scaffold pore size was found to be in the range 300–400 μm and the structure supported ingrowth of human bone tissue (Koch et al., 1998).

The Polytechnic Institute of Leiria developed a variation of FDM called BioExtruder for producing PCL scaffolds (Domingos et al., 2009; 2012). BioExtruder comprises two different deposition systems: one rotational system for multi-material deposition acted by a pneumatic mechanism and another one for a single material deposition that uses a screw to assist the deposition process.

Extrusion freeforming without material melting

New configurations for melt extrusion could open up the possibility for the use of a wider range of biomaterials, making the extrusion-based systems a more versatile and realizable alternative manufacturing process for composite scaffold materials. But limitations remain in terms of the high heat effect on raw biomaterial. Thus, researchers have made attempts to develop new configurations to process biomaterials without melting that can better preserve the bioactivities of the scaffold materials.

The PAM process is a technique developed by Vozzi and co-workers, (2002) which resembles FDM without the need for heating. PAM uses a pneumatic driven microsyringe to deposit biomaterial on a substrate. Material viscosity, deposition speed, tip diameter and the applied pressure correlate with the final deposited strand dimensions (Vozzi et al., 2003). Polymeric scaffolds with different polymer compositions such as PCL, PLLA, PLGA, PCL/PLLA, gelatin and alginate hydrogel scaffolds with three different geometries – square grids, hexagonal grids and octagonal grids were produced (Mariani et al., 2006; Tirella et al., 2008, 2009; Vozzi and Ahluwalia, 2007; Vozzi et al., 2004). Apart from TE scaffolds, PAM was used to deposit a polyurethane dielectric layer and a carbon black electrode layer above it (Tartarisco et al., 2009). Vozzi and his group used a modified system called piston assisted microsyringe (PAM2) for microfabrication of viscous, sol–gel or gelled inks (e.g., alginate solutions at different concentrations). PAM2 uses a stepper motor instead of compressed air to move the syringe plunger with a controlled speed (Tirella et al., 2012). PAM2 also has a temperature controlled syringe (TCS) module to control the temperature of processed materials using an aluminium jacket.

The key feature of the LDM process proposed by Xiong *et al.* (2002) is that it is a non-heating liquefying process. In it, material slurries are fed into the material supply that is connected to a screw pump nozzle using a soft pipe and the fabrication process is accomplished in a low temperature environment below 0°C in the freezer. The layer of deposited materials is frozen on the platform. After the forming process, the frozen scaffolds formed by the LDM system need to be freeze-dried for rather a long time (~38 h) to remove the solvent. The bone scaffolds made by this LDM system have good biocompatibility and bone conductive property as a molecular scaffold for bone morphogenic protein in the implantation experiments of repairing segment defects of rabbit radius (Yongnian *et al.*, 2003). Biomolecules can be applied in the LDM process to directly fabricate a bioactive scaffold. Incorporating multiple nozzles with different designs into the LDM technique gave rise to multinozzle low-temperature deposition and manufacturing (M-LDM) and multinozzle deposition manufacturing (MDM) (Liu *et al.*, 2009a, 2009b). The M-LDM system is proposed for fabricating scaffolds with heterogeneous materials and gradient hierarchical porous structures by the incorporation of more jetting nozzles into the system. The LDM process has been used to build multi-material (Liu *et al.*, 2009a), and different hydrogel scaffolds (Li *et al.*, 2009a, b). Cong Bang *et al.* (2008) developed a special LDM system based on rapid freeze prototyping (RFP) to produce scaffolds from chitosan solution.

3D bioplotting is a technique that was first developed by Landers and Mulhaupt (2000) at Freiburger group to produce scaffolds for soft tissue engineering purposes, and simplifying hydrogel manufacture. In this process, the material dispensing head normally moves in three dimensions, while the fabrication platform is stationary. Either a filtered air pressure (pneumatic nozzle) or a stepper motor (volume-driven injection nozzle) is used to plot a viscous material into a liquid (aqueous) plotting medium with a matching density. It is possible to perform either discontinuous dispensing of microdots or continuous dispensing of fine filaments. 3D bioplotting can process thermally sensitive biocomponents and cells since heating is not applied. Curing reactions can be performed by plotting in a co-reactive medium or by two-component dispensing using mixing nozzles. The filament thickness can be adjusted by varying the viscosity of the plotting solution, nozzle diameter and the applied pressure (Billiet *et al.*, 2012). Further surface treatment is normally applied to the scaffolds produced by 3D bioplotting as they mostly have smooth surfaces that are undesirable in terms of cell attachment. Geun Hyung and Joon Gon (2009) used a piezoelectric transducer (PZT) generating vibrations during plotting to make PCL scaffolds with a rough surface. Maher *et al.* (2009) developed a device based on bioplotting with the ability to heat the plotting materials and produced TE scaffolds using a variety of materials including poly (ethylene glycol) (PEG), gelatin, alginic acid and agarose at various concentrations and viscosities.

In comparison with other extrusion-based SFF processes, 3D bioplotting can process a remarkably wide variety of different biomaterials, including polymer melts, thermoset resins, polymer solutions and pastes with high filler contents, and bioactive polymers such as proteins. The plotting of biomaterials such as melts of PLA, PLGA, PHBV biodegradable thermoplastic, PCL, poly (butylene terephthalate-block oligoethylene oxide), biopolymer solutions of agar and gelatin (Landers et al., 2002), natural polymers such as collagen and reactive biosystems involving fibrin formation and polyelectrolyte complexation are all possible. In particular, the processing of materials with low viscosities benefits from buoyancy compensation (Pfister et al., 2004). The work of the Freiburger group led to the commercialization of the first 3D bioplotting system by EnvisionTec GmbH (www.envisiontec.com) to meet the demand for 3D scaffolds with well-defined external and internal structures in tissue engineering and controlled drug release. The 3D-Bioplotter™ has the capacity to fabricate scaffolds using the widest range of materials from soft hydrogels and polymer melts to hard ceramics and metals.

Recently, Schuurman et al. (2011) used a hybrid bioplotting approach for fabrication of solid biodegradable material (polymers, ceramics) with cell-laden hydrogels that could combine favourable mechanical properties with cells positioned in defined locations at high densities. The resulting mechanical properties of the scaffolds were significantly improved and could be tailored within the same range as those of native tissues. Moreover, the approach allows the use of multiple hydrogels, and can thus build constructs containing multiple cell types or bioactive factors. Furthermore, since the hydrogel is supported by the thermoplastic material, a broader range of hydrogel types, concentrations and cross-link densities can be used compared to the deposition of hydrogels alone, thereby improving the conditions for encapsulated cells to proliferate and deposit new matrix (Melchels et al., 2012).

Khalil et al. (2005) developed a special multinozzle bioplotter which was capable of extruding biopolymer solutions and living cells for freeform construction of 3D tissue scaffolds. The deposition is not into plotting media but the process is biocompatible and occurs at room temperature and low pressures to reduce damage to cells. The system was capable of, simultaneously with scaffold construction, depositing a controlled amount of cells, growth factors or other bioactive compounds with precise spatial position to form complex cell-seeded tissue constructs. They fabricated some scaffolds based on sodium alginate solutions and PCL.

Ang et al. (2002) set up a special robotic bioplotting device called rapid prototyping robot dispensing (RPBOD) for the design and fabrication of chitosan-HA scaffolds. Their system consists of a computer-guided desktop robot and a one-component pneumatic dispenser. Mixtures of sodium hydroxide solution and ethanol at different ratios were used as plotting medium to produce chitosan-HA scaffolds.

Further, the RPBOD system was improved to include a new manufacturing method called the dual dispensing system as, besides the pneumatic dispenser, a mechanical dispenser driven by a stepper motor was set up to deposit plotting medium (NaOH). The dual dispensing method overcomes the high sensitivity to material concentration compared with the method of dispensing plotting materials into a fluid medium, as precipitation occurs when the dispensing material and the coagulant medium merge on the base or on the previous layer. There is therefore no precipitated lump forming at the nozzle and no movement of the fluid medium to affect the shape of the precipitated strands of the scaffold. The chitosan scaffolds built by researchers at National University of Singapore using RPBOD exhibit excellent uniformity, interconnectivity, sufficient strength, good reproducibility and calibration (Li *et al.*, 2005).

A variety of extrusion-based techniques has also been developed for processing ceramics. Robocasting is a ceramic processing technique in which a computer controls the robotic deposition of highly concentrated (typically 50–65 vol.% ceramic powder) colloidal ceramic slurries. The slurry is deposited layer by layer from a syringe using constant displacement at a controlled rate. Upon deposition, robocasting relies on a small amount of drying to induce rheological transition of the slurry. The slurry changes from a flowable pseudo-plastic state to a solid-like dilatant mass. This transition gives each layer the strength necessary to support subsequent layers of freshly deposited slurry. Robocasting is a binderless process with low toxicity in which drying is necessary to build 3D parts. The concept of robocasting relies essentially on the rheology of the slurry and also on the partial drying of the deposited layers. In order to make good parts, high solids loadings are necessary, so powder surface chemistry and interparticle forces must be controlled. In addition to the aforementioned factors, the parameters associated with the freeform process, including table speeds, deposition rate and nozzle size, should be controlled appropriately. Miranda *et al.* (2006; 2008) used a robocasting process to produce β-TCP scaffolds with designed 3D geometry and mesoscale porosity using concentrated β-TCP inks with appropriate viscoelastic properties. The deposition was done in a non-wetting oil bath to prevent non-uniform drying during assembly.

Direct-write assembly is an extrusion-based system developed by Lewis and co-workers (Smay *et al.*, 2002) whereby a wide range of inks can be patterned in both planar and 3D shapes with feature sizes as fine as 250 nm. Robocasting and direct-write assembly are essentially identical – the primary difference is the way in which ink is extruded. Robocasting relies on a constant displacement process, whereas direct ink writing relies on a constant pressure process. In this latter process, compressed air is employed to push inks with controlled rheological properties through an individual nozzle (diameter ranging from 1 to 500 µm). The key components of a direct-write assembly system are: compressed air supply, nozzle, three-axis translation stage and optical microscope for real-time monitoring. Direct-write assembly deposits inks on substrates at room temperature or a proper coagulation reservoir using a controlled-printing speed and pressure which depend on ink rheology and nozzle diameter. Due to viscoelastic ink characteristics, direct-write assembly enables self-supporting and spanning features. Ink rheology strongly depends on solid loading for nanoparticle inks so that viscosity decreases by decreasing solid loading. Concentrated inks with solid loadings of 70–85 wt.% are normally required for printing planar and spanning filaments. Using low viscosity inks (i.e., dilute inks) results in a significant lateral spreading during printing.

A wide range of inks including colloidal suspensions and gels, nanoparticle-filled inks, polymer melts, fugitive organic inks, hydrogels, sol–gel and polyelectrolyte inks have been processed using direct-write assembly. Lewis and co-workers have achieved minimum feature sizes ranging from 250 nm for sol-gel inks to 200 µm for ceramic colloidal inks. Writing with some inks such as polyelectrolyte inks needs to be performed into a reservoir-induced coagulation to enable 3D printing, whereas some other inks such as sol-gel inks can be directly printed in air providing excellent control over the deposition process (e.g., the ink flow can be started/stopped repeatedly during assembly).

In recent years, Lewis and co-workers have focused on extending direct-write assembly to biomedical applications. Using biocompatible inks they printed different 3D scaffolds and microvascular networks for tissue engineering and cell culture. Different 3D HA scaffolds with 250 µm road width (Michna *et al.*, 2005; Simon *et al.*, 2007), and 3D scaffolds composed of a gradient array of silk/HA filaments of 200 µm size were fabricated by direct-write assembly (Sun *et al.*, 2012). The 3D silk/HA scaffolds were used to support the growth of co-cultures of human bone marrow-derived mesenchymal stem cells (hMSCs) and human mammary microvascular endothelial cells (hMMECs) to assess *in vitro* formation of bone-like tissue. 3D microperiodic scaffolds of regenerated silk fibroin have been fabricated using direct-write assembly for tissue engineering (Ghosh *et al.*, 2008). Biocompatible silk optical waveguides as fine as 5 µm were produced by direct-write assembly of a concentrated silk fibroin ink through a micronozzle into a methanol-rich coagulation reservoir (Parker *et al.*, 2009). 3D microperiodic hydrogel scaffolds composed of 1 µm (Barry *et al.*, 2009) and 10 µm (Shepherd *et al.*, 2011) filaments were produced for guided cell growth by direct writing through a gold-coated deposition micronozzle of a PHEMA-based ink that is simultaneously photopolymerized via UV illumination.

Solvent-based extrusion freeforming is another technique developed by our group to produce bioceramic scaffolds (Grida and Evans, 2003). In this process, continuous flow of materials in the form of paste or particulate slurries is dispensed onto the surface using a 3D motion system incorporated with the nozzle. Solvent-based extrusion freeforming is a relatively simple process in which phase change is based on solvent evaporation. Paste with a high yield strength is prepared by blending polymer, ceramic and a solvent in specific ratios. Defects such as dilatancy, drying cracks and surface fracture which happen in water-based extrusion systems (Yang *et al.*, 2008c) can be eliminated by appropriate adjustment of polymer content. Low solvent and high ceramic contents result in low drying and sintering shrinkages, respectively. Typical solvent-based extrusion freeforming equipment is shown in Fig. 2.13a and the schematic arrangement is described in Fig. 2.13b. As seen in Fig. 2.13a, there are four axes, including X, Y, Z and an extrusion drive. The stainless steel/glass syringe is mounted in the Z-axis and the sample substrate is placed on the X-Y table. The extrusion pressure can be measured by a load cell which is mounted on the extrusion axis. The control program, compiled by Labview Software (National Instruments, US), controls four motors for the movement of X-, Y- and Z-axes and extrusion. The overall solvent-based extrusion freeforming process steps are: (1) preparation of paste, (2) deposition of fine filaments and (3) post-processing including drying, debinding and sintering. The paste is normally prepared using an ultrasonic probe for dispersion of powder, drying to increase viscosity and limited vacuum de-airing. Thermoplastic binder, polyvinylbutyral (PVB) and plasticizer, PEG (MWt = 600), in the ratio of 75 wt.% PVB and 25 wt.% PEG, are fully dissolved in the solvent, propan-2-ol and then the desired bioceramic is added to the solution and finally is stirred (for at least 2 h) to achieve a well-dispersed paste.

A range of bioceramic scaffolds have been fabricated by our group with different compositions in the HA/β-TCP (different HA/β-TCP ratios) and sintered from 1100°C to 1300°C in steps of 50°C. Scaffolds with different porosities and pore sizes were produced, with raster width down to 60 μm and interconnected pores with interstices from 50 to 500 μm (Yang *et al.*, 2008a, b, c). Other ceramic pastes such as alumina, alumina/silica, zirconia and alumina/graphite have been used successfully for fabrication of 3D lattice structures with fine filaments (Xuesong *et al.*, 2009b, 2010). In addition to bioceramics, 3D carbon scaffolds have been produced using two different paste compositions with different polymers in the paste (Lu *et al.*, 2012). Figure 2.14 shows a sample HA scaffold with filament diameter 70 μm produced by solvent-based extrusion freeforming.

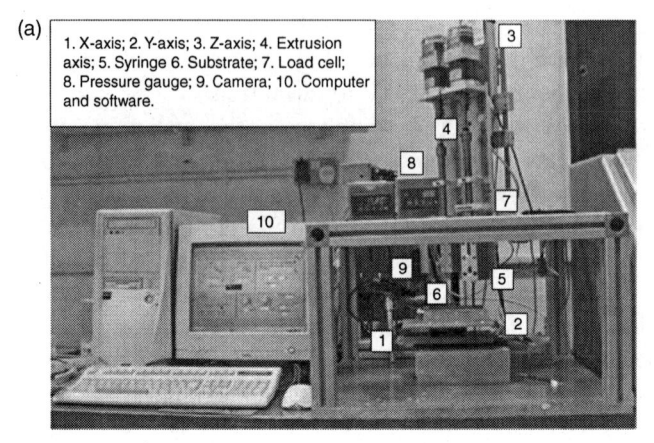

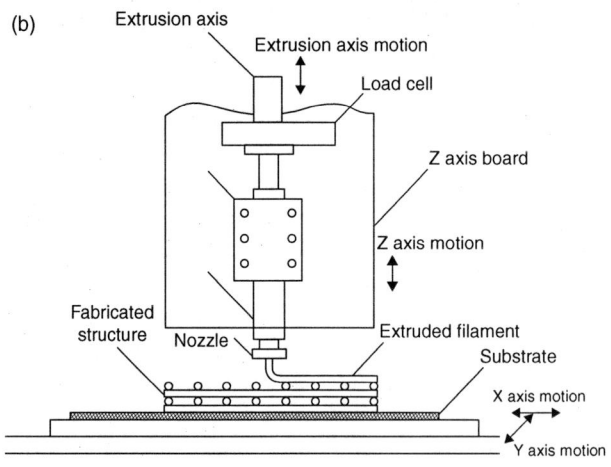

2.13 (a) Paste extrusion freeforming experimental set-up, (b) schematic of the extrusion axis (Xuesong *et al.*, 2009b).

Bioceramic scaffolds produced by solvent-based extrusion freeforming have the finest filament diameter amongst extrusion-based SFF techniques such as direct-write assembly or robocasting. However, producing bioceramic scaffolds with fine filaments (less than 100 µm) is still a challenging issue. A set of experiments was conducted to define the most significant factors that should be taken into consideration in order to reach fine filaments. Equipment accuracy (in particular levelling error) and appropriate adjustment of X-Y table velocity and nozzle path according to extrusion ram velocity have been determined as the most important factors for fine filament extrusion freeforming. The levelling error determines how large a sample can be fabricated. The relationship between X-Y table velocity

2.14 (a) Plane view of HA scaffold with filament diameter of 70 μm produced by solvent-based extrusion freeforming. (b) Fracture of a sintered filament near a weld area, indicating that welds formed in the slurry state are retained after sintering to provide a strong bridge between layers. 1–5 μm small pores on surface can be seen, and filaments are porous after sintering. Thus, both macroporosity of scaffold and microporosity of filaments can be controlled in solvent-based extrusion freeforming: macroporosity of the scaffold is computer controlled, and microporosity of filaments is controlled by varying sintering temperature (Yang *et al.*, 2008c).

and extrusion ram velocity under steady state extrusion conditions follows Equation [2.1]:

$$\frac{V_{ram}}{V_{paste}} = \left(\frac{D_{nozzle}}{D_{barrel}}\right)^2 \quad [2.1]$$

where, V_{ram} is extrusion ram velocity, V_{paste} is paste extrude velocity, D_{barrel} is syringe barrel diameter and D_{nozzle} the nozzle diameter. Based on Equation [2.1], the X-Y table velocity can be a thousand times higher than the extrusion ram velocity for extrusion of very fine filaments (smaller nozzle diameter), depending on the ratio of syringe barrel diameter to nozzle diameter. Hence, some control difficulties are introduced in order to make the X-Y table and extrusion work in harmony. Therefore, the ratio of barrel to nozzle diameter should be selected to adjust these two velocities (Xuesong *et al.*, 2009b). In short, X-Y table velocity should match the extrusion ram velocity and barrel diameter should be compatible with the nozzle diameter. Overfill and underfill of filament deposition can happen in acceleration and deceleration paths where extrusion is not in a steady state. Moreover, it was observed that air bubbles and particle agglomerates are two of the most common defects in fine filaments which are often produced in the paste preparation stage. The solvent content of paste is the most important parameter which determines the ability of the filament to span and retain planned height. Leakage paths and die swell can occur during extrusion and both perturb the ideal relationship between ram velocity and extrusion rate, hence hindering the control system. Low ram velocity can be selected to reduce die swell.

The extrusion pressure P for paste extrusion described by Benbow and Bridgewater (1993) and expressed by Equation [2.2]:

$$P = f + P_1 + P_2 = f + 2(\sigma_0 + \alpha V^m)\ln[D_0/D] + 4(\tau_0 + \beta V^n)[L/D] \quad [2.2]$$

where f is the friction between the plunger and the syringe wall; P_1 is the pressure drop in the die entry; P_2 is the pressure drop in the die land; V is the paste velocity in the die land; σ_0 is the yield stress extrapolated to zero velocity; α is a factor characterizing the effect of velocity; τ_0 is the initial wall stress; β is the wall velocity factor which accounts for the velocity-dependence of the wall shear stress; m and n are exponents for taking account of non-linear behaviour of the paste; D_0 is the diameter of the barrel; D is the diameter of the die land; and L is the length of the die land (see Fig. 2.15).

According to Equation [2.2], increasing the die land length (L) results in a remarkable increase of extrusion pressure so that extrusion pressure is slightly higher in processes such as PAM or direct-write assembly than solvent-based extrusion freeforming as they use needle like nozzles. The authors used nozzles with small die land length (~1 mm) to decrease the required extrusion pressure as much as possible. According to Equation [2.2], decreasing nozzle diameter results in increased extrusion pressure. The authors could decrease nozzle diameter and achieve fine filaments of 60 μm but decreasing nozzle diameter is not a matter to be taken lightly as it causes nozzle clogging.

From what we have discussed above, extrusion-based SFF processes can be employed as an efficient standard tool in tissue engineering. However, choosing the right extrusion-based process requires careful evaluation of the capabilities and limitations of each process. Table 2.2 provides a comparison between the key extrusion-based SFF techniques in terms of resolution, materials and their characteristics.

Scaffold-based tissue engineering is a proven approach in regenerative medicine but is still subject to some limitations and challenges including: (1) complications posed by host acceptance (immunogenicity, inflammatory response, mechanical mismatch); and (2) problems related to cell cultures (e.g., cell density, multiple cell types, specific localization) (Billiet et al., 2012). Cell-based printing techniques have been intensively investigated in recent years and many innovative approaches such as organ bioprinting (Mironov et al., 2007), laser writing of cells (Schiele et al., 2009), bio-electrospraying

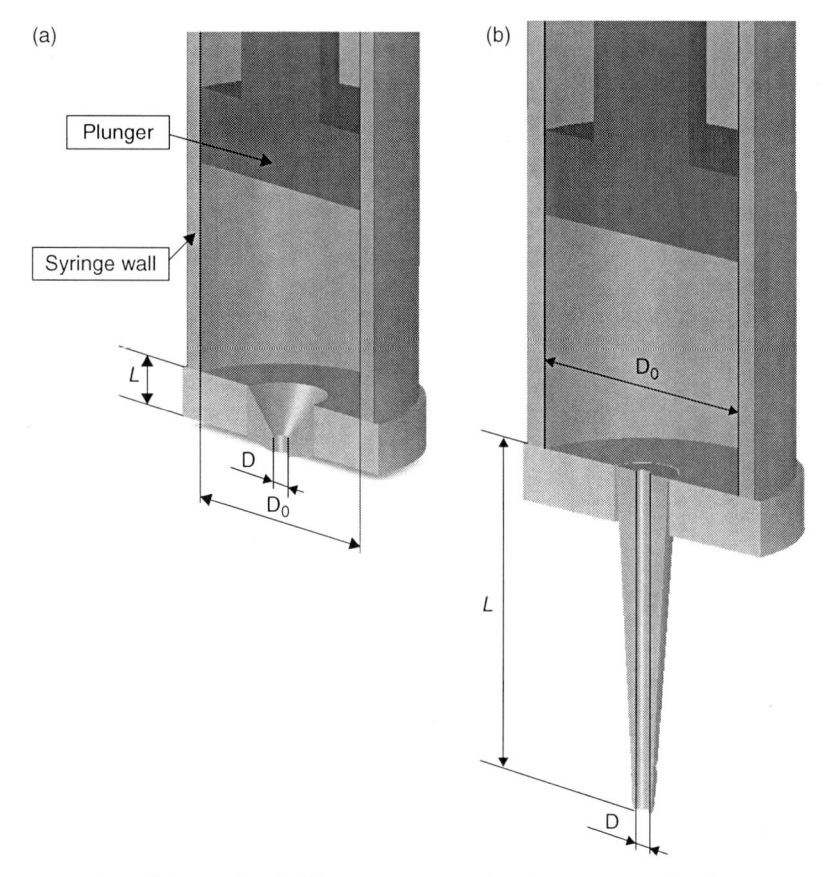

2.15 Schematic of different nozzle design for extrusion freeforming: (a) small nozzle length to decrease extrusion pressure, (b) needle like nozzle used in PAM and direct-write assembly techniques.

(Jayasinghe, 2007) and biological laser printing (BioLP) (Barron *et al.*, 2004) have surfaced to complement limitations in scaffold-based TE. Billiet and co-workers (2012) define organ bioprinting as 'the engineering of 3D living structures supported by the self-assembly/organizing capabilities of cells delivered through the application of SFF techniques based on laser, inkjet, or extrusion freeforming technologies'.

Inkjet and extrusion-based systems are utilized in a similar way to direct bioprinting: balls or continuous flows of bioinks are deposited in well-defined topological patterns into biopaper layers. The bioink building blocks typically have a spherical or cylindrical shape, and consist of single or multiple cell types. In a post-processing step, the construct is transferred to a bioreactor and the bioink spheres are fused. The biopaper, an inert and biocompatible hydrogel, can be removed after construction in post-processing

Table 2.2 Comparison of the key extrusion-based SFF techniques

Technique	Lateral resolution (μm)	Materials	Strengths	Drawbacks	References
FDM	250	PCL, PP-TCP, PCL-HA, PCL-TCP, PLGA, polybutylene terephthalate (PBT), etc.	Good mechanical strength, versatile in lay-down pattern design, no trapped particles or solvents	High temperature, need to produce filament material, rigid filament, limited material range, difficult to prepare structures with microscale porosity	Bose et al., 2003; Hung-Jen et al., 2009; Kalita et al., 2003; Tellis et al., 2008; Zein et al., 2002
RPBOD/dual dispensing	400	Chitosan-HA; chitosan	Enhanced range of materials can be used, can incorporate biomolecules	Low mechanical strength, precise control properties of material and medium, requires freeze drying	Ang et al., 2002; Li et al., 2005
PAM	5–10	PCL, PCL-PLLA, PLGA, PLLA, polyurethane elastomer (Polytek 74-20), alginate, gelatin and viscous inks (using PAM2)	Enhanced range of materials can be used, can incorporate biomolecules, high resolution, not subject to heat, can be used for multilayers	Small nozzle inhibits incorporation of particle, narrow range of printable viscosities, solvent is used, highly water-soluble materials cannot be used	Mariani et al., 2006; Tirella et al., 2009; Vozzi and Ahluwalia, 2007; Vozzi et al., 2004
Robocasting	100	Ceramic and organic inks	Enhanced range of materials can be used, high ceramic content, multi-material scaffold is possible	Precise control of ink properties is crucial	Cesarano, 1999; Miranda et al., 2008; 2006

Technique	Size (µm)	Materials	Applications/Features	Disadvantages	References
PED	250	PCL, PCL-HA	Bone regeneration and drug release applications: HA, titanium, TCP, PCL, PLGA, PLLA	Input material in pellet form	High temperature, rigid filament	Shor et al., 2009; Shor et al., 2007; Wang et al., 2004
3D bioplotting	45 (hydrogel)–250		Soft tissue fabrication/organ printing applications: Agar, chitosan, alginate, gelatine, collagen, fibrin, PU, silicone Hybrid scaffolds: PCL hybrid with alginate	Remarkably wide variety of different materials, can incorporate biomolecules, use of hydrogel materials (agar, gelatin, etc.)	Low mechanical strength, smooth surface which is not desired for appropriate cell attachment, low accuracy, slow processing, precise control of properties of plotting material and medium is required	Landers et al., 2002; Landers and Mulhaupt, 2000; Maher et al., 2009; Pfister et al., 2004
LDM/MDM/M-LDM	300	PLLA-TCP, PLGA, collagen, PLGA-collagen, chitosan, gelatin, alginate		Input material in grain form, preserve bioactivities of scaffold materials because of its non-heating liquefying processing of materials, can incorporate biomolecules	Solvent is used, requires freeze drying	Li et al., 2009a; Liu et al., 2009b; Yongnian et al., 2003
PEM	200–500	PLLA; PLLA-TCP		Input material is grains	High temperature, rigid filament	Xiong et al., 2001

(Continued)

Table 2.2 Continued

Technique	Lateral resolution (μm)	Materials	Strengths	Drawbacks	References
Direct write assembly	250 nm for sol-gel inks 1–20 μm for nanoparticle inks, hydrogel, and polyelectrolyte inks 200–250 for ceramic inks such as HA; silk-HA	Ceramic inks; fugitive organic inks; nanoparticle inks; polymer inks; sol-gel inks	A wide range of materials, very high resolution, non-heating process,	Precise control of ink properties is crucial	Lewis and Gratson, 2004; Shepherd et al., 2011; Sun et al., 2012
3D fibre deposition	250	Poly (ethylene glycol)-terephthalate/ polybutylene terephthalate) (PEGT-PBT), polyethyleneoxide terephthalate (PEOT) and polybutylene-terephthalate (PBT)	Input material in pellet form, preparation time is reduced	High temperature, rigid filament, difficult to prepare structures with microscale porosity	Moroni et al., 2006; Woodfield et al., 2004
Solvent-based extrusion freeforming	80	HA, HA-β-TCP, zirconia, alumina, alumina/silica	Simple process yet high resolution, low sintering shrinkage, controlled microporosity of filaments	Precise control of ink properties is crucial	Xuesong et al., 2009b, 2010, Yang et al., 2008a, 2008b, 2008c

step (Billiet *et al.*, 2012). Several extrusion-based systems such as the 3D bioplotter described earlier can be used as a bio-printer, if sterile conditions can be acquired. However, it should be noted that, technologically, bioprinting using SFF techniques is still in its infancy. Different living structures have been produced using hydrogel structures containing viable cells, but the designs have been simple and isotropic, and mechanical properties were not satisfactory (Melchels *et al.*, 2012). Figure 2.16 depicts some examples for which extrusion-based systems have been used in scaffold-based and scaffold-free tissue engineering.

2.6 Dry powder printing

Recently a promising printing method was developed by our research group to print dry powders precisely with the aims of metering, dispensing and multi-material solid freeforming. Metering and dispensing of small amounts of dry powder is an important demand in a wide range of processes and pharmaceutical industries. A variety of powders with different sizes/shapes need to be weighed out and dispensed in pharmaceutical industries for traditional solid-phase organic synthesis and routine analysis for large compound libraries (Islam and Gladki, 2008). In many cases, powders are metered in interim dose forms (within the range from hundreds of micrograms to tens of milligrams) in the combinatorial chemistry of drug development (Morissette *et al.*, 2004). However, manual weighing is often boring, time-consuming and the accuracy is insufficient. Dry powder printing is an efficient alternative to save time and increase the precision of metering and dispensing.

Furthermore, dry powder printing can be incorporated with powder-based SFF techniques such as SLS and 3D printing to bring new possibilities and complement their current challenges. Powder-based processes such as 3D printing and SLS have little further potential for enhancing the resolution for micromanufacturing applications: as the powder size gets increasingly smaller powder handling (recoating) becomes impossible. Moreover, producing true multi-material parts is still a challenge for powder-based SFF systems. Powder-based systems are currently able to build multi-material 3D parts in which material can change only in the vertical direction. Dry powder dispensing systems (especially ultrasonic nozzle dispensing systems) have demonstrated their great ability in precise placement of fine powders (Lu *et al.*, 2007; Yang and Evans, 2007). It is believed that employing a selective dry powder dispensing mechanism incorporated with powder-based systems would be an efficient way to solve the problem of fine powder handling as well as enhancing the capacity to produce multi-material parts with lateral material change. In this way, a higher level of material deposition control could be obtained which is very attractive

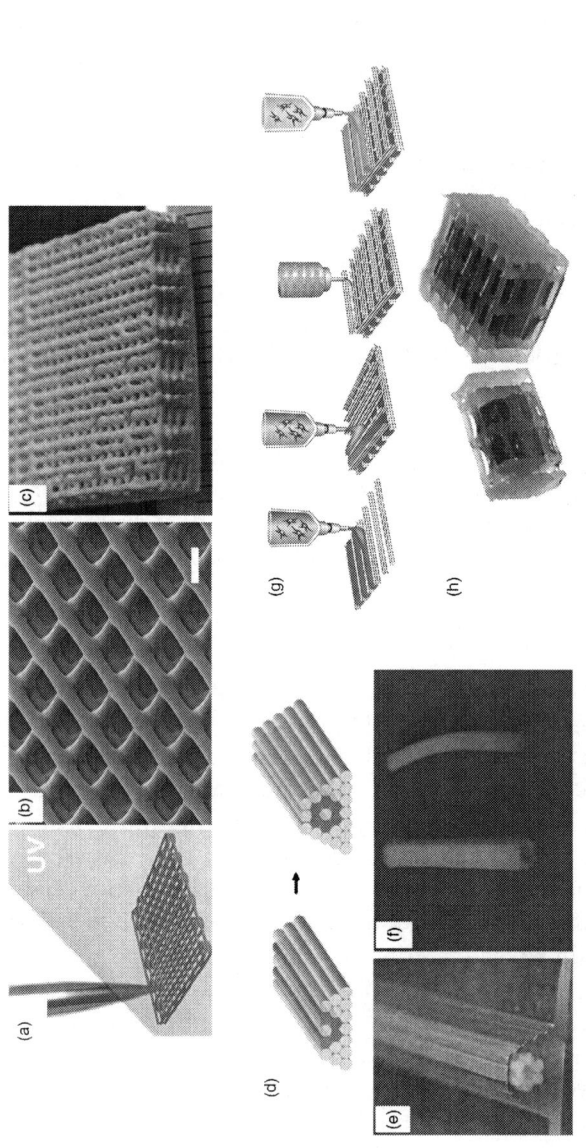

2.16 (a) Schematic illustration of direct-write assembly of a hydrogel-based ink through a gold-coated deposition micronozzle that is simultaneously photopolymerized via UV illumination. (b) SEM micrographs of 3D PHEMA hydrogel scaffold with 10 μm filament and 30 μm pitch produced using direct-write assembly (Shepherd *et al.*, 2011). (c) Multi-material scaffold (poly(lactic-co-glycolic acid), PLGA-collagen) fabricated via M-LDM (Liu *et al.*, 2009a). (d) Schematic illustration of bioprinting tubular structures using extrusion-based freeforming, layer-by-layer deposition of agarose hydrogel (light grey) cylinders and multicellular pig SMC cylinders (dark grey). (e) The printed construct. (f) Engineered pig SMC tubes of different diameters (left: 2.5 mm OD; right: 1.5 mm OD) resulted after 3 days of post-printed fusion and hydrogel removal. Agarose is not remodelled by the cells and can easily be removed after fusion of the bioink (Lee *et al.*, 2010). (g) Fabrication of solid biodegradable materials with cell-laden hydrogels: schematic illustration of a hybrid bioprinting process including alternating steps of printing biodegradable polymer and cell-laden hydrogel. (h) Layering of the dye-containing alginate results in specific confinement of the printed hydrogels (Schuurman *et al.*, 2011).

and a highly important issue in TE scaffold fabrication. Moreover, it is possible to incorporate growth factors and fabricate controlled-release scaffolds. Ongoing works in our lab are focussing on incorporating dry powder printing with SLS in order to extend the approach to the areas of multi-material and micromanufacturing. Figure 2.17 depicts an example of our preliminary results in high resolution dry powder printing for electronics applications. Dry powder printing enabled us to print lines as fine as ~70 μm. The results prove that high resolution, true 3D multi-material parts can be produced efficiently using the proposed approach.

By application of an acoustic energy driven system containing a glass capillary as funnel and a delicate computer control system, the vibrations from a piezoelectric disc can precisely initiate and halt the flow of powder from a fine nozzle. Once the ultrasonic vibration is switched off, arrest of powder flow is brought about by the formation of domes in the capillary due to wall–particle and particle–particle friction. The powder flow rate can be adjusted by varying the frequency and amplitude of the vibration (Yang and Evans, 2004, 2005). In the following section, processing of nanobiomaterials using dry powder printer are discussed in more detail.

Figure 2.18 shows the experimental device set up for printing different nanobiomaterial dry powders. The device includes a computer, an analogue waveform generator (National Instruments Corporation Ltd., Berkshire, UK), a power amplifier (PB58A, Apex Co., USA), a piezoelectric ring (SPZT-4 A3544C-W, MPI Co., Switzerland), a glass nozzle (made from a capillary tube), a purpose-built water vessel and a microbalance (2100 mg ± 0.1 μg, Sartorius AG, Germany). The system generates a voltage signal, which can be varied by different waveforms (e.g., square, sine, triangle and sawtooth), frequencies and amplitudes (Shoufeng and Evans, 2004; Xuesong *et al.*, 2006). The PZT excited by the high-frequency signal (>20 kHz) transmits the vibration through water to the capillary. The inner diameter of the water tank is 40 mm, with an inserted feed tube made of a 10 mm (inner diameter) glass capillary. The upper section functions as a hopper for the powder sample. The piezoelectric ceramic ring was attached to the bottom of the glass tank with an adhesive commonly used in ultrasonic cleaning tank construction (9340 GRAY Hysol Epoxi-Patch Structural Adhesive, DEXTER Co., Seabrook, USA). The microbalance is employed to verify and record the dose mass.

Different nanobio dry powders have been processed and characterized based on particle size, density, shape and angle of repose (Table 2.3). Angle of repose is tested as a relative measure of friction and cohesiveness of the powders. Generally a powder with an angle of repose greater than about 40°

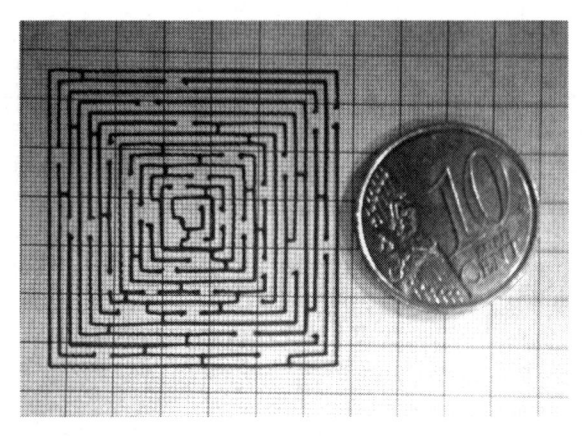

2.17 Fine features printed by dry powder printing.

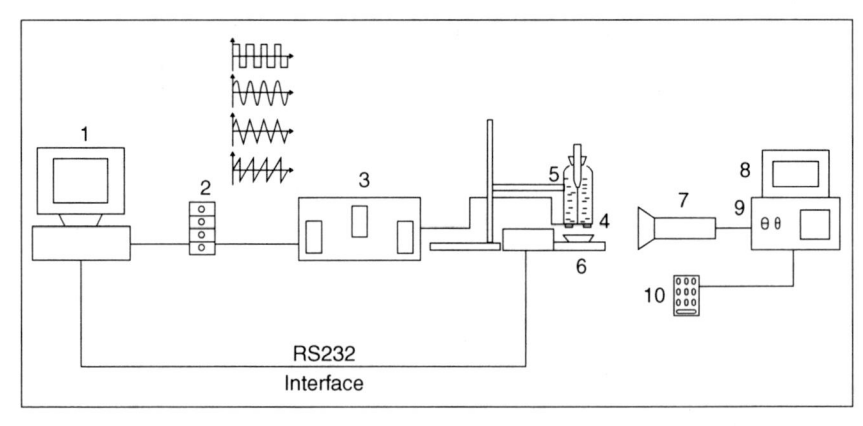

2.18 Experimental arrangement of the ultrasonic vibration controlled micro-feeding system: 1. Computer; 2. D/A card; 3. Electrical controller; 4. Piezoelectric ring; 5. Ultrasonic nozzle; 6. Microbalance; 7. High-speed camera; 8. Video monitor; 9. Image motion analyser; 10. Control panel (Xuesong *et al.*, 2009a).

is classified as cohesive and non-free flowing, which is difficult to dispense in conventional dry powder handling methods.

Further, some results from nanosize HA powder, CAPTAL® R Sintering grade HA (Batch P201), CAPTAL® S Sintering grade HA (Batch P221S BM168) and (β-TCP (Batch P228S) will be presented and discussed in more detail to give an outline of the effects of process parameters.

Figure 2.19 shows scanning electron microscope (SEM) images of different nanobiomaterial powders. The CAPTAL® S has a larger particle size (500 nm–3 μm) due to the sintering and grinding process used during powder manufacture. Merck HA powder and β-TCP have a near-spherical

Table 2.3 Different nanobiomaterials which have been tested with dry powder printing

Powder	Particle size	Particle density (kg/m^3)	Repose angle	Manufacturing company
CAPTAL® S	500 nm–3 μm	–	51.5	Merck KGaA, Germany
CAPTAL® R	20 nm width–200 nm Length	–	60.3	Merck KGaA, Germany
Merck HA	50–300 nm	–	55.2	Merck KGaA, Germany
β-TCP	100–500 nm	–	58.2	Plasma Biotal Limited, UK
TiO$_2$	180 nm	4150	38	Tioxide UK Ltd.
MgO	100 nm	3580	53	PI-KEM Ltd, UK

Source: Li and Yang, 2012.

shape. The CAPTAL® S has an angular shape whereas CAPTAL® R has a long needle shape.

Measurement of angle of repose indicated that all processed powders are cohesive. In particular, the needle shape of CAPTAL® R particles hinders their flow.

All of the nanobiomaterial powder was extruded out as discrete rods due to the strong agglomerations (Fig. 2.20(a)). In the dispensing process, some powder cracks might be formed randomly in the nozzle, as shown in Fig. 2.20(a). The movement difficulties of sticky nanopowder lead to uneven packing densities in different positions of the nozzle (Fig. 2.20(b)). The unpredictable crack would distinctly bring deviations to the amount of dispensed powders for each dose. To reduce the chance of crack forming, the preferential higher amplitude is used.

Dosage can be controlled by adjusting some process parameters but the most efficient way is to change the time of vibration (T_V). The dosage can be varied simply from a few micrograms to several grams by changing the time of vibration (T_V) from 0.01 s to a few seconds. Figure 2.21 shows dependency of dose mass on vibration time for the processed nanobiomaterials.

Nozzle size has a significant effect on dispensing of nanobiomaterial powders as well. To reach uniform powder dispensing, nozzle size should be selected appropriately according to particle size/shape and flowability of the nanobiopowder used. Generally, if the nozzle size is too small the powder is dispensed sporadically, and if the nozzle is too big the powder may be over-run. Dose mass change was studied in different nanobiopowders when nozzle size is changed from 1 to 0.8 mm while other process parameters were kept constant. The results show that CAPTAL® S has the highest mean dose mass (and flow rate) at the same flow condition (nozzle size of 1 mm); as seen from the SEM images (Fig. 2.19) it has the biggest particle size, rounded particle shape and less agglomeration which give better flowability. On the other hand, Merck HA nanopowder has the lowest mean dose mass at the same

2.19 SEM images of different nanobiomaterial powders (Li and Yang, 2012): (a) Merck® hydroxyapatite, (b) CAPTAL® S hydroxyapatite, (c) CAPTAL® R hydroxyapatite, (d) β-tricalcium phospate.

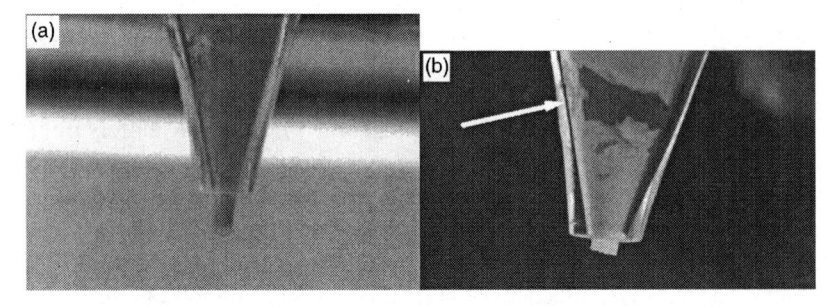

2.20 (a) High speed camera image of Merck HA nanopowder flow (Li and Yang, 2012), (b) uneven packing of β-TCP nanopowders in the nozzle.

flow condition (nozzle size of 0.8 mm) as it has the smallest particle size, which forms very strong agglomeration and has the lowest flowability.

2.7 Conclusion

In this chapter different SFF techniques suitable for nanobiomaterial processing were reviewed comprehensively. Nanofillers are utilized in SFF

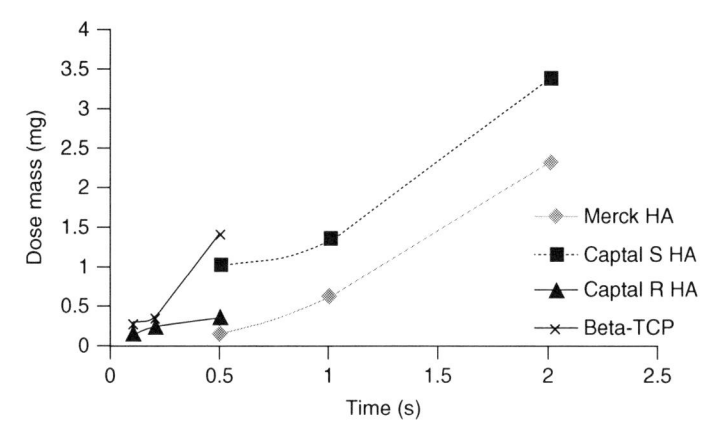

2.21 Mean dose mass of Merck R® HA, CAPTAL® S HA, CAPTAL® R HA and β-TCP nanopowders against vibration time at 750 V amplitude in 1.0 mm nozzle (Li and Yang, 2012).

techniques such as SLS, SL, inkjet printing, etc., to control characteristics such as bioactivity, electrical, optical and mechanical properties of 3D printed parts for medical applications. Tissue engineering scaffold fabrication procedures, including material, micro-macrostructure design and manufacturing, were described briefly. Moreover, various extrusion-based SFF processes were classified and their applications in scaffolding investigated extensively. Direct-write assembly and 3D bioplotting are two approaches with great potential as they can process a wide range of biomaterials. Dry powder printing opens up the possibility of producing high resolution, multi-material parts and is a more versatile and realizable alternative manufacturing process for composite scaffold materials.

2.8 References

Ahn, S. H., Kim, S. G., Chu, W. S. and Jung, W. K. (2007a). Evaluation of mechanical and electrical properties of nanocomposite parts fabricated by nanocomposite deposition system (NCDS). *Journal of Materials Processing Technology*, **187–188**, 331–334.

Ahn, S. H., Lee, C. S., Kim, S. G. and Kim, H. J. (2007b). Measurement of anisotropic compressive strength of rapid prototyping parts. *Journal of Materials Processing Technology*, **187–188**, 627–630.

Allahverdi, M., Danforth, S. C., Jafari, M. and Safari, A. (2001). Processing of advanced electroceramic components by fused deposition technique. *Journal of the European Ceramic Society*, **21**, 1485–1490.

Ang, T. H., Sultana, F. S. A., Hutmacher, D. W., Wong, Y. S., Fuh, J. Y. H., Mo, X. M., Loh, H. T., Burdet, E. and Teoh, S. H. (2002). Fabrication of 3D chitosan-hydroxyapatite scaffolds using a robotic dispensing system. *Materials Science and Engineering C-Biomimetic and Supramolecular Systems*, **20**, 35–42.

Barron, J. A., Wu, P., Ladouceur, H. D. and Ringeisen, B. R. (2004). Biological laser printing: A novel technique for creating heterogeneous 3-dimensional cell patterns. *Biomedical Microdevices*, **6**, 139–147.

Barry, R. A., Shepherd, R. F., Hanson, J. N., Nuzzo, R. G., Wiltzius, P. and Lewis, J. A. (2009). Direct-write assembly of 3D hydrogel scaffolds for guided cell growth. *Advanced Materials*, **21**, 1–4.

Bartolo, P. J. (2011). *Stereolithography: Materials, Processes and Applications*, New York, Springer.

Bartolo, P. J. and Bidanda, B. (2008). *Bio-Materials and Prototyping Applications in Medicine*, New York, USA, Springer.

Bartolo, P. J. and Gaspar, J. (2008). Metal filled resin for stereolithography metal part. *Cirp Annals-Manufacturing Technology*, **57**, 235–238.

Benbow, J. J. and Bridgwater, J. (1993). *Paste Flow and Extrusion*, UK, Oxford University Press.

Bertsch, A., Jiguet, S. and Renaud, P. (2004). Microfabrication of ceramic components by microstereolithography. *Journal of Micromechanics and Microengineering*, **14**, 197–203.

Billiet, T., Vandenhaute, M., Schelfhout, J., Van Vlierberghe, S. and Dubruel, P. (2012). A review of trends and limitations in hydrogel-rapid prototyping for tissue engineering. *Biomaterials*, **33**, 6020–6041.

Bin, D., Min, W., Wen You, Z., Wai Lam, C., Zhao Yang, L. and Lu, W. W. (2010). Three-dimensional nanocomposite scaffolds fabricated via selective laser sintering for bone tissue engineering. *Acta Biomaterialia*, **6**, 4495–4505.

Bose, S., Darsell, J., Kintner, M., Hosick, H. and Bandyopadhyay, A. (2003). Pore size and pore volume effects on alumina and TCP ceramic scaffolds. *Materials Science and Engineering C-Biomimetic and Supramolecular Systems*, **23**, 479–486.

Cesarano, J. (1999). A review of robocasting technology. In: Dimos, D., Danforth, S. C. and Cima, M. J. (eds.) *Solid Freeform and Additive Fabrication*. Warrendale, Materials Research Society.

Cheah, C. M., Chua, C. K., Leong, K. F., Cheong, C. H. and Naing, M. W. (2004). Automatic algorithm for generating complex polyhedral scaffold structures for tissue engineering. *Tissue Engineering*, **10**, 595–610.

Chen, M. H., Chen, C. R., Hsu, S. H., Sun, S. P. and Su, W. F. (2006). Low shrinkage light curable nanocomposite for dental restorative material. *Dental Materials*, **22**, 138–145.

Chen, P. C., Chen, H. T., Qiu, J. and Zhou, C. W. (2010). Inkjet printing of single-walled carbon nanotube/RuO2 nanowire supercapacitors on cloth fabrics and flexible substrates. *Nano Research*, **3**, 594–603.

Chu, W. S., Kim, S. G., Kim, H. J., Lee, C. S. and Ahn, S. H. (2007). Fabrication of bio-composite drug delivery system using rapid prototyping technology. In: Kim, Y. H., Cho, C. S., Kang, I. K., Kim, S. Y. and Kwon, O. H. (eds.) *ASBM7: Advanced Biomaterials VII*. Stafa-Zurich, Trans Tech Publications Ltd.

Chua, C. K., Leong, K. F. and Lim, C. S. (2004). *Rapid Prototyping: Principles and Applications*, Singapore, World Scientific Publishing.

Chung, H. and Das, S. (2008). Functionally graded Nylon-11/silica nanocomposites produced by selective laser sintering. *Materials Science and Engineering A-Structural Materials Properties Microstructure and Processing*, **487**, 251–257.

Cong Bang, P., Kah Fai, L., Tze Chiun, L. and Kerm Sin, C. (2008). Rapid freeze prototyping technique in bio-plotters for tissue scaffold fabrication. *Rapid Prototyping Journal*, **14**, 246–253.

De Gans, B. J. and Schubert, U. S. (2003). Inkjet printing of polymer micro-arrays and libraries: Instrumentation, requirements, and perspectives. *Macromolecular Rapid Communications*, **24**, 659–666.

Domingos, M., Chiellini, F., Gloria, A., Ambrosio, L., Bartolo, P. and Chiellini, E. (2012). Effect of process parameters on the morphological and mechanical properties of 3D Bioextruded poly(epsilon-caprolactone) scaffolds. *Rapid Prototyping Journal*, **18**, 56–67.

Domingos, M., Dinucci, D., Cometa, S., Alderighi, M., Bartolo, P. J. and Chiellini, F. (2009). Polycaprolactone scaffolds fabricated via bioextrusion for tissue engineering applications. *International Journal of Biomaterials*, **2009**, 239643.

Dong-Woo, C., Jong Young, K., Jin Woo, L., Seung-Jae, L., Eui Kyun, P. and Shin-Yoon, K. (2007). Development of a bone scaffold using HA nanopowder and microstereolithography technology. *Microelectronic Engineering*, **84**, 1762–1765.

Doreau, F., Chaput, C. and Chartier, T. (2000). Stereolithography for manufacturing ceramic parts. *Advanced Engineering Materials*, **2**, 493–496.

Dorj, B., Park, J. H. and Kim, H. W. (2012). Robocasting chitosan/nanobioactive glass dual-pore structured scaffolds for bone engineering. *Materials Letters*, **73**, 119–122.

Evans, J. R. G., Edirisinghe, M. J., Coveney, P. V. and Eames, J. (2001). Combinatorial searches of inorganic materials using the ink jet printer: Science, philosophy and technology. *Journal of the European Ceramic Society*, **21**, 2291–2299.

Fang, Z., Starly, B. and Sun, W. (2005). Computer-aided characterization for effective mechanical properties of porous tissue scaffolds. *Computer-Aided Design*, **37**, 65–72.

Ferrari, M. (2005). Cancer nanotechnology: Opportunities and challenges. *Nature Reviews Cancer*, **5**, 161–171.

Fong, H. (2004). Electrospun nylon 6 nanofibre reinforced BIS-GMA/TEGDMA dental restorative composite resins. *Polymer*, **45**, 2427–2432.

Fuller, S. B., Wilhelm, E. J. and Jacobson, J. M. (2002). Ink-jet printed nanoparticle microelectromechanical systems. *Journal of Microelectromechanical Systems*, **11**, 54–60.

Gaspar, J., Bartolo, P. J. and Duarte, F. M. (2008). Cure and rheological analysis of reinforced resins for stereolithography. *In:* Marques, A. T., Silva, A. F., Baptista, A. P. M., Sa, C., Alves, F., Malheiros, L. F. and Vieira, M. (eds.) *Advanced Materials Forum IV*. Stafa-Zurich: Trans Tech Publications Ltd.

Geun Hyung, K. and Joon Gon, S. (2009). 3D polycarprolactone (PCL) scaffold with hierarchical structure fabricated by a piezoelectric transducer (PZT)-assisted bioplotter. *Applied Physics A: Materials Science and Processing*, **94**, 781–785.

Ghosh, S., Parker, S. T., Wang, X. Y., Kaplan, D. L. and Lewis, J. A. (2008). Direct-write assembly of microperiodic silk fibroin scaffolds for tissue engineering applications. *Advanced Functional Materials*, **18**, 1883–1889.

Greco, A., Licciulli, A. and Maffezzoli, A. (2001). Stereolithography of ceramic suspensions. *Journal of Materials Science*, **36**, 99–105.

Greulich, M., Greul, M. and Pintat, T. (1995). Fast, functional prototypes via multiphase jet solidification. *Rapid Prototyping Journal*, **1**, 20–25.

Grida, I. and Evans, J. R. G. (2003). Extrusion freeforming of ceramics through fine nozzles. *Journal of the European Ceramic Society*, **23**, 629–635.

Griffith, M. L., Chu, T. M., Wagner, W. and Halloran, J. W. (1995). Ceramic stereolithography for investment casting and biomedical applications. *Solid Freeform Fabrication Conference*, University of Texas at Austin, USA, pp. 31–38.

Griffith, M. L. and Halloran, J. W. (1996). Freeform fabrication of ceramics via stereolithography. *Journal of the American Ceramic Society*, **79**, 2601–2608.

Haizhong, Z., Jian, Z., Shiqiang, L., Gaochao, W. and Zhifeng, X. (2006). Effect of core-shell composite particles on the sintering behavior and properties of nanoAl 2O 3/polystyrene composite prepared by SLS. *Materials Letters*, **60**, 1219–1223.

Hao, L., Savalani, M. M., Zhang, Y., Tanner, K. E. and Harris, R. A. (2006). Selective laser sintering of hydroxyapatite reinforced polyethylene composites for bioactive implants and tissue scaffold development. *Proceedings of the Institution of Mechanical Engineers Part H-Journal of Engineering in Medicine*, **220**, 521–531.

Hasirci, V., Vrana, E., Zorlutuna, P., Ndreu, A., Yilgor, P., Basmanav, F. B. and Aydin, E. (2006). Nanobiomaterials: A review of the existing science and technology, and new approaches. *Journal of Biomaterials Science-Polymer Edition*, **17**, 1241–1268.

Heo, S. J., Kim, S. E., Wei, J., Hyun, Y. T., Yun, H. S., Kim, D. H., Shin, J. W. and Shin, J. W. (2009). Fabrication and characterization of novel nano and micro-HA/PCL composite scaffolds using a modified rapid prototyping process. *Journal of Biomedical Materials Research Part A*, **89A**, 108–116.

Hinczewski, C., Corbel, S. and Chartier, T. (1998). Ceramic suspensions suitable for stereolithography. *Journal of the European Ceramic Society*, **18**, 583–590.

Ho, H. C. H., Cheung, W. L. and Gibson, L. (2002). Effects of graphite powder on the laser sintering behaviour of polycarbonate. *Rapid Prototyping Journal*, **8**, 233–242.

Hollister, S. J. (2006). Porous scaffold design for tissue engineering (*Nature Materials*, **4**, 518, 2005). Erratum in *Nature Materials*, **5**, 590.

Hollister, S. J. and Kikuchi, N. (1994). Homogenization theory and digital imaging – A basis for studying the mechanics and design principles of bone tissue. *Biotechnology and Bioengineering*, **43**, 586–596.

Hollister, S. J., Levy, R. A., Chu, T. M., Halloran, J. W. and Feinberg, S. E. (2000). An image-based approach for designing and manufacturing craniofacial scaffolds. *International Journal of Oral and Maxillofacial Surgery*, **29**, 67–71.

Hollister, S. J., Maddox, R. D. and Taboas, J. M. (2002). Optimal design and fabrication of scaffolds to mimic tissue properties and satisfy biological constraints. *Biomaterials*, **23**, 4095–4103.

Hon, K. K. B., Li, L. and Hutchings, I. M. (2008). Direct writing technology-Advances and developments. *Cirp Annals-Manufacturing Technology*, **57**, 601–620.

Hoque, M. E., San, W. Y., Wei, F., Li, S. M., Huang, M. H., Vert, M. and Hutmacher, D. W. (2009). Processing of polycaprolactone and polycaprolactone-based copolymers into 3D scaffolds, and their cellular responses. *Tissue Engineering Part A*, **15**, 3013–3024.

Hosseini Zori, M. and Soleimani-Gorgani, A. (2012). Ink-jet printing of micro-emulsion TiO2 nanoparticles ink on the surface of glass. *Journal of the European Ceramic Society*, **32**(16), 4271–4277.

Huang, Z. M., Zhang, Y. Z., Kotaki, M. and Ramakrishna, S. (2003). A review on polymer nanofibres by electrospinning and their applications in nanocomposites. *Composites Science and Technology*, **63**, 2223–2253.

Hung-Jen, Y., Ching-Shiow, T., Shan-Hui, H. and Ching-Lin, T. (2009). Evaluation of chondrocyte growth in the highly porous scaffolds made by fused deposition manufacturing (FDM) filled with type II collagen. *Biomedical Microdevices*, **11**, 615–624.

Islam, N. and Gladki, E. (2008). Dry powder inhalers (DPIs) – A review of device reliability and innovation. *International Journal of Pharmaceutics*, **360**, 1–11.

Jain, P. K., Pandey, P. M. and Rao, P. V. M. (2010). Selective laser sintering of clay-reinforced polyamide. *PolymerComposites*, **31**, 732–743.

Jayasinghe, S. N. (2007). Bio-electros prays: The development of a promising tool for regenerative and therapeutic medicine. *Biotechnology Journal*, **2**, 934–937.

Jin Woo, L., Geunseon, A., Dae Shick, K. and Dong-Woo, C. (2009). Development of nano and microscale composite 3D scaffolds using PPF/DEF-HA and microstereolithography. *Microelectronic Engineering*, **86**, 1465–1467.

Jinku, K., Mcbride, S., Tellis, B., Alvarez-Urena, P., Young-Hye, S., Dean, D. D., Sylvia, V. L., Elgendy, H., Joo, O. and Hollinger, J. O. (2012). Rapid-prototyped PLGA/beta-TCP/hydroxyapatite nanocomposite scaffolds in a rabbit femoral defect model. *Biofabrication*, **4**, 025003 (11 pp.).

Kalita, S. J., Bose, S., Hosick, H. L. and Bandyopadhyay, A. (2003). Development of controlled porosity polymer-ceramic composite scaffolds via fused deposition modeling. *Materials Science and Engineering C-Biomimetic and Supramolecular Systems*, **23**, 611–620.

Khalil, S., Nam, J. and Sun, W. (2005). Multi-nozzle deposition for construction of 3D biopolymer tissue scaffolds. *Rapid Prototyping Journal*, **11**, 9–17.

Ko, S. H., Chung, J., Hotz, N., Nam, K. H. and Grigoropoulos, C. P. (2010). Metal nanoparticle direct inkjet printing for low-temperature 3D micro metal structure fabrication. *Journal of Micromechanics and Microengineering*, **20**(12), 125010 (7pp).

Ko, S. H., Pan, H., Grigoropoulos, C. P., Luscombe, C. K., Frechet, J. M. J. and Poulikakos, D. (2007). All-inkjet-printed flexible electronics fabrication on a polymer substrate by low-temperature high-resolution selective laser sintering of metal nanoparticles. *Nanotechnology*, **18**.

Koch, K. U., Biesinger, B., Arnholz, C. and Jansson, V. (1998). *Time – Compression Technologies '98 Conferences*. Rapid News Publications, Chester, UK.

Kordas, K., Mustonen, T., Toth, G., Jantunen, H., Lajunen, M., Soldano, C., Talapatra, S., Kar, S., Vajtai, R. and Ajayan, P. M. (2006). Inkjet printing of electrically conductive patterns of carbon nanotubes. *Small*, **2**, 1021–1025.

Kosmala, A., Wright, R., Zhang, Q. and Kirby, A. (2011). Synthesis of silver nano particles and fabrication of aqueous Ag inks for inkjet printing. *Materials Chemistry and Physics*, **129**, 1075–1080.

Landers, R., Hubner, U., Schmelzeisen, R. and Mulhaupt, R. (2002). Rapid prototyping of scaffolds derived from thermoreversible hydrogels and tailored for applications in tissue engineering. *Biomaterials*, **23**, 4437–4447.

Landers, R. and Mulhaupt, R. (2000). Desktop manufacturing of complex objects, prototypes and biomedical scaffolds by means of computer-assisted design combined with computer-guided 3D plotting of polymers and reactive oligomers. *Macromolecular Materials and Engineering*, **282**, 17–21.

Lee, G. and Barlow, J. W. (2003). Selective laser sintering of bioceramic materials for implants. Proceedings of the Solid Freeform Fabrication Symposium, Austin, TX. 376–380, 9–11 August 1993.

Lee, J. H., Ha, W. S., Chu, W. S., Park, C. W., Ahn, S. H. and Chi, S. C. (2008). Preparation of micro-fabricated biodegradable polymeric structures using NCDS. *Archives of Pharmacal Research*, **31**, 125–132.

Lee, Y. B., Polio, S., Lee, W., Dai, G. H., Menon, L., Carroll, R. S. and Yoo, S. S. (2010). Bio-printing of collagen and VEGF-releasing fibrin gel scaffolds for neural stem cell culture. *Experimental Neurology*, **223**, 645–652.

Lejeune, M., Chartier, T., DOSSOU-Yovo, C. and Noguera, R. (2009). Ink-jet printing of ceramic micro-pillar arrays. *Journal of the European Ceramic Society*, **29**, 905–911.

Lewis, J. A. and Gratson, G. M. (2004). Direct writing in three dimensions. *Materials Today*, **7**, 32–39.

Li, G., Wei, F., Hutmacher, D. W., Yoke San, W., Han Tong, L. and Fuh, J. Y. H. (2005). Direct writing of chitosan scaffolds using a robotic system. *Rapid Prototyping Journal*, **11**, 90–97.

Li, S. J., Xiong, Z., Wang, X. H., Yan, Y. N., Liu, H. X. and Zhang, R. J. (2009a). Direct fabrication of a hybrid cell/hydrogel construct by a double-nozzle assembling technology. *Journal of Bioactive and Compatible Polymers*, **24**, 249–265.

Li, S. J., Yan, Y. N., Xiong, Z., Weng, C. Y., Zhang, R. J. and Wang, X. H. (2009b). Gradient hydrogel construct based on an improved cell assembling system. *Journal of Bioactive and Compatible Polymers*, **24**, 84–99.

Li, Z. and Yang, S. (2012). Nanobiomaterials library synthesis for high-throughput screening using a dry powder printing method. *Nano LIFE*, **2**, 1250006 (11 pp.).

Lin, C. Y., Hsiao, C. C., Chen, P. Q. and Hollister, S. J. (2004a). Interbody fusion cage design using integrated global layout and local microstructure topology optimization. *Spine*, **29**, 1747–1754.

Lin, C. Y., Kikuchi, N. and Hollister, S. J. (2004b). A novel method for biomaterial scaffold internal architecture design to match bone elastic properties with desired porosity. *Journal of Biomechanics*, **37**, 623–636.

Lin, L., Shen, Y., Zhang, J. and Fang, M. (2009). Microstructure and mechanical properties analysis of β-tricalcium phosphate/carbon nanotubes scaffold based on rapid prototyping. *Journal of Shanghai University*, **13**, 349–351.

Liu, H. N. and Webster, T. J. (2011). Enhanced biological and mechanical properties of well-dispersed nanophase ceramics in polymer composites: From 2D to 3D printed structures. *Materials Science and Engineering C-Materials for Biological Applications*, **31**, 77–89.

Liu, L., Xiong, Z., Yan, Y. N., Zhang, R. J., Wang, X. H. and Jin, L. (2009a). Multinozzle low-temperature deposition system for construction of gradient tissue engineering scaffolds. *Journal of Biomedical Materials Research Part B-Applied Biomaterials*, **88B**, 254–263.

Liu, L., Xiong, Z., Zhang, R. J., Jin, L. and Yan, Y. N. (2009b). A novel osteochondral scaffold fabricated via multi-nozzle low-temperature deposition manufacturing. *Journal of Bioactive and Compatible Polymers*, **24**, 18–30.

Lu, X. S., Chen, L. F., Amini, N., Yang, S. F., Evans, J. R. G. and Guo, Z. X. (2012). Novel methods to fabricate macroporous 3D carbon scaffolds and ordered surface mesopores on carbon filaments. *Journal of Porous Materials*, **19**, 529–536.

Lu, X. S., Yang, S. F. and Evans, J. R. G. (2007). Dose uniformity of fine powders in ultrasonic microfeeding. *Powder Technology*, **175**, 63–72.

Magdassi, S. (2010). *The Chemistry of Silver Ink*, Singapore, World Scientific Publishing.

Maher, P. S., Keatch, R. P., Donnelly, K., Mackay, R. E. and Paxton, J. Z. (2009). Construction of 3D biological matrices using rapid prototyping technology. *Rapid Prototyping Journal*, **15**, 204–210.

Mariani, M., Rosatini, F., Vozzi, G., Previti, A. and Ahluwalia, A. (2006). Characterization of tissue-engineered scaffolds microfabricated with PAM. *Tissue Engineering*, **12**, 547–557.

Mattioli-Belmonte, M., Vozzi, G., Whulanza, Y., Seggiani, M., Fantauzzi, V., Orsini, G. and Ahluwalia, A. (2012). Tuning polycaprolactone-carbon nanotube composites for bone tissue engineering scaffolds. *Materials Science and Engineering C-Materials for Biological Applications*, **32**, 152–159.

Melchels, F. P. W., Domingos, M. A. N., Klein, T. J., Malda, J., Bartolo, P. J. and Hutmacher, D. W. (2012). Additive manufacturing of tissues and organs. *Progress in Polymer Science*, **37**, 1079–1104.

Mengyan, L., Mondrinos, M. J., Gandhi, M. R., Ko, F. K., Weiss, A. S. and Lelkes, P. I. (2005). Electrospun protein fibres as matrices for tissue engineering. *Biomaterials*, **26**, 5999–6008.

Michna, S., Wu, W. and Lewis, J. A. (2005). Concentrated hydroxyapatite inks for direct-write assembly of 3-D periodic scaffolds. *Biomaterials*, **26**, 5632–5639.

Miranda, P., Pajares, A., Saiz, E., Tomsia, A. P. and Guiberteau, F. (2008). Mechanical properties of calcium phosphate scaffolds fabricated by robocasting. *Journal of Biomedical Materials Research Part A*, **85A**, 218–227.

Miranda, P., Saiz, E., Gryn, K. and Tomsia, A. P. (2006). Sintering and robocasting of beta-tricalcium phosphate scaffolds for orthopaedic applications. *Acta Biomaterialia*, **2**, 457–466.

Mironov, V., Prestwich, G. and Forgacs, G. (2007). Bioprinting living structures. *Journal of Materials Chemistry*, **17**, 2054–2060.

Morissette, S. L., Almarsson, O., Peterson, M. L., Remenar, J. F., Read, M. J., Lemmo, A. V., Ellis, S., Cima, M. J. and Gardner, C. R. (2004). High-throughput crystallization: Polymorphs, salts, co-crystals and solvates of pharmaceutical solids. *Advanced Drug Delivery Reviews*, **56**, 275–300.

Moroni, L., De Wijn, J. R. and Van Blitterswijk, C. A. (2006). 3D fibre-deposited scaffolds for tissue engineering: Influence of pores geometry and architecture on dynamic mechanical properties. *Biomaterials*, **27**, 974–985.

Mott, M. and Evans, J. R. G. (1999). Zirconia/alumina functionally graded material made by ceramic ink jet printing. *Materials Science and Engineering: A*, **271**, 344–352.

Mott, M., Song, J. H. and Evans, J. R. G. (1999). Microengineering of ceramics by direct inkjet printing. *Journal of the American Ceramic Society*, **82**, 1653–1658.

Nguyen, K. T. and West, J. L. (2002). Photopolymerizable hydrogels for tissue engineering applications. *Biomaterials*, **23**, 4307–4314.

Nobusa, Y., Yomogida, Y., Matsuzaki, S., Yanagi, K., Kataura, H. and Takenobu, T. (2011). Inkjet printing of single-walled carbon nanotube thin-film transistors patterned by surface modification. *Applied Physics Letters*, **99**, 183106 (3pp).

O'Connell, C., Okimoto, H., Takenobu, T. and Iwasa, Y. (2008). Inkjet printing of carbon nanomaterials. *22nd Annual Rice Quantum Institute Summer Research Colloquium*. Houston, TX, 1 August 2008.

Park, C. W., Rhee, Y. S., Park, S. H., Danh, S. D., Ahn, S. H., Chi, S. C. and Park, E. S. (2010). In vitro/in vivo evaluation of NCDS-micro-fabricated biodegradable implant. *Archives of Pharmacal Research*, **33**, 427–432.

Parker, S. T., Domachuk, P., Amsden, J., Bressner, J., Lewis, J. A., Kaplan, D. L. and Omenetto, F. G. (2009). Biocompatible silk printed optical waveguides. *Advanced Materials*, **21**, 2411–2415.

Pfister, A., Landers, R., Laib, A., Hubner, U., Schmelzeisen, R. and Mulhaupt, R. (2004). Biofunctional rapid prototyping for tissue-engineering applications: 3D bioplotting versus 3D printing. *Journal of Polymer Science Part a-Polymer Chemistry*, **42**, 624–638.

Price, R. L., Waid, M. C., Haberstroh, K. M. and Webster, T. J. (2003). Selective bone cell adhesion on formulations containing carbon nanofibres. *Biomaterials*, **24**, 1877–1887.

Qin, X. Y., Kim, J. G. and Lee, J. S. (1999). Synthesis and magnetic properties of nanostructured gamma-Ni-Fe alloys. *Nanostructured Materials*, **11**, 259–270.

Rimell, J. T. and Marquis, P. M. (2000). Selective laser sintering of ultra high molecular weight polyethylene for clinical applications. *Journal of Biomedical Materials Research*, **53**, 414–420.

Sachlos, E. and Czernuszka, J. T. (2003). Making tissue engineering scaffolds work: Review on the application of solid freeform fabrication technology to the production of tissue engineering scaffolds. *European Cells and Materials*, **5**, 29–40.

Sandler, J., Werner, P., Shaffer, M. S. P., Demchuk, V., Altstadt, V. and Windle, A. H. (2002). Carbon-nanofibre-reinforced poly(ether ether ketone) composites. *Composites Part A-Applied Science and Manufacturing*, **33**, 1033–1039.

Sandoval, J. H., Soto, K. F., Murr, L. E. and Wicker, R. B. (2007). Nanotailoring photocrosslinkable epoxy resins with multi-walled carbon nanotubes for stereolithography layered manufacturing. *Journal of Materials Science*, **42**, 156–165.

Schiele, N. R., Koppes, R. A., Corr, D. T., Ellison, K. S., Thompson, D. M., Ligon, L. A., Lippert, T. K. M. and Chrisey, D. B. (2009). Laser direct writing of combinatorial libraries of idealized cellular constructs: Biomedical applications. *Applied Surface Science*, **255**, 5444–5447.

Schmidt, M., Pohle, D. and Rechtenwald, T. (2007). Selective laser sintering of PEEK. *Cirp Annals-Manufacturing Technology*, **56**, 205–208.

Schuurman, W., Khristov, V., Pot, M. W., Van Weeren, P. R., Dhert, W. J. A. and Malda, J. (2011). Bioprinting of hybrid tissue constructs with tailorable mechanical properties. *Biofabrication*, **3**, 021001.

Seerden, K. A. M., Reis, N., Evans, J. R. G., Grant, P. S., Halloran, J. W. and Derby, B. (2001). Ink-jet printing of wax-based alumina suspensions. *Journal of the American Ceramic Society*, **84**, 2514–2520.

Shepherd, J. N. H., Parker, S. T., Shepherd, R. F., Gillette, M. U., Lewis, J. A. and Nuzzo, R. G. (2011). 3D microperiodic hydrogel scaffolds for robust neuronal cultures. *Advanced Functional Materials*, **21**, 47–54.

Shor, L., Guceri, S., Chang, R., Gordon, J., Kang, Q., Hartsock, L., An, Y. H. and Sun, W. (2009). Precision extruding deposition (PED) fabrication of polycaprolactone (PCL) scaffolds for bone tissue engineering. *Biofabrication*, **1**, 015003 (5pp).

Shor, L., Guceri, S., Wen, X. J., Gandhi, M. and Sun, W. (2007). Fabrication of three-dimensional polycaprolactone/hydroxyapatite tissue scaffolds and osteoblast-scaffold interactions in vitro. *Biomaterials*, **28**, 5291–5297.

Shoufeng, Y. and Evans, J. R. G. (2004). A dry powder jet printer for dispensing and combinatorial research. *Powder Technology*, **142**, 219–222.

Sigmund, O. 1994. Materials with prescribed constitutive parameters – an inverse homogenization problem. *International Journal of Solids and Structures*, **31**, 2313–2329.

Simon, J. L., Michna, S., Lewis, J. A., Rekow, E. D., Thompson, V. P., Smay, J. E., Yampolsky, A., Parsons, J. R. and Ricci, J. L. (2007). In vivo bone response to 3D periodic hydroxyapatite scaffolds assembled by direct ink writing. *Journal of Biomedical Materials Research Part A*, **83A**, 747–758.

Simpson, R. L., Wiria, F. E., Amis, A. A., Chua, C. K., Leong, K. F., Hansen, U. N., Chandrasekaran, M. and Lee, M. W. (2008). Development of a 95/5 poly(L-lactide-co-glycolide)/hydroxylapatite and beta-tricalcium phosphate scaffold as bone replacement material via selective laser sintering. *Journal of Biomedical Materials Research Part B-Applied Biomaterials*, **84B**, 17–25.

Smay, J. E., Gratson, G. M., Shepherd, R. F., Cesarano, J. and Lewis, J. A. (2002). Directed colloidal assembly of 3D periodic structures. *Advanced Materials*, **14**, 1279–1283.

Song, J. W., Kim, J., Yoon, Y. H., Choi, B. S., Kim, J. H. and Han, C. S. (2008). Inkjet printing of single-walled carbon nanotubes and electrical characterization of the line pattern. *Nanotechnology*, **19**, 095702 (6pp).

Sun, L., Parker, S. T., Syoji, D., Wang, X., Lewis, J. A. and Kaplan, D. L. (2012). Direct-write assembly of 3D silk/hydroxyapatite scaffolds for bone co-cultures. *Advanced Healthcare Materials*, **1**, 729–735.

Tan, K. H., Chua, C. K., Leong, K. F., Cheah, C. M., Gui, W. S., Tan, W. S. and Wiria, F. E. (2005). Selective laser sintering of biocompatible polymers for applications in tissue engineering. *Bio-Medical Materials and Engineering*, **15**, 113–124.

Tartarisco, G., Gallone, G., Carpi, F. and Vozzi, G. (2009). Polyurethane unimorph bender microfabricated with Pressure Assisted Microsyringe (PAM) for biomedical applications. *Materials Science and Engineering C-Materials for Biological Applications*, **29**, 1835–1841.

Teixeira, A. I., Abrams, G. A., Bertics, P. J., Murphy, C. J. and Nealey, P. F. (2003). Epithelial contact guidance on well-defined micro- and nanostructured substrates. *Journal of Cell Science*, **116**, 1881–1892.

Tellis, B. C., Szivek, J. A., Bliss, C. L., Margolis, D. S., Vaidyanathan, R. K. and Calvert, P. (2008). Trabecular scaffolds created using micro CT guided fused deposition modeling. *Materials Science and Engineering C-Biomimetic and Supramolecular Systems*, **28**, 171–178.

Tianming, W. and Derby, B. (2005). Ink-jet printing and sintering of PZT. *Journal of the American Ceramic Society*, **88**, 2053–2058.

Tirella, A., DE Maria, C., Criscenti, G., Vozzi, G. and Ahluwalia, A. (2012). The PAM(2) system: A multilevel approach for fabrication of complex three-dimensional microstructures. *Rapid Prototyping Journal*, **18**, 299–307.

Tirella, A., Orsini, A., Vozzi, G. and Ahluwalia, A. (2009). A phase diagram for microfabrication of geometrically controlled hydrogel scaffolds. *Biofabrication*, **1**, 045002 (7pp).
Tirella, A., Vozzi, G. and Ahluwalia, A. (2008). *Biomimicry of PAM Microfabricated Hydrogel Scaffold*, Springfield, Soc Imaging Science & Technology.
Tolochko, N. K., Laoui, T., Khlopkov, Y. V., Mozzharov, S. E., Titov, V. I. and Ignatiev, M. B. (2000). Absorptance of powder materials suitable for laser sintering. *Rapid Prototyping Journal*, **6**, 155–160.
Van Cleynenbreugel, T., Van Oosterwyck, H., Vander Sloten, J. and Schrooten, J. (2002). Trabecular bone scaffolding using a biomimetic approach. *Journal of Materials Science-Materials in Medicine*, **13**, 1245–1249.
Vasir, J. K., Reddy, M. K. and Labhasetwar, V. D. (2005). Nanosystems in drug targeting: Opportunities and challenges. *Current Nanoscience*, **1**, 47–64.
Vozzi, G. and Ahluwalia, A. (2007). Microfabrication for tissue engineering: Rethinking the cells-on-a scaffold approach. *Journal of Materials Chemistry*, **17**, 1248–1254.
Vozzi, G., Flaim, C., Ahluwalia, A. and Bhatia, S. (2003). Fabrication of PLGA scaffolds using soft lithography and microsyringe deposition. *Biomaterials*, **24**, 2533–2540.
Vozzi, G., Flaim, C. J., Bianchi, F., Ahluwalia, A. and Bhatia, S. (2002). Microfabricated PLGA scaffolds: A comparative study for application to tissue engineering. *Materials Science and Engineering C-Biomimetic and Supramolecular Systems*, **20**, 43–47.
Vozzi, G., Previti, A., Ciaravella, G. and Ahluwalia, A. (2004). Microfabricated fractal branching networks. *Journal of Biomedical Materials Research Part A*, **71A**, 326–333.
Wake, M. C., Patrick, C. W. and Mikos, A. G. (1994). Pore morphology effects on the fibrovascular tissue-growth in porous polymer substrates. *Cell Transplantation*, **3**, 339–343.
Wang, F., Shor, L., Darling, A., Khalil, S., Sun, W., Guceri, S. and Lau, A. (2004). Precision extruding deposition and characterization of cellular poly-epsilon-caprolactone tissue scaffolds. *Rapid Prototyping Journal*, **10**, 42–49.
Wen You, Z., Siu Hang, L., Min, W., Wai Lam, C. and Wing Yuk, I. (2008). Selective laser sintering of porous tissue engineering scaffolds from poly(L-lactide)/carbonated hydroxyapatite nanocomposite microspheres. *Journal of Materials Science: Materials in Medicine*, **19**, 2535–2540.
Wenjuan, C., Wensheng, L., Yakun, Z., Guanhua, L., Tianxin, W. and Long, J. (2010). Gold nanoparticle ink suitable for electric-conductive pattern fabrication using in ink-jet printing technology. *Colloids and Surfaces A: Physicochemical and Engineering Aspects*, **358**, 35–41.
Williams, J. M., Adewunmi, A., Schek, R. M., Flanagan, C. L., Krebsbach, P. H., Feinberg, S. E., Hollister, S. J. and Das, S. (2005). Bone tissue engineering using polycaprolactone scaffolds fabricated via selective laser sintering. *Biomaterials*, **26**, 4817–4827.
Wiria, F. E., Leong, K. F., Chua, C. K. and Liu, Y. (2007). Poly-epsilon-caprolactone/hydroxyapatite for tissue engineering scaffold fabrication via selective laser sintering. *Acta Biomaterialia*, **3**, 1–12.
Won-Shik, C., Sung-Geun, K., Woo-Kyun, J., Hyung-Jung, K. and Sung-Hoon, A. (2007). Fabrication of micro parts using nano composite deposition system. *Rapid Prototyping Journal*, **13**, 276–283.

Woodfield, T. B. F., Malda, J., De Wijn, J., Peters, F., Riesle, J. and Van Blitterswijk, C. A. (2004). Design of porous scaffolds for cartilage tissue engineering using a three-dimensional fibre-deposition technique. *Biomaterials*, **25**, 4149–4161.

Xiong, Z., Yan, Y. N., Wang, S. G., Zhang, R. J. and Zhang, C. (2002). Fabrication of porous scaffolds for bone tissue engineering via low-temperature deposition. *Scripta Materialia*, **46**, 771–776.

Xiong, Z., Yan, Y. N., Zhang, R. J. and Sun, L. (2001). Fabrication of porous poly(L-lactic acid) scaffolds for bone tissue engineering via precise extrusion. *Scripta Materialia*, **45**, 773–779.

Xu, C. Y., Inai, R., Kotaki, M. and Ramakrishna, S. (2004). Aligned biodegradable nanofibrous structure: A potential scaffold for blood vessel engineering. *Biomaterials*, **25**, 877–886.

Xuesong, L., Shoufeng, Y. and Evans, J. R. G. (2006). Studies on ultrasonic microfeeding of fine powders. *Journal of Physics D (Applied Physics)*, **39**, 2444–2453.

Xuesong, L., Shoufeng, Y. and Evans, J. R. G. (2009a). Microfeeding with different ultrasonic nozzle designs. *Ultrasonics*, **49**, 514–521.

Xuesong, L., Yoonjae, L., Shoufeng, Y., Yang, H., Evans, J. R. G. and Parini, C. G. (2009b). Fine lattice structures fabricated by extrusion freeforming: Process variables. *Journal of Materials Processing Technology*, **209**, 4654–4661.

Xuesong, L., Yoonjae, L., Shoufeng, Y., Yang, H., Evans, J. R. G. and Parini, C. G. (2010). Solvent-based paste extrusion solid freeforming. *Journal of the European Ceramic Society*, **30**, 1–10.

Yang, H. Y., Thompson, I., Yang, S. F., Chi, X. P., Evans, J. R. G. and Cook, R. J. (2008a). Dissolution characteristics of extrusion freeformed hydroxyapatite-tricalcium phosphate scaffolds. *Journal of Materials Science-Materials in Medicine*, **19**, 3345–3353.

Yang, H. Y., Yang, S. F., Chi, X. P., Evans, J. R. G., Thompson, I., Cook, R. J. and Robinson, P. (2008b). Sintering behaviour of calcium phosphate filaments for use as hard tissue scaffolds. *Journal of the European Ceramic Society*, **28**, 159–167.

Yang, S. and Evans, J. R. G. (2004). Acoustic control of powder dispensing in open tubes. *Powder Technology*, **139**, 55–60.

Yang, S. and Evans, J. R. G. (2005). On the rate of descent of powder in a vibrating tube. *Philosophical Magazine*, **85**, 1089–1109.

Yang, S. and Evans, J. R. G. (2007). Metering and dispensing of powder; the quest for new solid freeforming techniques. *Powder Technology*, **178**, 56–72.

Yang, S. F., Hongyi, Y., Xiaopeng, C., Evans, J. R. G., Thompson, I., Cook, R. J. and Robinson, P. (2008c). Rapid prototyping of ceramic lattices for hard tissue scaffolds. *Materials and Design*, **29**, 1802–1809.

Yang, S. F., Leong, K. F., Du, Z. H. and Chua, C. K. (2001). The design of scaffolds for use in tissue engineering. Part 1. Traditional factors. *Tissue Engineering*, **7**, 679–689.

Yang, S. F., Leong, K. F., Du, Z. H. and Chua, C. K. (2002). The design of scaffolds for use in tissue engineering. Part II. Rapid prototyping techniques. *Tissue Engineering*, **8**, 1–11.

Ye, L., Zeng, X. C., Li, H. J. and Ai, Y. (2010). Fabrication and biocompatibility of nano non-stoichiometric apatite and poly(epsilon-caprolactone) composite scaffold by using prototyping controlled process. *Journal of Materials Science-Materials in Medicine*, **21**, 753–760.

Yexin, G., Xuening, C., Joung-Hyun, L., Monteiro, D. A., Hongjun, W. and Lee, W. Y. (2012). Inkjet printed antibiotic- and calcium-eluting bioresorbable nanocomposite micropatterns for orthopedic implants. *Acta Biomaterialia*, **8**, 424–431.

Yongnian, Y., Zhuo, X., Yunyu, H., Shenguo, W., Renji, Z. and Chao, Z. (2003). Layered manufacturing of tissue engineering scaffolds via multi-nozzle deposition. *Materials Letters*, **57**, 2623–2628.

Yoshimoto, H., Shin, Y. M., Terai, H. and Vacanti, J. P. (2003). A biodegradable nanofibre scaffold by electrospinning and its potential for bone tissue engineering. *Biomaterials*, **24**, 2077–2082.

Zein, I., Hutmacher, D. W., Tan, K. C. and Teoh, S. H. (2002). Fused deposition modeling of novel scaffold architectures for tissue engineering applications. *Biomaterials*, **23**, 1169–1185.

Zhang, Y. Z., Lim, C. T., Ramakrishna, S. and Huang, Z. M. (2005). Recent development of polymer nanofibres for biomedical and biotechnological applications. *Journal of Materials Science-Materials in Medicine*, **16**, 933–946.

Zhao, X. L., Evans, J. R. G., Edirisinghe, M. J. and Song, J. H. (2002). Direct inkjet printing of vertical walls. *Journal of the American Ceramic Society*, **85**, 2113–2115.

Zhuo, X., Yongnian, Y., Shenguo, W., Renji, Z. and Chao, Z. (2002). Fabrication of porous scaffolds for bone tissue engineering via low-temperature deposition. *Scripta Materialia*, **46**, 771–776.

Zong, X. H., Bien, H., Chung, C. Y., Yin, L. H., Fang, D. F., Hsiao, B. S., Chu, B. and Entcheva, E. (2005). Electrospun fine-textured scaffolds for heart tissue constructs. *Biomaterials*, **26**, 5330–5338.

3
Rapid prototyping techniques for the fabrication of biosensors

K. PATAKY and J. BRUGGER, Ecole Polytechnique Fédérale de Lausanne, Switzerland

DOI: 10.1533/9780857097217.75

Abstract: This chapter discusses the use of rapid prototyping techniques in the fabrication of biosensors. First it discusses the use of micromoulding, extrusion, 3D printing, stereolithography, and xurography in the prototyping of microfluidic biosensors. This is followed by a discussion of functionalization and patterning methods for adding bioelement and transducer materials to biosensor surfaces. Finally, the chapter discusses future trends in biosensors and the continued yet evolving role of rapid prototyping methods in their fabrication.

Key words: biosensors, rapid prototyping, 3D printing, inkjet, biomaterial, microfabrication.

3.1 Introduction

The first biosensor was described in 1962 by Clark and Lyons as an 'enzyme electrode' and consisted of glucose oxidase trapped by a dialysis membrane at an electrode surface (Clark and Lyons, 1962). Despite numerous technological developments since then, this electrochemical detection concept is still in use (Newman and Setford, 2006). In fact, glucose oxidase is still frequently used to demonstrate a new biosensor design or fabrication technique and features in many of the works discussed in this chapter. The use of horseradish peroxidase (HRP) is also common – most probably owing to its frequent use in enzyme-linked immunosorbent assays (ELISA).

As microelectromechanical systems (MEMS) and semiconductor fabrication processes have become more commonly used and as life-sciences research has grown, biosensors have been developed that operate on a range of different physical principles such as fluorescence detection, impedance measurements, gravimetric detection, and plasmonic detection (Cooper and Cass, 2004; Fritz, 2008; Noh et al., 2011). In spite of these different detection methods, all biosensors operate on a common principle. A bioelement interacts with a component in a sample, and a transducer recognizes a change in the bioelement and outputs an electrical signal (Mohanty and Kougianos,

2006) (see Fig. 3.1). Rather than focusing on the specific physical methods behind biosensors, this chapter will focus on the role of rapid prototyping (RP) techniques in their fabrication and the technical aspects of their use. For more information we refer the reader to the following references: (Cui *et al.*, 2001; Drummond *et al.*, 2003; Cooper and Cass, 2004; Fritz, 2008; Grieshaber *et al.*, 2008; Noh *et al.*, 2011).

As new diagnostic tests, biomaterials, and transducer materials are developed, they may not yet be compatible with the existing fabrication processes used in biosensor fabrication (Delaney *et al.*, 2009). Consequently, RP techniques have begun to play a significant role in the functionalization of prototype biosensors. Also, as biosensors are miniaturized and integrated into the lab on a chip platform, or incorporate high-cost reagents, there is an increased role for microfluidics in sample handling. Thus, RP techniques are also being increasingly used for producing the microfluidic systems associated with biosensors.

One key feature of most of the RP techniques described in this chapter is that they are all data-driven, enabling researchers to evaluate different biosensor parameters rapidly. For the most part, they are also serial manufacturing techniques in that material is added or removed from one point at a time.

Though various groups have published their research on biosensors fabricated by using highly novel fabrication methods (Stokes *et al.*, 2008; Bompart *et al.*, 2012; Ongaro and Ugo, 2012), this chapter will focus on commonly-cited RP techniques. The first section will focus on RP techniques used for fabricating microfluidics for biosensor applications. It will describe micromoulding, xurography, computer numerical control (CNC) milling, laser machining, and finally 3D printing and extrusion processes. The second section will focus on the functionalization of biosensors and will discuss modular surface chemistries, inkjet printing, and microspotting. Finally, this chapter will discuss future trends in these areas.

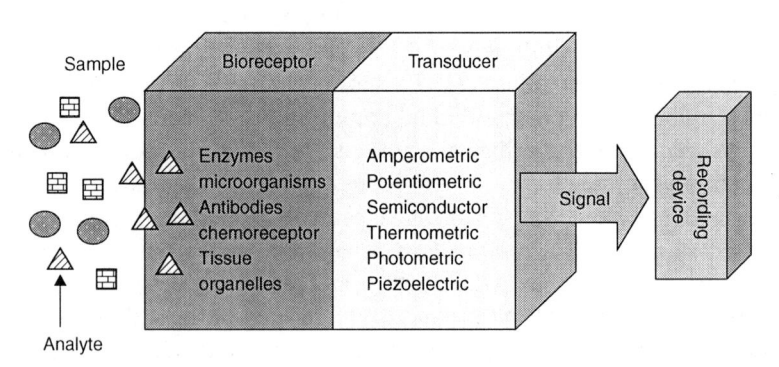

3.1 Schematic outlining common bioelement (bioreceptor) recognition strategies and transformation or transduction methods to produce an electrical output signal (Newman and Setford, 2006).

3.2 Rapid prototyping (RP) of microfluidic systems

This section will begin with a discussion of RP of microfluidic systems for biosensor applications. As 'lab on a chip' systems become more commonplace, biosensors are frequently implemented within microfluidic systems – to minimize analyte and reagent volumes or because the sensing technique relies on a unique physical microfluidic effect (Beebe *et al.*, 2002). In a typical microfluidic biosensor, the analyte is introduced through an injection port and flows along one or more microfluidic channels. The analyte may undergo various reactions before arriving at the actual biosensing element (Beebe *et al.*, 2002; Noh *et al.*, 2011). For examples of the types of biosensors that have been developed from microfluidic systems, we refer the reader to Noh *et al.* (2011).

One of the most widely cited approaches to RP microfluidics is a silicone micromoulding technique popularized by the Whitesides Research Group at Harvard University (Duffy *et al.*, 1998). Briefly, a computer aided design (CAD) model of a microfluidic system is designed and transferred to a high-resolution plotter or photomask writer. This photomask is then used to pattern a thick photosensitive resist thus creating a mould (Natarajan *et al.*, 2008). Finally, polydimethylsiloxane (PDMS) is cast over the mould to produce the microfluidic channels. The microfluidic device is sealed by mechanically fixing or chemically bonding this PDMS structure against a flat surface such as a microscope slide (Duffy *et al.*, 1998; Natarajan *et al.*, 2008). In order to produce high-resolution microfluidics this technique typically relies on semiconductor or MEMS fabrication facilities that may not be accessible to many groups. However, when lower resolutions are acceptable, manufacturing can be undertaken outside of a clean room.

Moreira *et al.* (2009) report producing a photomask on an overhead transparency using a high-resolution photo-plotter (8000 dpi) such as might be found at a commercial printing service. The authors used this photomask to produce a microfluidic electrochemical biosensor with channels of 50 and 100 μm diameter using the PDMS micromoulding technique.

The following example demonstrates the value of RP in optimizing microfluidic biosensors. Frey *et al.* (2010) produced a glucose and lactate biosensor incorporating moulded PDMS channels (see Fig. 3.2). In their device, an injected analyte diffuses through an injected buffer layer before being analysed by a multiplexed enzyme electrode array downstream. Computational fluid dynamics simulations by the authors showed that varying both the PDMS channel dimensions along with the analyte and buffer flow rates affected the uniformity of the analyte concentrations across an electrode array downstream. Thus by modifying channel widths, the dilution ratio of the sample could be held constant at both the glucose and lactate electrodes (see Fig. 3.2).

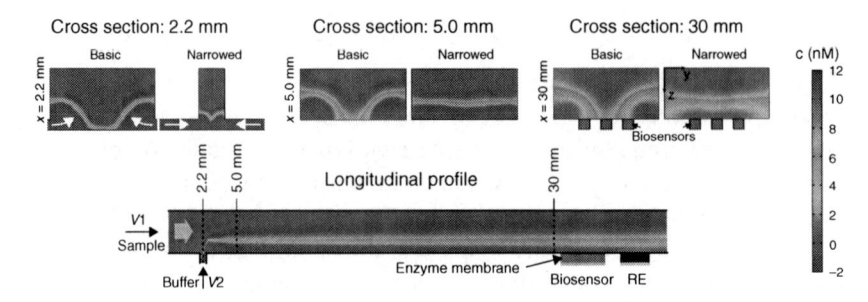

3.2 Simulations of sample concentration at positions 2.2, 5.0, and 30 mm along the microfluidic channel. Cross sections demonstrate the effect of altering injection port geometry on the uniformity of sample dilution (Frey *et al.*, 2010). (*Source*: Modified figure provided by O. Frey.)

Though this example was specific to PDMS moulding, it neatly demonstrates the value of varying channel diameters in prototyping microfluidic biosensors. With this context in mind, the following examples describe how various RP techniques have been used to prototype microfluidic biosensors.

Xurography, from the Greek word for razor (xuron), is a RP technique in which a small cutting blade is traced over a thin polymeric film according to a CAD pattern (Bartholomeusz, 2005). While this technique has been used for RP macroscale structures, Bartholomeusz *et al.* (2005) investigated the use of this technique for prototyping microfluidics. Using a Graphtec FC5100A-75 plotter (Graphtec Inc., Irvine, CA) the authors produced microfluidics from a variety of polymeric films including vinyl, polyester, and thermal adhesives. The authors report that the technique is suitable for films ranging from 25 to 1000 μm thickness, and that resolutions of 18 μm have been obtained in films 25 μm thick (Bartholomeusz, 2005). As xurography relies on tracing a cutting blade across a surface, corners present a potential discontinuity and the authors recommend optimizing the corner cutting algorithm for different films (Bartholomeusz, 2005).

In a practical application, Kim *et al.* (2009) reported using xurography to produce a microfluidic electrolyte sensor and DNA ultrafiltration system . Briefly, 120–180 μm of PDMS was spin-coated onto double-sided tape to produce a thin composite. This composite was patterned by xurography, and then the PDMS side was affixed to a larger PDMS block containing injection ports. The contact adhesive of the tape only leaked at 586 ± 34 kPa and enabled the microfluidic network to be sealed to a variety of substrates including polycarbonate (Kim *et al.*, 2009). The authors produced an electrochemical sensor by bonding the microfluidic channels to a glass slide presenting platinum microelectrodes and a screen-printed Ag/AgCl reference electrode (Kim *et al.*, 2009).

As CNC machining processes have evolved, the resolutions obtainable by milling are now suitable for prototyping microfluidic networks though they are potentially unsuitable for production due to sidewall roughness (Lee *et al.*, 2001). Resolutions as low as 50 μm have been reported for milling microfluidic channels (Lee *et al.*, 2001; Mecomber *et al.*, 2006; Shackman *et al.*, 2007).

Henry *et al.* (2009) report producing a microfluidic distribution system for a biosensor by milling polycarbonate sheets. The biosensor was configured as an immunosensor or a genosensor by grafting either antibodies or complementary DNA segments to the electrode surface. Henry and co-workers report good detection of the cancer marker carcinoembryonic antigen at 10 ng/mL, and a good linear response range for DNA concentrations from 3 to 20 nM (Henry *et al.*, 2009).

Laser machining or ablation of thin polymer films is one of the more commonly used RP techniques for producing microfluidic biosensors. In this technique a laser (typically CO_2) is focused to a spot on a polymer surface, and the material at the spot site evaporates. The substrate is then moved under the laser spot at a controlled rate to produce a pattern by ablation. A patterned polymer sheet can then easily be applied between two substrates to produce a microfluidic system (Hasenbank *et al.*, 2008). Hasenbank *et al.* (2008) report producing a biosensor flow cell by laser machining of a mylar film. This work will be discussed in more detail later in the chapter.

Jin *et al.* (2012) report producing a microfluidic biosensor for nucleic acid detection in polymethyl methacrylate (PMMA) by laser machining. Using a 30 W desktop laser (Epilog Laser, Golden, CO) the authors produced 500 μm wide channels in a 400 μm thick PMMA sheet. The biosensor was operated on the principle of $[Ru(bipy)_3]^{2+}$ electrochemiluminescence under intercalation in double-stranded DNA – indicating hybridization (Lee *et al.*, 2007). In the operation of their biosensor, magnetic microbeads conjugated to reporter DNA segments were combined with a sample containing unknown DNA segments and $[Ru(bipy)_3]Cl_2$ doped nanoparticles. This analyte was micropipetted into the channel, with flow being driven by an absorbent pad at the far end of the device. The magnetic microparticles were separated from the flow by a magnet placed at an observation port on the device at which luminescence could be observed under a microscope (Jin *et al.*, 2012).

Though this microfluidic biosensor was relatively simple, consisting only of the sample channel and a buffer rinsing channel, it is a prime example of minimizing reagent volumes by the use of microfluidic analyte handling. The approach of Jin *et al.* also demonstrates how RP techniques can make microfluidic biosensors available to researchers without specialized microfabrication equipment, and without extensive experience with microfluidics and electronic instrumentation.

In addition to producing microchannels by ablation, Edwards (2010) describes the use of laser milling for spot welding to produce functional valve elements in a microfluidic system. The author produced a stacked microfluidic electrochemical biosensor consisting of a laminated stack of alternating poly (ethylene terephthalate) (PET) and acrylic adhesive layers that were bonded to a substrate on which an electrode array had been patterned (Edwards, 2010). A check valve and a thermally-actuated one-shot valve were integrated by using a 60 W CO_2 laser to tack weld points between layers in the laminate. The details of the tack points and device are shown in Fig. 3.3

In addition to ablation and welding, lasers can also be used in additive techniques such as in stereolithography and two-photon polymerization. For details of these two techniques, we refer the reader to the paper by Gittard and Narayan (2010). An interesting application of such additive processes is in producing cantilevers for cell force measurements. Chan et al. (2012) report using Model 250/50 stereolithography apparatus (3D Systems, Rock Hill, SC) to produce PEG-based cantilevers for cardiomyocyte contractility measurements. The authors were able to control the stiffness of the cantilevers by altering the molecular weight of the gels and by changing the cantilever dimensions.

As a RP technique, extrusion typically refers to the expression of a viscous liquid through an orifice, which subsequently undergoes a phase transition in order to retain its shape. A prototype is patterned from a CAD file by moving either the orifice or the substrate as the material is extruded. For most microfluidic prototyping applications, the extruded material is a microcrystalline wax that serves as a mould for an epoxy cast (Bey-Oueslati et al., 2006). The extruded wax pattern is then melted and removed, leaving a complex microfluidic network with channel diameters in the range of 100–250 μm (Bey-Oueslati et al., 2006).

By comparison, 3D printing involves the layer-by-layer inkjet printing of a photosensitive resin that is cured to produce a shape based on a CAD model. Bonyár et al. (2010, 2012) report using an Eden 250 printer (Objet Geometries, Israel) to produce masters for two moulded PDMS microfluidic devices: a surface plasmon resonance biosensor, and a cell lysis chamber (see Fig. 3.4). The authors state that from an initial CAD model, they can produce a fluidic device in 3–5 h: truly a testament to the flexibility of the technique.

Beyond 3D printing, it is worth noting the use of inkjet printing in the RP of paper-based microfluidic systems. Paper-based biochemical tests are widely used, with two of the most common examples being home pregnancy and influenza tests. These tests operate by being dipped into a liquid sample, which is then wicked upwards along the test strip, carrying analytes with it to reagent regions along the strip. Inkjet printing has been used to produce paper-based immunoassay tests such as might be used for blood-typing (Delaney et al., 2009).

In these paper-based tests, the analyte is drawn through the paper by uncontrolled capillary action. In cases where a potential for cross-talk exists between analysis points, patterned paper-based microfluidic biosensors offer a means of effectively isolating these designated analysis points

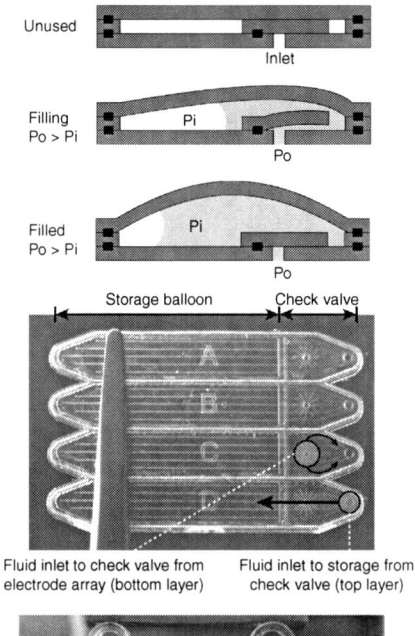

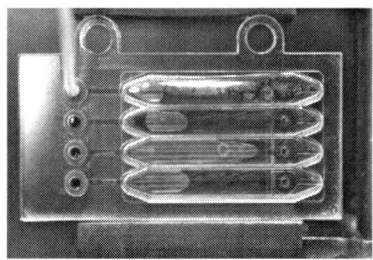

3.3 A microfluidic network produced by laser micromachining. Top – schematic of check-valve showing laser weld points. Middle – Photograph of four parallel sample handling devices (unused). Bottom – Photograph of filled channels; C did not fill due to incomplete sealing (Edwards, 2010). (*Source*: Modified figure provided by T. Edwards.)

from one another by creating a wetting contrast (Martinez *et al.*, 2010). Li *et al.* report using a modified Canon desktop inkjet printer to create hydrophobic boundaries on hydrophilic filter paper (Li *et al.*, 2010a). To produce the hydrophobic patterns, the authors printed a heptane solution of an alkenyl-dimer that binds to cellulose. This was followed by a second printing step in which a colorimetric NO_2^- indicator (citric acid, sulfanilamide, and N-(1-naphthyl)-ethylenediamine) was printed in a detection zone (Li *et al.*, 2010a). In another work, the authors report how paper microfluidics can be adapted for quantitative analysis of NO_2^- concentration (Li *et al.*, 2010b).

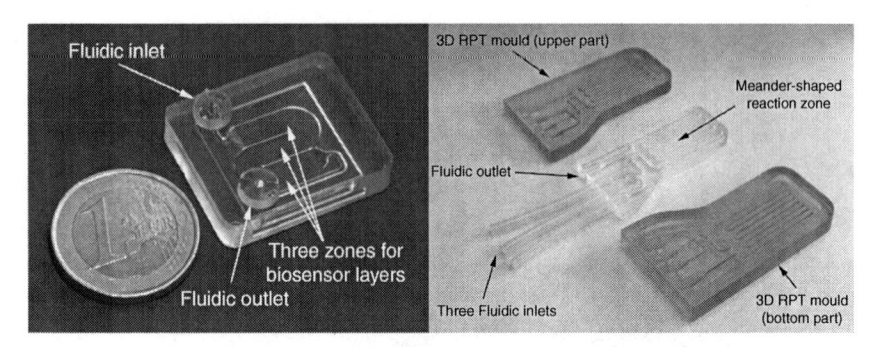

3.4 Left – PDMS flow cell for plasmonic biosensor. Right – PDMS cell lysis chamber, (centre) cast from 3D printed moulds shown *L* and *R* (Bonyár *et al.*, 2010). (*Source*: Modified figure provided by A. Bonyár.)

3.3 Functionalization

While microfluidics provide for sample handling, the bioelement and transducer ultimately form the functional heart of a biosensor (Mohanty and Kougianos, 2006). Thus the development of new functional materials and biomolecules offer the potential to create new biosensors and improve the performance of existing platforms.

The challenge is to evaluate the potential of these new materials before they are adapted to conventional manufacturing processes. For example, incorporating carbon nanotubes (CNTs) in an electrode improves electrical performance in biosensing applications (Wang, 2005). Various methods exist to pattern CNT electrodes, but many involve micromachining. Venkatanarayanan *et al.* (2012) describe a method of producing a CNT electrode by inkjet printing a solution of CNT in dimethylfolmamide (DMF) onto $FeCl_3$ printed upon a printed nafion planarization layer. In another example, Crowley *et al.* (2008) report the inkjet printing of an ammonia sensor using doped polyaniline nanoparticles. Bulk forms of polyaniline are difficult to process and it would have been more complicated to produce such a sensor by conventional means.

As we shall see in this section, RP enables the evaluation of new materials with minimal modification. Most of the RP techniques that will be discussed here are based on solution processing and in some cases a material can even be patterned in the solution in which it was synthesized. By comparison, if a new material is modified to be photolithographically patterned, it is likely to be mixed into a polymer matrix with a photosensitizer and cross-linking agent, and then further optimized for roughness, resolution, and process compatibility (Reichmanis *et al.*, 1991; Shirai and Tsunooka, 1998; Smith *et al.*, 2004). Excessive adulteration of a bioelement or transducer material affects its suitability for biosensing.

Perhaps the most evident ways of applying a new material in a biosensor are dip-coating and spin-coating. While these are not necessarily considered

RP techniques, dip-coating is discussed here to establish the baseline of what is possible and then to contextualize the benefits of RP techniques.

The uniformity of a dried dip- or spin-coated film is difficult to optimize as Marangoni and capillary flows redistribute solutes unevenly (Hurd, 1994; Berteloot et al., 2008; Doumenc and Guerrier, 2010). Bietsch et al. (2004) report dip-coating an array of cantilever biosensors into an aligned array of capillaries, resulting in a simple form of micropatterning. The authors report that aligning the capillaries was time-consuming and that both sides were coated due to the lack of an underlying patterning effect (Bietsch et al., 2004).

Selective surface reactive chemistries, such as self-assembled monolayer (SAM) systems, provide a means of controlling uniformity in dip-coating. When a SAM forms on a surface it occupies all available binding sites and forms a 2D crystal at the surface (Love et al., 2005). Examples of these selectively reactive surface chemistries are the thiol/noble-metal system (Veiseh et al., 2007), the silane/SiO_2 system (Brzoska et al., 1994; Nusz et al., 2006; Jedlicka et al., 2010), and the phosphate/metal-oxide system (Brovelli et al., 1999; Tosatti et al., 2002). These SAM layers can then be chemically functionalized with biomolecules with various biochemical systems such as the N-hydroxy-succinimidyl (NHS), carboxylic acid, or biotin-streptavidin systems (Roberts et al., 2002; Orth et al., 2003). The biotin-streptavidin system is frequently used because biotin (also known as vitamin B7) can be easily bonded to proteins, causing them to link with one of four biotin binding sites on the streptavidin molecule (Orth et al., 2003).

Xu et al. (2008) functionalized a series of silicon photonic wire biosensors with amine-reactive silanes, such that the sensors could later be functionalized with biotin to measure streptavidin binding events. Haes and Van Duyne (2002) report using a similar method to biotinylate a plasmonic Ag nanoparticle biosensor capable of detecting streptavidin binding events. In both cases the biotin-streptavidin binding served as a surrogate to demonstrate the potential of their biosensor to detect other binding events such as antibody–antigen interactions.

Technically it would still be possible to create a multiplexed microstructured biosensor by dip-coating; however, each element would have to be coated independently and then assembled mechanically. McBride et al. (2008) report microfabricating microcantilever sensors that were designed to be detached from one another so that they could be coated independently. While this approach does work, it clearly has limitations with large numbers of biosensors.

Inkjet printing features frequently as a RP technique for biosensor fabrication. As a liquid-based, non-contact patterning technique, it is suited to handling most biomaterials without damaging biosensor components. Also, with the emergence of several research and production inkjet manufacturers in recent years, researchers now have access to the method without having to develop their own in-house experience or equipment from scratch.

Setti and co-workers report fabricating both HRP (Setti *et al.*, 2007) and glucose oxidase (Setti *et al.*, 2005) biosensors by thermal inkjet printing – representing two of the most common types of biosensors. The authors used a Canon i905D (Setti *et al.*, 2007) and Olivetti I-Jet (Olivetti, Aosta, Italy) (Setti *et al.*, 2005) thermal inkjet to deposit conductive (poly(3,4-ethylenedioxythiophene)/poly(styrenesulfonate) (PEDOT/PSS) onto an indium tin oxide (ITO) electrode. The relevant enzyme was then printed on the PEDOT/PSS, and the entire structure was sealed in an acetate-cellulose membrane by dip-coating (Setti *et al.*, 2005, 2007).

One of the earliest examples of using inkjet printing for biosensor fabrication is a work by Kimura *et al.* (1989). The biosensor consisted of an array of ion-sensitive field effect transistors (ISFETs) inkjet patterned with a solution of either bovine serum albumin (BSA)-urease or BSA-glucose oxidase. Using a custom 50 µm diameter piezoelectric inkjet nozzle, the authors printed clusters of 50 droplets of a BSA-enzyme solution at different positions on each ISFET. This BSA-enzyme deposit was cross-linked by placing the functionalized devices into a vapour chamber containing a 25% solution of glutaraldehyde. The biosensor was capable of detecting glucose and urea concentrations ranging from 1–1000 mg/dL (Kimura *et al.*, 1989).

Inkjet manufacturers, such as Fujifilm Dimatix Inc., MicroFab Technologies Inc., and Microdrop Technologies GmbH, offer standard inkjet models with diameters near 50 µm, and many are now marketing inkjet heads that print 1 pL droplet volumes. In extreme cases, droplet diameters of several microns have been reported (Basaran *et al.*, 2013). These resolutions are well-suited to creating multiplexed biosensors.

Kirk *et al.* used a piezoelectric non-contact Scienion S3 Flexarrayer (BioDot, Irvine, CA) to multiplex an array of resonant photonic ring biosensors to study the interaction of various proteins and carbohydrates on the biosensor surfaces. Resonant photonic biosensors are extremely sensitive and operate on the basis of an evanescent field interacting with molecules on the surface of the resonator in a manner somewhat similar to surface plasmon resonant biosensors (Blättler *et al.*, 2009; Mortazavi *et al.*, 2011). Shifts in resonant frequency are measured and correlate with the intensity of interaction of an analyte with the resonant ring or biomolecules at its surface.

The authors report using an 80 µm nozzle to print BSA-conjugated mannose (BSA-mannose), BSA-lactose, BSA-galactose, RNase B, and oligo ethylene-glycol on different resonant sensor rings. The multiplexed biosensor was capable of detecting various lectins in ranges from 200 nM to 1 µM (Kirk *et al.*, 2011) (see Fig. 3.5).

While the patterning resolution of inkjet printing has benefits for multiplexing, it also provides a significant level of control on constructing well ordered coatings at larger length-scales – like printing a photograph. In essence, each dried droplet can be thought of as a building-block of a bioelement or transducer material. While individual droplet compositions, drying properties, and surface interactions can affect the uniformity of individual droplets (Derby 2010), *en masse* the droplets create a macroscopically uniform quality that researchers have used to enhance biosensor performance. This surface uniformity principle is demonstrated in Fig. 3.6.

In another example, O'Toole *et al.* (2009) report inkjet printing a pH sensitive wireless vapour sensor. A colorimetric pH sensitive polymer film was printed between a diode and a photodetector using a Fujifilm Dimatix model DMP-2822 inkjet printer (Fujifilm Dimatix Inc., Santa Clara, CA). The printed material was a solution of bromophenol blue (the pH indicator), a buffering agent (tetrahexylammonium bromide), and ethyl cellulose (matrix) in 1-butanol. The authors optimized the substrate temperature (40°C) and droplet spacing (30 μm) to create ordered uniform pH sensitive films. They report the relative standard deviation of ten inkjet-printed sensors as being

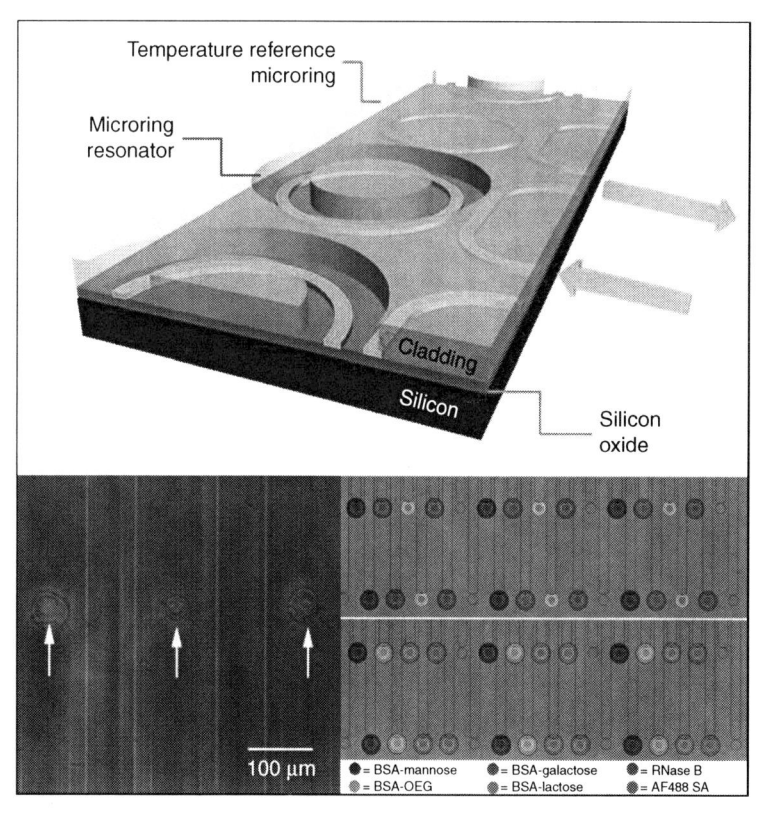

3.5 Top – Schematic of microring resonator biosensor. Bottom – On the left a fluorescence micrograph showing AF488 conjugated streptavidin patterned on microrings; on the right a schematic detailing multiplexing patterns (Kirk *et al.*, 2011).

over 12 times lower than ten equivalent reference sensors produced by drop casting and still 71.7% lower than comparable sensors produced by spin-coating (Pacquit et al., 2006; O'Toole et al., 2009). The sensor output of a series of inkjet-printed sensors vs drop-cast sensors is shown in Fig. 3.7.

Inkjet printing presents a unique challenge when patterning resolutions approach the drop sizes in question. Due to various surface transport, pinning, and coalescence effects caused by droplet drying, printed droplets tend to present a 'coffee-cup' morphology (Derby 2010). This heterogeneous distribution can even be different within a droplet for different ink components such as solvent and surfactant (Filenkova et al., 2011).

Ness et al. (2012) have presented a unique means of creating a uniform biomolecule covering on silicon photonic microcantilevers whose width approaches the printed droplet diameter. After printing a solution of biotinylated BSA onto the cantilevers, the authors placed the devices in a humidity-controlled incubator, causing the droplets to re-swell and coat the cantilevers uniformly (Ness et al., 2012) (see Fig. 3.8). When an ink 'bulb' was present at the base of the cantilever, the authors report that the ink transported salt and other residues off the cantilever, leaving a uniform coating which would doubtless improve sensing.

In another example, Carter et al. (2006) printed a pH-sensitive epoxy onto the tip of an optical fibre, with the droplets retaining a lens-like shape. When epoxy inks are formulated with low fractions of high vapour-pressure solvents, printed droplets tend not to display a 'coffee-cup' morphology and instead retain a somewhat spherical shape (Jacot-Descombes et al., 2012). This effect has been used to produce optical coupling elements by inkjet printing in the past (Cox et al., 2001; Chen et al., 2003).

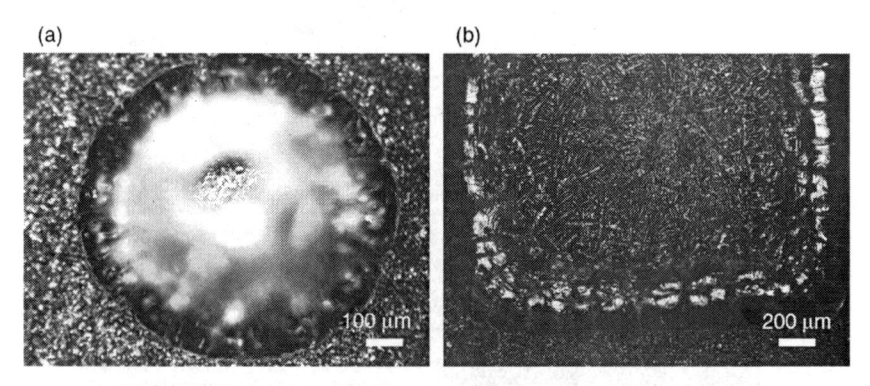

3.6 (a) 4000 droplets of glucose oxidase solution inkjet printed at a single point. The equivalent volume (0.44 µL) could also have been micropipetted by hand. (b) 4000 droplets of glucose oxidase inkjet printed in a 20 × 20 array with ten droplets per site. (*Source*: Reprinted with permission from Wang *et al.* 2008. Copyright 2008, The Electrochemical Society.)

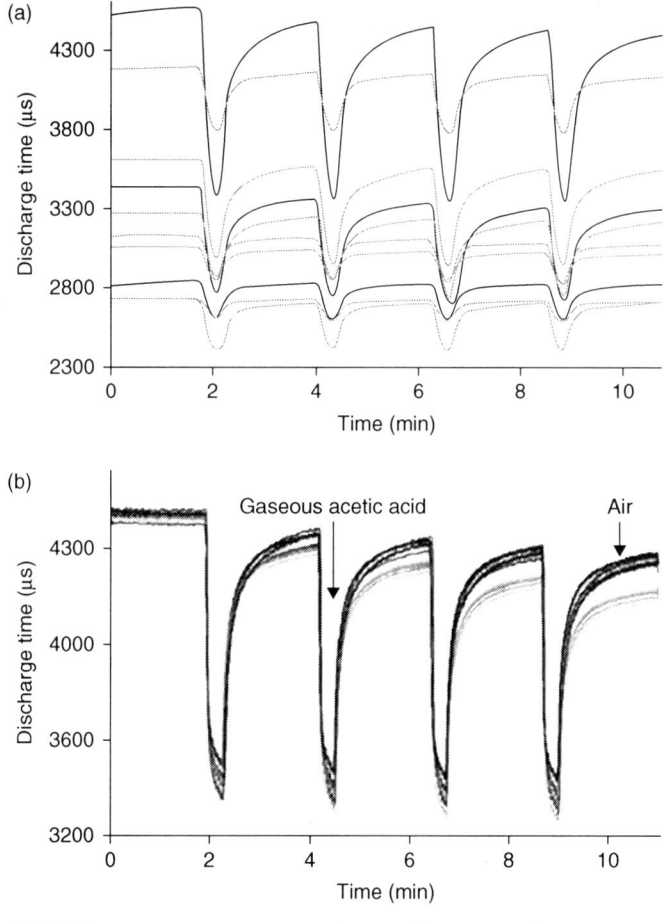

3.7 (a) Sensor response curves for ten drop-cast pH sensors under gaseous acetic acid injection. (b) The response curves of ten inkjet-printed pH sensors showing greatly enhanced uniformity (O'Toole et al., 2009).

To eliminate enzyme aggregates, Turcu et al. (2005) report printing a solution of glucose oxidase in Vinnapas polymer. In the absence of the Vinnipas matrix, the glucose oxidase formed surface aggregates and would dissolve from the surface in aqueous solution (Turcu et al., 2005). The authors suggest that formulating a polymer–enzyme matrix is a trade-off between substrate diffusion in the matrix and the mechanical stability of the printed structure (Turcu et al., 2005).

In addition to enabling multiplexing, the principle of an inkjet droplet being a building block permits control over the surface density of a bioelement at specific sites. In creating a simple biosensor array, Hasenbank et al. (2008) report using a 50 μm diameter MicroFab inkjet head to pattern a biotinylated PEG-thiol molecule into 300 μm squares on a gold substrate before back-

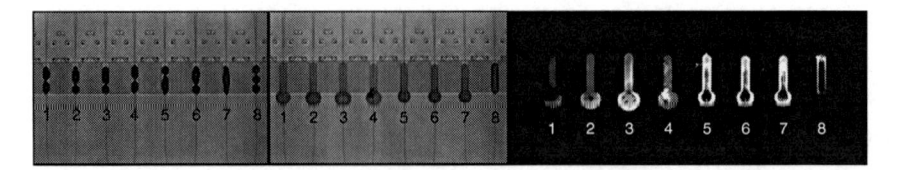

3.8 Left – Micrograph of droplets of biotinylated BSA printed on microcantilevers in varying states of coalescence. Middle – Coalesced and uniformly distributed droplets after incubation in a humidity chamber. Note the beads that have formed near the bases of the cantilevers. Right – Fluorescence image of microcantilevers after incubation in a humidity chamber (Ness *et al.*, 2012).

filling the remaining gold sites. Using the Au-thiol system imparted an extra level of control (as discussed previously) and decreased non-specific biotin adsorption outside the printed patterns. By increasing the number of printing passes at each square from one to eight, the authors were able to increase the density of biotin molecules. When the squares were functionalized with streptavidin and then further with biotinylated HRP or BSA, this directly translated into increased surface protein densities (Hasenbank *et al.*, 2008).

Turcu *et al.* (2005) report fabricating glucose oxidase gradients by dispensing a glucose oxidase and Vinnapas polymer solution with a custom inkjet print head (Laurell *et al.*, 1999; Turcu *et al.*, 2005). The gradient was produced by changing the number of droplets printed at a given point. The authors used scanning electrochemical microscopy to demonstrate the increase in activity caused by increasing droplet number.

An additional advantage of inkjet printing and other RP techniques over dip- or spin-coating is that there is minimal material waste which is an important consideration when the materials being tested are costly or can be synthesized only in small quantities. Crowley estimates that materials waste in inkjet printing is roughly 2%, compared with 95% in spin-coating (Crowley *et al.*, 2008). The waste in an inkjet printer typically corresponds to the dead volume of the device (i.e., cartridge and fluid lines) and the material used to optimize the printing parameters. Some inkjet manufacturers have introduced low-waste inkjet heads for this reason.

A common low-waste head design being adopted consists of a micropipette with an inkjet-style aperture and piezo actuator at one end. Several microlitres of a liquid are aspirated into the pipette by vacuum and then ejected using the piezo actuator. Boero *et al.* (2011) used such a non-contact pipetting system – the sciFLEXARRAYER™ DW (Scienion, Germany) – to print multi-walled CNT electrodes functionalized with glucose oxidase, glutamate oxidase, or lactate oxidase. The authors report droplet diameters of 50–80 μm, and that the system ejected a minimum of 400 pL at each site. As CNTs can aggregate during solvent evaporation, the authors optimized droplet spacing and number to produce uniform electrodes, similar to the methods mentioned above for inkjet printing. The resulting amperometric biosensors had detection limits of 216 μM for glutamate and 115 μM for both glucose and lactate (Boero *et al.*, 2011).

Densmore *et al.* (2009) report using a similar micropipette spotter – a Nano-Plotter NP2.1 from GeSiM (Dresden, Germany) – to functionalize sensing sites on a photonic wire biosensor. A 450 pL droplet containing rabbit or goat antibodies was ejected into a 130 μm diameter sensor cup at each sensor site. Anti-rabbit and anti-goat antibodies were reliably detected by the respective sensor sites at concentrations of 200 nM.

Though these micropipette spotters operate on a similar principle to inkjet printing, they are typically not optimized to eject small diameter droplets. Most manufacturers claim that the minimum volumes ejected by their devices are several hundred picolitres, whereas an optimized high-resolution inkjet printer can eject droplets of less than 10 pL. Despite their somewhat limited use, the larger dispensing volumes of micropipette spotters would appear to be adequate for most bioelement patterning applications, except where high-density multiplexing is required.

Another source of material loss is non-specific adsorption of biomaterials to inkjet surfaces. However, Delehanty and Ligler (2003) found that this can be minimized by pretreating the inkjet with a solution of BSA. Where the goal is merely ensuring the same protein concentration in the printed ink as in the source solution, the inkjet can be flushed with the source solution for several minutes before printing (Nishioka *et al.*, 2004).

3.4 Biomaterials compatibility

One issue in biomaterials printing that has been investigated by various means is whether any damage occurs to molecules during printing. DNA and RNA consist of long nucleotide chains and show rheological similarities to polyelectrolyte. In one of the earliest discussions of microarray fabrication by inkjet printing, Okamoto *et al.* (2000) found no evidence of DNA damage. Presumably this is because DNA and RNA function is determined by covalently bonded nucleotide sequences. Protein and enzyme function is more complicated as it relies on complex three-dimensional folding that is maintained by weaker electron interactions.

Cook *et al.* (2010) investigated the effect of inkjet printing on glucose oxidase activity and structure using a 60 μm MicroFab MJ-AT-01–60 inkjet head (MicroFab Inc., Plano, TX) . The authors found that activity decreased as a function of increasing ejection voltage. No significant changes in protein mass or folding were identified by size-exclusion chromatography, analytical ultra-centrifugation, or circular dichroism, suggesting that protein changes are somewhat occult. Nonetheless, even at the maximum ejection voltage tested, enzyme activity was reduced only by 30%. A related work by Wang *et al.* (2008) using dynamic light scattering also showed no change in molecular mass for inkjet-printed glucose oxidase.

Di Risio and Yan (2007) investigated the effect of several ink viscosity modifiers on the activity of HRP. The modifiers included PEG, ethylene glycol, glycerol, polyvinyl alcohol, carboxymethyl cellulose. The authors found that HRP activity was reduced with increasing molecular weight of the additive and increasing concentration. They suggested that the reduction in activity was caused by impaired diffusion of the enzyme and substrates (Uribe and Sampedro, 2003; Di Risio and Yan, 2007).

3.5 Conclusion and future trends

A growing area in biosensor research is the development of screening tests for cancer. These rely on detecting a protein or other biomarker that is overexpressed by cancer cells (Herold and Rasooly, 2012). In order to detect these biomarkers specifically, a biosensor must be functionalized with some sort of highly specific bioelement (e.g., an antibody). Also, rather than focusing on a specific biomarker, it is now recognized that combined biomarker expression profiles should improve the sensitivity and specificity of cancer screening tests (Herold and Rasooly, 2012). We expect that this inherent multiplexing requirement will create an ongoing role for RP functionalization techniques.

Because biomarkers (e.g., cytokines) may be present only in a sample at extremely low concentrations, much of the development work consists of improving biosensor sensitivities and detection limits (Arlett *et al.*, 2011; Brolo, 2012). As a reference these would be detections in the pM or fM range (Arlett *et al.*, 2011), whereas many of the devices reported in this chapter operate in the nM or μM range. In cases where these new constraints imply evaluating new transducer materials (i.e., graphene or CNT arrays), there would appear to be a continued role for RP techniques. A recent article by Arlett *et al.* (2011) provides an excellent overview of the minimum detection limits of various sensing modalities. It also discusses the thermodynamic issues that result in the trade-off between the sampling time and detection limit (Arlett *et al.*, 2011).

Another growing area in biosensor research is in diagnosing infectious disease to guide treatment (Hays and Leeuwen, 2012). Again, similarly to cancer, it is expected that biomarker profiles are the key to improving sensitivity and specificity in diagnosis (Zhou *et al.*, 2012). An interesting application in pathogen detection is the miniaturization of biosensors for integration with in-dwelling medical devices to screen for nascent infections (Li and Narayan, 2012). Again, RP techniques would seem to be suited to the development of such miniaturized sensors.

A recent trend in RP is that many companies now offer mail-in services where a user can upload a CAD file, and a prototype is then manufactured and dispatched to the client. Examples of service providers include: ProtoCAM (www.protocam.com), Linear Mold Inc. (www.linearmold.com), and Alpha Prototypes Inc. (www.aphaprototypes.com). Depending on the resolutions they seek, a research lab should be able have moulds manufactured to cast their own microfluidics and sample handling systems.

It is expected that as the market for RP increases, new materials – such as elastomers and improved photosensitive glasses – will be developed offering new resolutions and new manufacturing capabilities. Similarly, RP equipment is expected to become more user-friendly, requiring less optimization by research labs.

3.6 Sources of further information and advice

For further information on RP techniques and technology in general, we refer the reader to *Rapid Prototyping: Principles and Applications* (Chua et al., 2010). *Microdrop Generation* by Eric R. Lee (2010) provides an excellent introduction to inkjet printing in general and contains several simple device recipes for groups wishing to manufacture on their own. A review by Derby (2010) provides insights into ink formulation and inkjet physics.

The Society for Imaging Science and Technology holds an annual conference on digital printing and digital fabrication (www.imaging.org). The IEEE Engineering in Medicine and Biology Society is active in biosensor research (www.embs.org). Elsevier, in conjunction with the journal *Biosensors and Bioelectronics*, hosts a biennial congress on biosensors (http://www.biosensors-congress.elsevier.com).

The following companies produce RP equipment and may offer RP services: Stratasys Inc. (www.stratasys.com), 3D Systems Inc. (www.3dsystems.com), envisionTEC GmbH (www.envisiontec.com), Makerbot Industries LLC (www.makerbot.com). The following companies produce inkjet printing or spotter systems: Fujifilm Dimatix Inc. (www.fujifilmusa.com), microdrop Technologies GmbH (www.microdrop.de), MicroFab Technologies Inc. (www.microfab.com), Scienion AG (www.scienion.com).

3.7 References

Arlett, J.L., Myers, E.B. and Roukes, M.L. (2011). Comparative advantages of mechanical biosensors. *Nature Nanotechnology*, **6**(4), pp. 203–215.

Bartholomeusz, D. (2005). Xurography: rapid prototyping of microstructures using a cutting plotter. *Journal of Microelectromechanical Systems*, **14**(6), pp. 1364–1374.

Basaran, O.A., Gao, H. and Bhat, P.P. (2013). Nonstandard inkjets. *Annual Review of Fluid Mechanics*, **45**(1), pp. 85–113.

Beebe, D.J., Mensing, G.A and Walker, G.M. (2002). Physics and applications of microfluidics in biology. *Annual Review of Biomedical Engineering*, **4**, pp. 261–286.

Berteloot, G. Pham, C.T., Daerr, A., Lequeux, F. and Limat, L. (2008). Evaporation-induced flow near a contact line: consequences on coating and contact angle. *EPL (Europhysics Letters)*, **83**(1), p. 14003.

Bey-Oueslati, R. Palm, S.J., Therriault, D. and Martel, S. (2006). High speed direct-write for rapid fabrication of three dimensional microfluidic devices. *International Journal of Heat and Technology*, **26**(1), pp. 125–131.

Bietsch, A. Zhang, J., Hegner, M., Lang, H.P. and Gerber, C. (2004). Rapid functionalization of cantilever array sensors by inkjet printing. *Nanotechnology*, **15**(8), pp. 873–880.

Blättler, T.M. Senn, P., Textor, M., Vörös, J. and Reimhult, E. (2009). Microarray spotting of nanoparticles. *Colloids and Surfaces A: Physicochemical and Engineering Aspects*, **346**, pp. 61–65.

Boero, C., Carrara, S. and De Micheli, G. (2011). New technologies for nanobiosensing and their applications to real-time monitoring. In *2011 IEEE Biomedical Circuits and Systems Conference (BioCAS)*. IEEE, pp. 357–360.

Bompart, M., Haupt, K. and Ayela, C. (2012). Micro and nanofabrication of molecularly imprinted polymers. *Topics in Current Chemistry*, **325**, pp. 83–110.

Bonyár, A. Sántha, H., Ring, B., Varga, M., Gábor Kovács, J. and Harsányi, G. (2010). 3D Rapid Prototyping Technology (RPT) as a powerful tool in microfluidic development. *Procedia Engineering*, **5**, pp. 291–294.

Bonyár, A. Sántha, H., Varga, M. and Ring, B. (2012). Characterization of rapid PDMS casting technique utilizing molding forms fabricated by 3D rapid prototyping technology (RPT). *International Journal of Materials Forming*, (Online First Article).

Brolo, A. (2012). Plasmonics for future biosensors. *Nature Photonics*, **6**, pp. 709–713.

Brovelli, D. Hähner, G., Ruiz, L., Hofer, R., Kraus, G., Waldner, A., Schloesser, J., Oroszlan, P., Ehrat, M. and Spencer, N.D. (1999). Highly oriented, self-assembled alkanephosphate monolayers on tantalum (V) oxide surfaces. *Langmuir*, **15**(13), pp. 4324–4327.

Brzoska, J., Azouz, I. and Rondelez, F. (1994). Silanization of solid substrates: a step toward reproducibility. *Langmuir*, **10**(11), pp. 4367–4373.

Carter, J.C., Alvis, R.M., Brown, S.B., Langry, K.C., Wilson, T.S., McBride, M.T., Myrick, M.L., Cox, W.R., Grove, M.E. and Colston, B.W. (2006). Fabricating optical fiber imaging sensors using ink jet printing technology: a pH sensor proof-of-concept. *Biosensors and Bioelectronics*, **21**(7), pp. 1359–1364.

Chan, V. Jeong, J.H., Bajaj, P., Collens, M., Saif, T., Kong, H. and Bashir, R. (2012). Multi-material bio-fabrication of hydrogel cantilevers and actuators with stereolithography. *Lab on a Chip*, **12**(1), pp. 88–98.

Chua C.K., Leong K.F. and Lim, C.S. (2010). *Rapid Prototyping: Principles and Applications* 3rd ed. World Scientific Publishing Co. Pte. Ltd., pp. 512.

Chen, T. Hayes, D.J. and Wallace, D.B. (2003). Inkjet printing for optical/electrical interfacing of VCSEL and PD arrays. In *SPIE: International Symposium on Microelectronics*. Boston, MA, pp. 864–868.

Clark, L. and Lyons, C. (1962). Enzyme electrode. *Annals of the New York Academy of Sciences*, **102**, pp. 29–45.

Cook, C.C., Wang, T. and Derby, B. (2010). Inkjet delivery of glucose oxidase. *Chemical Communications (Cambridge, England)*, **46**, pp. 5452–5454.

Cooper, J. and Cass, A. (2004). Biosensors. *The Practical Approach Series*, Volume 268, Oxford, Oxford University Press.

Cox, W. Chen, T., Hayes, D.J. and Grove, M.E. (2001). Low-cost fiber collimation for MOEMS switches by ink-jet printing. In M. E. Motamedi and R. Goering, eds. *SPIE Proc. Vol. 4561 MOEMS and Miniaturized Systems II*. pp. 93–101.

Crowley, K. O'Malley, E., Morrin, A., Smyth, M.R. and Killard, A.J. (2008). An aqueous ammonia sensor based on an inkjet-printed polyaniline nanoparticle-modified electrode. *The Analyst*, **133**(3), pp. 391–399.

Cui, Y. Wei, Q., Park, H. and Lieber, C.M. (2001). Nanowire nanosensors for highly sensitive and selective detection of biological and chemical species. *Science*, **293**, pp. 1289–1292.

Delaney, J.T., Smith, P.J. and Schubert, U.S. (2009). Inkjet printing of proteins. *Soft Matter*, **5**, pp. 4866–4877.

Delehanty, J. and Ligler, F. (2003). Method for printing functional protein microarrays. *Biotechniques*, **34**(2), pp. 380–385.

Densmore, A. Vachon, M., Xu, D.X., Janz, S., Ma, R., Li, Y.H., Lopinski, G., Delâge, A., Lapointe, J., Luebbert, C.C., Liu, Q.Y., Cheben, P. and Schmid, J. H. (2009). Silicon photonic wire biosensor array for multiplexed real-time and label-free molecular detection. *Optics Letters*, **34**(23), pp. 3598–3600.

Derby, B. (2010). Inkjet printing of functional and structural materials: fluid property requirements, feature stability, and resolution. *Annual Review of Materials Research*, **40**(1), pp. 395–414.

Doumenc, F. and Guerrier, B. (2010). Drying of a solution in a meniscus: a model coupling the liquid and the gas phases. *Langmuir : The ACS Journal of Surfaces and Colloids*, **26**(17), pp. 13959–13967.

Drummond, T.G., Hill, M.G. and Barton, J.K. (2003). Electrochemical DNA sensors. *Nature Biotechnology*, **21**(10), pp. 1192–1199.

Duffy, D.C. McDonald, J.C., Schueller, O.J., and Whitesides, O.J. (1998). Rapid prototyping of microfluidic systems in poly(dimethylsiloxane). *Analytical Chemistry*, **70**(23), pp. 4974–4984.

Edwards, T.L. (2010). A system of parallel and selective microchannels for biosensor sample delivery and containment. In *Sensors, 2010 IEEE*, pp. 1460–1463, 1–4 Nov. 2010, Kona, HI, 10.1109/icsens.2010.5690370, 1930-0395

Filenkova, A. Acosta, E., Brodersen, P.M., Sodhi, R.N.S. and Farnood, R. (2011). Distribution of inkjet ink components via ToF-SIMS imaging. *Surface and Interface Analysis*, **43**(1–2), pp. 576–581.

Frey, O. Talaei, S., van der Wal, P.D., Koudelka-Hep, M. and de Rooi, N.F. (2010). Continuous-flow multi-analyte biosensor cartridge with controllable linear response range. *Lab on a Chip*, **10**, pp. 2226–2234.

Fritz, J. (2008). Cantilever biosensors. *The Analyst*, **133**(7), pp. 855–863.

Gittard, S. and Narayan, R. (2010). Laser direct writing of micro-and nano-scale medical devices. *Expert Review of Medical Devices*, **7**(3), pp. 343–356.

Grieshaber, D. MacKenzie, R., Voeroes, J. and Reimhult, E. (2008). Electrochemical biosensors—sensor principles and architectures. *Sensors*, **8**(3), pp. 1400–1458.

Haes, A.J. and Van Duyne, R.P. (2002). A nanoscale optical biosensor: sensitivity and selectivity of an approach based on the localized surface plasmon resonance spectroscopy of triangular silver nanoparticles. *Journal of the American Chemical Society*, **124**(35), pp. 10596–10604.

Hasenbank, M.S. Edwards, T., Fu, E., Garzon, R., Kosar, T.F., Look, M., Mashadi-Hossein, A. and Yager, P. (2008). Demonstration of multi-analyte patterning using piezoelectric inkjet printing of multiple layers. *Analytica Chimica Acta*, **611**(1), pp. 80–88.

Hays, J. and Leeuwen, W. Van (2012). *The Role of New Technologies in Medical Microbiological Research and Diagnosis*, The Role of New Technologies in

Medical Microbiological Research and Diagnosis (Bentham Science Publishers, Bussum, The Netherlands, 2012).

Henry, O.Y. Fragoso, A., Beni, V., Laboria, N., Sánchez, J.L.A., Latta, D., Von Germar, F., Drese, K., Katakis, I. and O'Sullivan, C.K. (2009). Design and testing of a packaged microfluidic cell for the multiplexed electrochemical detection of cancer markers. *Electrophoresis*, **30**, pp. 3398–3405.

Herold, K. and Rasooly, A. (2012). *Biosensors and Molecular Technologies for Cancer Diagnostics*, Sensors series (CRC Press/Taylor & Francis, Boca Raton, FL, 2012).

Hurd, A. (1994). Evaporation and surface tension effects in dip coating. *Advances in Chemistry Series*, pp. 433–450.

Jacot-Descombes, L. Gullo, M.R., Cadarso, V.J. and Brugger, J. (2012). Fabrication of epoxy spherical microstructures by controlled drop-on-demand inkjet printing. *Journal of Micromechanics and Microengineering*, **22**(7), p. 074012.

Jedlicka, S., Rickus, J. and Zemlyanov, D. (2010). Controllable surface expression of bioactive peptides incorporated into a silica thin film matrix. *The Journal of Physical Chemistry*, **114**(1), pp. 342–344.

Jin, S. Dai, M., He, F., Wang, Y., Ye, B.-C. and Nugen, S.R. (2012). Development and characterization of a capillary-flow microfluidic device for nucleic acid detection. *Microsystem Technologies*, **18**(6), pp. 731–737.

Kim, J., Surapaneni, R. and Gale, B.K. (2009). Rapid prototyping of microfluidic systems using a PDMS/polymer tape composite. *Lab on a Chip*, **9**, pp. 1290–1293.

Kimura, J., Kawana, Y. and Kuriyama, T. (1989). An immobilized enzyme membrane fabrication method using an ink jet nozzle. *Biosensors*, **4**, pp. 41–52.

Kirk, J.T. Fridley, G.E., Chamberlain, J.W., Christensen, E.D., Hochberg, M. and Ratner, D.M. (2011). Multiplexed inkjet functionalization of silicon photonic biosensors. *Lab on a Chip*, **11**, pp. 1372–1377.

Laurell, T., Wallman, L. and Nilsson, J. (1999). Design and development of a silicon microfabricated flow-through dispenser for on-line picolitre sample handling. *Journal of Micromechanics and Microengineering*, **9**, pp. 369–376.

Lee, E. (2010). *Microdrop Generation*, Nano- and Microscience, Engineering, Technology and Medicine (CRC Press, Boca Raton, FL, 2010).

Lee, J.-G. Yun, K., Lim, G.-S., Lee, S.E., Kim, S. and Park, J.-K. (2007). DNA biosensor based on the electrochemiluminescence of Ru(bpy)3(2+) with DNA-binding intercalators. *Bioelectrochemistry*, Amsterdam, The Netherlands, **70**(2), pp. 228–234.

Lee, L. Madou, M.J., Koelling, K.W. and Daunert, S. (2001). Design and fabrication of CD-like microfluidic platforms for diagnostics: polymer-based microfabrication. *Biomedical Microdevices*, **3**(4), pp. 339–351.

Li, C. and Narayan, R.K. (2012). Development of a novel catheter for early diagnosis of bacterial meningitis caused by the ventricular drain. In *2012 IEEE 25th International Conference on Micro Electro Mechanical Systems (MEMS)*. Paris, pp. 120–123, Jan. 29-Feb. 2, Paris.

Li, X., Tian, J., Garnier, G. and Shen, W. (2010a). Fabrication of paper-based microfluidic sensors by printing. *Colloids and Surfaces. B, Biointerfaces*, **76**, pp. 564–570.

Li, X., Tian, J. and Shen, W. (2010b). Quantitative biomarker assay with microfluidic paper-based analytical devices. *Analytical and Bioanalytical Chemistry*, **396**(1), pp. 495–501.

Love, J. Estroff, L.A., Kriebel, J.K., Nuzzo, R.G. and Whitesides, G.M. (2005). Self-assembled monolayers of thiolates on metals as a form of nanotechnology. *Chemical Reviews*, **105**(4), pp. 1103–1169.

Martinez, A.W. Phillips, S.T., Whitesides, G.M. and Carrilho, E. (2010). Diagnostics for the developing world: microfluidic paper-based analytical devices. *Analytical Chemistry*, **82**(1), pp. 3–10.

McBride, K.W. Snow, D.E., Walters, S., Jernigan, Z., Weeks, B.L. and Dallas, T. (2008). Decoupling functionalization from sensor array assembly using detachable cantilevers. *Scanning*, **30**(2), pp. 203–207.

Mecomber, J.S. Stalcup, A.M., Hurd, D., Halsall, H. B., Heineman, W.R., Seliskar, C.J., Wehmeyer, K.R. and Limbach, P.A. (2006). Analytical performance of polymer-based microfluidic devices fabricated by computer numerical controlled machining. *Analytical Chemistry*, **78**(3), pp. 3853–3858.

Mohanty, S. and Kougianos, E. (2006). Biosensors: a tutorial review. *Potentials IEEE*, **25**(2), pp. 35–40.

Moreira, N.H. de Almeida, A.L.D.J., Piazzeta, M.H.D.O., de Jesus, D.P., Deblire, A., Gobbi, A.L. and da Silva, J.A.F. (2009). Fabrication of a multichannel PDMS/glass analytical microsystem with integrated electrodes for amperometric detection. *Lab on a Chip*, **9**, pp. 115–121.

Mortazavi, D. Kouzani, A.Z., Kaynak, A. and Duan, W. (2011). Nano-plasmonic biosensors: a review. In *The 2011 IEEE/ICME International Conference on Complex Medical Engineering*. Harbin, China: IEEE, pp. 31–36, May 22–25.

Natarajan, S., Chang-Yen, D.A and Gale, B.K. (2008). Large-area, high-aspect-ratio SU-8 molds for the fabrication of PDMS microfluidic devices. *Journal of Micromechanics and Microengineering*, **18**(4), p. 045021.

Ness, S.J. Kim, S., Woolley, A.T. and Nordin, G.P. (2012). Single-sided inkjet functionalization of silicon photonic microcantilevers. *Sensors and Actuators B: Chemical*, **161**, pp. 80–87.

Newman, J.D. and Setford, S.J. (2006). Enzymatic biosensors. *Molecular Biotechnology*, **32**(3), pp. 249–268.

Nishioka, G., Markey, A.A. and Holloway, C.K. (2004). Protein damage in drop-on-demand printers. *Journal of the American Chemical Society*, **126**(50), pp. 16320–16321.

Noh, J., Kim, H.C. and Chung, T.D. (2011). Biosensors in microfluidic chips. *Topics in Current Chemistry*, **304**, pp. 117–152.

Nusz, G. Johannes, E., Allen, N.S. and Hallen, H.D. (2006). Self-assembled monolayer coating of biological probes to avoid protein adhesion. *Nano Biotechnology*, **2**, pp. 61–65.

Okamoto, T., Suzuki, T. and Yamamoto, N. (2000). Microarray fabrication with covalent attachment of DNA using bubble jet technology. *Nature Biotechnology*, **18**(4), pp. 438–441.

Ongaro, M. and Ugo, P. (2012). Bioelectroanalysis with nanoelectrode ensembles and arrays. *Analytical and Bioanalytical Chemistry*, **405**, pp. 3715–3729.

Orth, R.N., Clark, T.G. and Craighead, H.G. (2003). Biosensors : avidin-biotin micropatterning methods for biosensor. *Biomedical Microdevices*, **5**(3), pp. 29–34.

O'Toole, M. Shepherd, R., Wallace, G.G. and Diamond, D. (2009). Inkjet printed LED based pH chemical sensor for gas sensing. *Analytica Chimica Acta*, **652**, pp. 308–314.

Pacquit, A. Lau, K.T., McLaughlin, H., Frisby, J., Quilty, B. and Diamond, D. (2006). Development of a volatile amine sensor for the monitoring of fish spoilage. *Talanta*, **69**, pp. 515–520.

Reichmanis, E. Houlihan, F.M., Nalamasu, O. and Neenan, T.X. (1991). Chemical amplification mechanisms for microlithography. *Chemistry of Materials*, **4**(218), pp. 394–407.

Di Risio, S. and Yan, N. (2007). Piezoelectric ink-jet printing of horseradish peroxidase: effect of ink viscosity modifiers on activity. *Macromolecular Rapid Communications*, **28**, pp. 1934–1940.

Roberts, M.J., Bentley, M.D. and Harris, J.M. (2002). Chemistry for peptide and protein PEGylation. *Advanced Drug Delivery Reviews*, **54**(4), pp. 459–476.

Setti, L. Fraleoni-Morgera, A., Ballarin, B., Filippini, A., Frascaro, D. and Piana, C. (2005). An amperometric glucose biosensor prototype fabricated by thermal inkjet printing. *Biosensors and Bioelectronics*, **20**(10), pp. 2019–2026.

Setti, L. Fraleoni-Morgera, A., Mencarelli, I., Filippini, A., Ballarin, B. and Dibiase, M. (2007). An HRP-based amperometric biosensor fabricated by thermal inkjet printing. *Sensors and Actuators B: Chemical*, **126**(1), pp. 252–257.

Shackman, J.G., Munson, M.S. and Ross, D. (2007). Gradient elution moving boundary electrophoresis for high-throughput multiplexed microfluidic devices. *Analytical Chemistry*, **79**(2), pp. 565–571.

Shirai, M. and Tsunooka, M. (1998). Photoacid and photobase generators: prospects and their use in the development of polymeric photosensitive systems. *Bulletin of the Chemical Society of Japan*, **71**(11), pp. 2483–2507.

Smith, M.D., Byers, J.D. and Mack, C.A. (2004). The lithographic impact of resist model parameters. In J. L. Sturtevant ed. *Proc. SPIE 5376, Advances in Resist Technology and Processing XXI*. Santa Clara, CA, pp. 322–332.

Stokes, R.J. Dougan, J., Irvine, E., Haaheim, J., Stiles, P.L., Levesque, T. and Graham, D. (2008). Fabrication of biosensor arrays via DPN and detection by surface enhanced resonance Raman scattering. In M.I. Stockman ed. *Proc. SPIE 7032, Plasmonics: Metallic Nanostructures and Their Optical Properties VI*. San Diego, CA, p. 70320W–70320W–9.

Tosatti, S. Michel, R., Textor, M. and Spencer, N.D. (2002). Self-assembled monolayers of dodecyl and hydroxy-dodecyl phosphates on both smooth and rough titanium and titanium oxide surfaces. *Langmuir*, **18**(9), pp. 3537–3548.

Turcu, F. Hartwich, G., Schäfer, D. and Schuhmann, W. (2005). Ink-jet microdispensing for the formation of gradients of immobilised enzyme activity. *Macromolecular Rapid Communications*, **26**(4), pp. 325–330.

Uribe, S. and Sampedro, J.G. (2003). Measuring solution viscosity and its effect on enzyme activity. *Biological Procedures Online*, **5**(1), pp. 108–115.

Veiseh, M. Veiseh, O., Martin, M.C., Asphahani, F. and Zhang, M. (2007). Short peptides enhance single cell adhesion and viability on microarrays. *Langmuir*, **23**(8), pp. 4472–4479.

Venkatanarayanan, A. Crowley, K., Lestini, E., Keyes, T.E., Rusling, J.F. and Forster, R.J. (2012). High sensitivity carbon nanotube based electrochemiluminescence sensor array. *Biosensors and Bioelectronics*, **31**, pp. 233–239.

Wang, J. (2005). Carbon-nanotube based electrochemical biosensors: a review. *Electroanalysis*, **17**(1), pp. 7–14.

Wang, T., Cook, C. and Derby, B. (2008). Inkjet printing glucose oxidase for biosensor applications. *ECS Transactions*, **16**(11), pp. 15–20.

Xu, D. Densmore, A., Delâge, A., Waldron, P., McKinnon, R., Janz, S., Lapointe, J., Lopinski, G., Mischki, T., Post, E., Cheben, P. and Schmid, J.H. (2008). Folded cavity SOI microring sensors for high sensitivity and real time measurement of biomolecular binding. *Optics Express*, **16**(19), pp. 15137–15148.

Zhou, Q. Kwa, T., Liu, Y. and Revzin, A. (2012). Cytokine biosensors: the future of infectious disease diagnosis? *Expert Review of Anti-Infective Therapy*, **10**(10), pp. 1079–1081.

4
Rapid prototyping technologies for tissue regeneration

V. TRAN and X. WEN,
Virginia Commonwealth University, USA

DOI: 10.1533/9780857097217.97

Abstract: Tissue engineering and regenerative medicine hold a great promise for restoring functional tissues and organs and yet is still limited by its inability to reproduce macro- and micro-structures of native tissues and organs. The limitation may be overcome by fully mimicking those structures with computer-aided design (CAD) and fabrication. To date, rapid prototyping technologies with layer-by-layer construction provide a very powerful tool to fabricate intricate 3D scaffold and/or cell/tissue constructs with precisely controlled macro- and micro-features. Moreover, fine features, needed to surpass the oxygen diffusion problem at 100–200 μm, can be easily achieved with rapid prototyping technologies compared to traditional scaffold fabrication approaches. Over the last two decades, more than 20 rapid prototyping devices have been developed and used in laboratories for biomedical applications. In this review, these devices are categorized into laser-assisted based, extrusion or dispensing-based, and inkjet-based rapid prototyping technologies. Depending on specific technologies and types of materials used, rapid prototyping technologies may be used in engineering hard and soft tissues or even whole organs. These applications will be discussed along with their advantages, shortcomings, and future trends.

Key words: biomaterials, tissue engineering, regenerative medicine, biomimetics, rapid prototyping.

4.1 Introduction

Tissue engineering and regenerative medicine are interdisciplinary fields, drawing on expertise from material science, chemical engineering, mechanical engineering, computer science, life science, medical science, and so on (Lalan *et al.*, 2001; Langer and Vacanti, 1993). These fields have attracted a great deal of attention in recent years because of their potential to address several major clinical challenges, such as shortages of donor organs and tissues, chronic rejection, cell morbidity, etc. (Yeong *et al.*, 2004). Tissue engineering and regenerative medicine were envisioned to regenerate functional tissues and organs, build *in vitro* testing systems, and partially reduce the needs of using live animals in research (Langer and Vacanti, 1993). Today, the scope of tissue engineering and regenerative medicine has expanded to include their promise as a tool for preventative medicine as well. Traditional tissue engineering and regenerative medicine approaches include (1) cell transplantation technology, which allows direct injection of individual cells or microtissues into the damaged tissues and relies on the host tissue for vascular supply, engraftment, and reorganization; (2) bioactive encapsulation technology, which allows the delivery of bioactive molecules into the lesion sites to promote regeneration; and (3) the implantation of cell-loaded scaffolds or engineered tissues into the lesion sites. Of these technologies, the scaffold-based tissue engineering technology has been one major research area over the last 20 years, due to the possibility of using scaffolds to mimic the extracellular structures in the native tissue. In addition, scaffolds also provide 3D culture systems for many other research areas, such as drug screening, cancer research, developmental biology, and so on (Hutmacher et al., 2010). Since the major roles of scaffolds are to mimic the extracellular matrix (ECM) in the tissue for accommodating cells and guiding their growth and functional regeneration in three dimensions, its three-dimensional feature provides the necessary support for cells to attach, proliferate, migrate, and maintain their differentiated function, and its architecture defines the ultimate shape and histological structure of the new tissues and organs. Therefore the technologies used to fabricate scaffolds may have significant impact on the outcome of 3D tissue and organ growth. Often, scaffolds manufactured by conventional methods, such as solvent casting, particulate leaching, gas forming, freeze drying, emulsification, or phase separation, possess inherent hindrances in porosity control, pore size, interconnectivity, inhomogeneities of pore networks, and scaffold geometry. Intensive use of solvents during polymer processing may leave solvent residuals which are toxic to cells. In addition, it is difficult for these methods to load bioactive molecules into constructs (Peltola et al., 2008). These challenges limit the use of these methods of scaffold fabrication for tissue engineering applications.

Emerging as an alternative to the conventional scaffold fabrication technologies, rapid prototyping has been employed in the biomedical area since the 1990s (Webb, 2000). Rapid prototyping is also recognized under many different names, such as desktop processing, layer technology, freeform fabrication, additive manufacturing, or bottom-up. The field emerged in the late 1980s when personal computer technology evolved, mainly for producing models and prototypes in a layer-by-layer manner (Upcraft and Fletcher, 2003; Wüst et al., 2011)). A typical rapid prototyping technique begins with a CAD rendering of a product in the industrial standardized format, standard tessellation language (STL). The CAD design is based on computer tomography (CT) or magnetic resonance imaging (MRI) data scanned from a patient in a 3D format. The design then is run through a slicing software to obtain numeric data for individual layers. The data is fed to a controller of the rapid prototyping device that fabricates the scaffold from its bottom, proceeding upward. Each layer is bonded to the previous layer to form a 3D object. Post processing, such as surface finishing and cleaning, may be needed, depending on the technology used. Usually a rapid prototyping device consists of a stock (raw material) container, a product (build) container, a stock delivery system, a fabrication motion station, and a rapid prototyping tool (Fig. 4.1).

In contrast to conventional scaffold fabrication methods in which a mold or a cast is necessary, rapid prototyping technologies offer virtually any design of constructs with intricate and precisely controllable geometries and microstructures as well as macroscopic features. Rapid prototyping is a magnet in biomedical applications, especially in prostheses and biomedical prototype fabrication, as it has a number of advantages: no tooling, minimal set-up time, minimal design costs, fewer process steps, and mass customization, to name a few. In particular, its ability to match anatomical

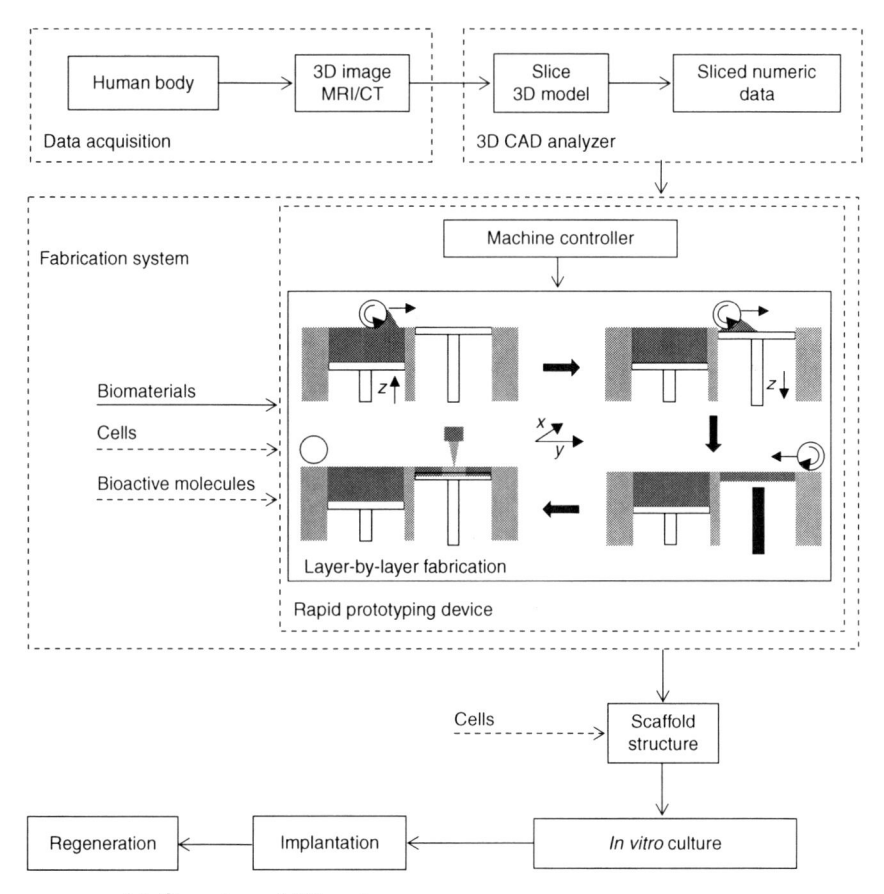

4.1 Overview of RP systems.

and histological features of a patient and the demands of the surgeon by means of CT and MRI 3D imaging, while reducing waiting time, makes it very efficient. Rapid prototyping (RP) systems provide reproducible, high throughput and scalable products while keeping precise control of the micro- and macrostructures of the constructs.

4.2 Rapid prototyping (RP) technologies in tissue regeneration

The earliest applications of rapid prototyping technologies were in the craniomaxillofacial areas. Their primary use was to fabricate models for surgeons to plan surgical procedures in order to achieve optimal outcomes. One such use was to print skull models from patients for surgical planning on reshaping craniosynostosis (David et al., 2004). Auricular prostheses have also much benefited from rapid prototyping technologies since 1999 through stereolithography (SLA) (Goiato et al., 2011). In addition to manufacturing implants, rapid prototyping has gradually become a powerful tool for supporting tissue and organ regeneration. These tissue regeneration applications of RP encompass three main directions.

4.2.1 Scaffold fabrication

Biocompatible scaffolds with microstructures serve as an artificial ECM for anchor-dependent cells to organize into 3D tissue-like structures and for the delivery of stimuli to cells to produce their own matrix as the scaffolds degrade. Scaffolds with appropriate 3D microstructures and biomechanical properties may be used for supporting cell attachment, growth, proliferation, and functional differentiation (Hutmacher et al., 2007; Karageorgiou and Kaplan, 2005; Peltola et al., 2008). 3D scaffolds may provide important guidance cues for tissue regeneration. The guidance cues could be geometrical, biomechanical, and/or biochemical. Therefore, the ideal scaffolds may provide all these cues closely mimicking native tissues to support their regeneration (Bártolo et al., 2011).

Geometrically, a tissue-resembling structure is likely to guide cells to form tissue-like structure (Leong et al., 2003). In order to do so, cells need an interconnective and open-pore structure for ingrowth (Yang et al., 2001). The interconnectivity of the pore structure further assists mass transport, diffusion of nutrients and waste products and other processes during tissue healing and normal physiological function. For scaffolds fabricated with traditional techniques, such as salt leaching, a maximum penetration depth of 240 μm of a mineralized matrix and a 2 mm penetration depth from the surface observed in poly(L-lactide–co-glycolide) (PLGA) foams and poly-β-caprolactone- tricalcium phosphate (PCL-TCP) scaffolds, respectively, suggested limited cell infiltration and nutrient supply (Ishaug *et al.*, 1997; Zhang *et al.*, 2009). The interconnectivity and porosity of the scaffolds fabricated using rapid prototyping technology are believed to achieve much greater penetration depth due to their ability to control the pore structure and degree in the whole scaffold. Porous constructs yield larger surface areas for cells to proliferate. In general, scaffolds with porosity as high as 90% facilitate better cell infiltration, ingrowth, and ECM deposition (Zeltinger *et al.*, 2001). Another important consideration for porous scaffolds is pore size and distribution, which is particularly important because most tissues require more than one cell type and can also be well controlled through rapid prototyping. Not surprisingly, cell-substrate behaviors are distinctive due to their differences in cell size and cell-matrix interactions (Salem *et al.*, 2002; Leong *et al.*, 2008). When designing a scaffold, pore size must thus take into account cell size to attain optimal cell growth and bioactivity. For example, a pore size of 5 μm or larger is necessary for neovascularization whereas 20 μm or larger is essential for ingrowth of hepatocytes whose size is about 20–40 μm, and 100–350 μm for bone regeneration to house and bridge osteoblasts of 20–30 μm size (Yang *et al.*, 2001). Pore size, porosity, and interconnectivity have also been demonstrated to influence osteogenic signal expression and differentiation (Kim *et al.*, 2010).

In addition to geometrical cues, mechanical cues have been investigated for scaffolds to match host tissue to prevent stress shielding and support mechanotransduction between scaffolds and cells, as well as physiological loadings *in vivo* and neotissue growth (Leong *et al.*, 2008). In load bearing tissues like bone, for instance, bone turnover or remodeling shows clear evidence of how the mechanical environment influences the tissue to adapt to external stresses (in Leong *et al.*, 2008). Cells are believed to be able to sense the mechanical properties of the adhesion substrates and adjust their shape and structure accordingly through the regulation of integrin binding as well as the assembly of focal adhesion apparatus and the cytoskeleton (Wang *et al.*, 2001). The mechanical properties of scaffolds include their stiffness and strength. It appears that ECM stiffness affects its responsiveness to external loads, consequently changing interstitial fluid flow and cell–cell distance and signaling (Griffith and Swartz, 2006). Scaffold stiffness can be characterized as a function of porosity and permeability. Increasing scaffold stiffness may reduce its porosity and permeability resulting in improper ingrowth of neotissues, while increasing scaffold porosity may render a higher permeability but compromise its stiffness (Chen *et al.*, 2011). As a consequence, optimal scaffold structures are crucial for the desired biological performance of scaffolds. It was confirmed by computational modeling and simulation that scaffold elastic tensor should match or be slightly higher than the elastic properties of the host bone due to the change of the construct stiffness in parallel with neotissue ingrowth

(Sturm et al., 2010). Serving as temporary substrates, biodegradable scaffolds should also maintain their integrity throughout the tissue regeneration process. Yet bioengineers are still trying to balance the rate of scaffold material with tissue modeling. The degradation of scaffolds in terms of their mechanical strength, and hence structural stability, is one of the major challenges of tissue regeneration in general and has been well documented in literature (Rezwan et al., 2006).

Biomolecular cues are also of importance in tissue healing. Tissue engineering scaffolds can be loaded with growth factors and other signaling molecules that stimulate cells to differentiate onto the desired pathway or with drugs to deliver to the defective area to promote healing. Signaling molecules can be attached onto scaffolds by chemical conjugation, by physical adsorption, or by entrapment within scaffolds primarily by mixing of the scaffold material with the signaling and therapeutic molecules before processing (Sokolsky-Papkov et al., 2007). One issue with conventionally manufactured scaffolds is the difficulty in preserving function and conformation of such proteins. Eliminating the use of solvent and temperature dependent processes in some rapid prototyping techniques presents a potential solution for growth factor loaded scaffolds.

To fully capitalize on the advantages of a scaffold-based approach, several other criteria should be considered for functional regeneration of tissues or organs. Surface properties, such as surface chemistry and surface topography that regulate cell attachment, along with ECM deposition, must be appropriately controlled throughout the RP manufacturing process. Rapid prototyping techniques allow a consistency in the microarchitecture obtained owing to the accuracy of the processes. This advantage allows for the ease of repeating the experiments with highly reproducible structures. Finally, the promise and premise of rapid prototyping in scaffold fabrication lie in its resolution. Later in this review while discussing each type of rapid prototyping techniques, it will be seen that resolution ranging from a few microns to nanometers undoubtedly favors the use of RP in tissue regeneration applications.

4.2.2 Hydrogel-based printing

Despite the anatomic resemblance of porous scaffolds to natural tissues, cells cannot be readily entrapped within these scaffolds during the fabrication process. Degradable hydrogels, therefore, have been considered an excellent candidate for this purpose. In this capacity, hydrogels provide a supportive 3D environment for cell survival, proliferation, migration, and differentiation. A hydrogel can also be a good carrier for signaling molecules or therapeutic drugs to direct tissue healing or new tissue formation. Hydrogels have a wide utility as injectable systems, desirable for minimally invasive surgery. However, most hydrogels do not possess sufficient mechanical strength for load bearing tissues, even though their mechanical strength can be substantially reinforced through crosslinkers and elastic motifs. Nevertheless, hydrogels can assist in load bearing applications for critical size defects as a promoting factor while facilitating delivery and distribution of cells within porous scaffolds (Lee et al., 2001; Lisignoli et al., 2002; Weinand et al., 2006).

Degradable hydrogels have potential for engineering tissues to a large extent due to their high water content, structural similarity to ECM, flexibility, and high permeability to oxygen and metabolites (Peltola et al., 2008). Accordingly, soft tissues like skeletal muscle (Rowley and Mooney, 2002), adipose tissues (von Heimburg et al., 2001), or neural tissues have been intensively investigated with relatively encouraging results. Moreover, hydrogels have been very widely employed in cartilage engineering (Elisseeff, 2004; Temenoff and Mikos, 2000; Xu et al., 2004) due to the macromolecular structural similarity to cartilage. Another notable application of hydrogel is its capacity of carrying large numbers of cells for cell transplantation. Thus, cell loaded hydrogels have great potential in cell transplantation (Nicodemus and Bryant, 2008; Zhong et al., 2010)

Ultimately, the ideal hydrogel structure for tissue printing will require at least three criteria: (1) compatibility with cells and tissues to maintain their viability and function, (2) shape preservation post printing, and (3) easy handling. The liquid-like state of hydrogel during extrusion causes difficulties in controlling the microstructure and macrostructure of the printed constructs, as well as in maintaining cell positions within the gel. Gelation rate, gel deposition velocity, and pressure are decisive parameters in attaining a desired hydrogel shape (Fedorovich et al., 2008). These, along with other process parameters, play significant roles in controlling the gel structure and cell deposition location.

4.2.3 Live structure printing

Perhaps the most appealing application of RP in tissue regeneration is live structure printing. Live structure printing, such as cell printing, tissue printing, and even organ printing, were only recently conceptualized but have very quickly grasped the attention of researchers as a potential solution for tissue vascularization foremost and organ printing furthermore (Boland et al., 2003; Mironov et al., 2003). The principles of using live structure printing for tissue regeneration lie in cellular self-assembly mechanisms, defined as autonomous organization from an initial state into a final structure without external intervention (Whitesides and Grzybowski, 2002). To date, two self-assembly processes characterized and utilized in live structure printing technology are tissue fusion and cell sorting (Pérez-Pomares and Foty,

2006; Suwińska *et al.*, 2008) (Fig. 4.2). Tissue fusion is a process in which two or more isolated cell populations contact and adhere to become one unit (Pérez-Pomares and Foty, 2006). Tissue fusion results in the disappearance of the gap between neighboring microtissues. Cell sorting takes place when one cell population coalesces to form discrete islands within the other cell population (Pérez-Pomares and Foty, 2006). These cellular assemblies are

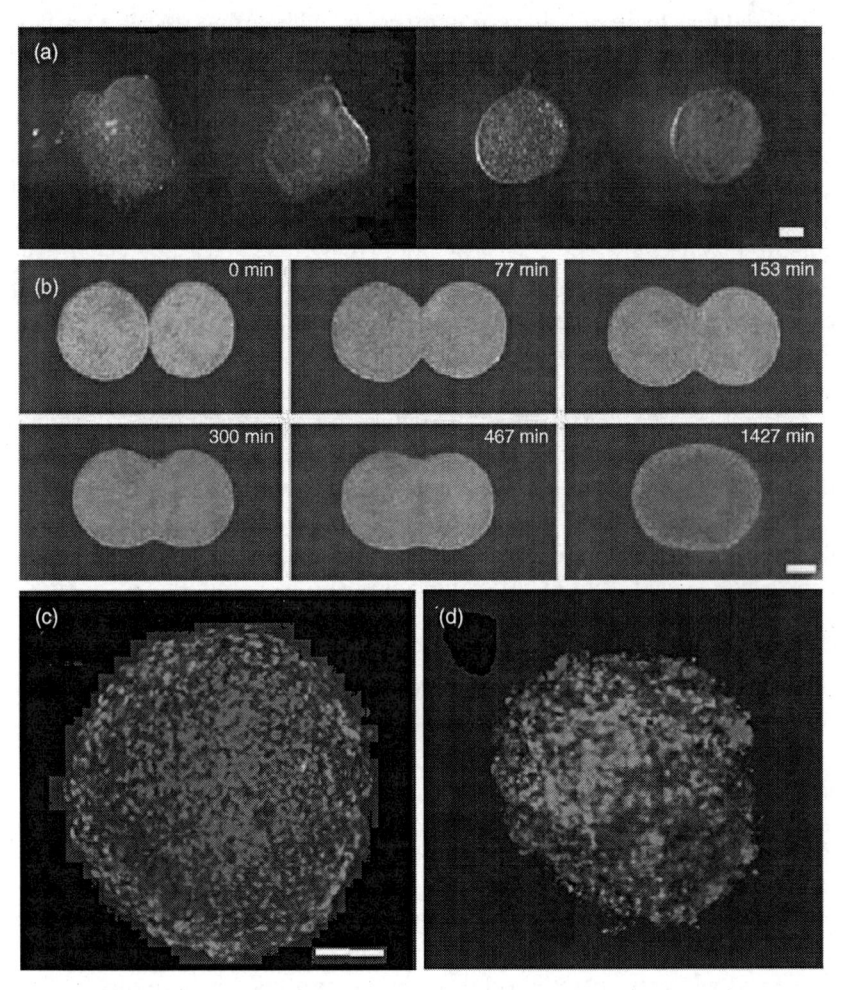

4.2 Fusion of tissue and cell aggregates indicates liquidity properties. (a) An irregular tissue fragment, excised from a day 5 chick embryo, rounds up into a spheroid in about 24 h. (b) Fusion of two spheroids in hanging drop in culture medium. (c) Cell sorting of E-cad expressing cell line E8a, paired with cell line LP1, expressing P-cad at a higher level, segregated externally (Jakab *et al.*, 2008). (d) LE-Dex cell line expressing E-cad at higher level than that of P-cad-expressing line LP1 segregated internally (Duguay *et al.*, 2003). Scale bar 100 μm.

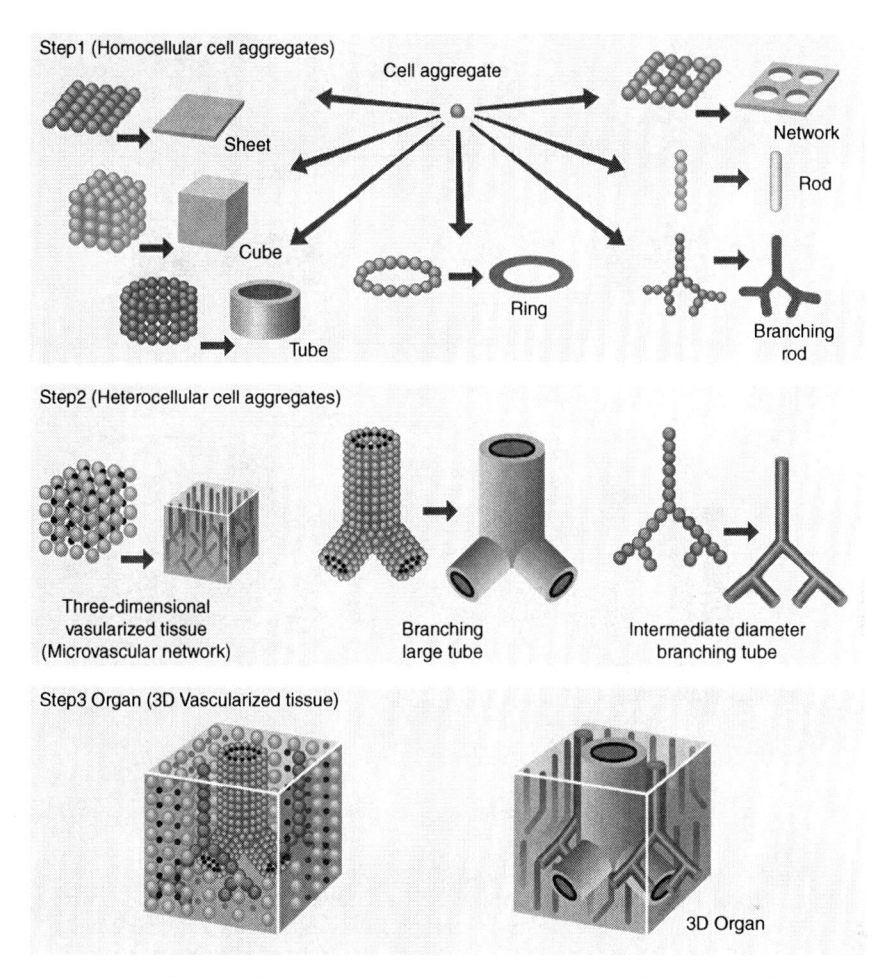

4.3 Organization of microvascular networks from building blocks (Bártolo *et al.*, 2011).

presumably driven by the similarity of surface receptors or surface tension to achieve equilibrium; without them, incorrect anatomic configurations might be generated. These driving forces indicate a fluid nature of self-assembly processes, as in the early development of embryonic tissues. That being said, these mechanisms may act at an early stage of development or only for a period of time, depending on tissue type (Jakab *et al.*, 2010). To date, the fabrication of vascular networks and multicellular living constructs has yielded some exciting results (Boland *et al.*, 2003; Cui and Boland, 2009; Jakab *et al.*, 2008; Wu and Ringeisen, 2010).

To print 3D live tissue or organ-like structures, two basic elements are needed: first, bio-ink or cell aggregates as building blocks prepared from cell suspensions of either a single cell type or multiple cell types; and

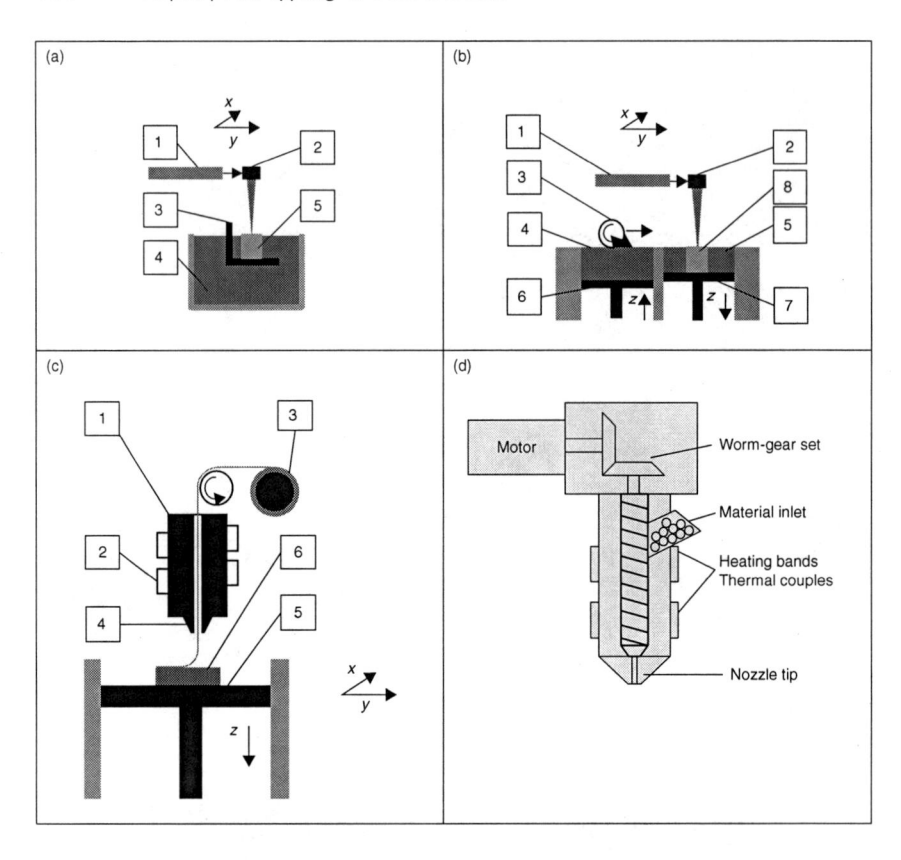

4.4 Main types of rapid prototyping systems. (a) SLA: 1 Laser; 2 Scanner; 3 Elevator; 4 Resin vat; 5 Fabricated object. (b) SLS: 1 Laser; 2 Scanner; 3 Roller; 4 Powder supply container; 5 Sintering container; 6 Powder delivery piston; 7 Fabrication piston; 8 Fabricated object. (c) FDM: 1 Liquefier; 2 Temperature control; 3 Nozzle; 4 Filament supply; 5 Elevator; 6 Fabricated object. (d) PED (Wang *et al*., 2004). (e) 3DP: 1 Inkjet nozzle; 2 Powder delivery piston; 3 Fabrication piston; 4 Powder supply container; 5 Fabrication container; 6 Roller; 7 Fabricated object. (f) BAT (Smith *et al*., 2004). (g) IP: 1 Printhead; 2 Elevator; 3 Uncrosslinked hydrogel; 4 Fabricated object. (h) Piezoelectric Inkjet Head (Arai *et al*., 2011).

second, bio-paper or a hydrogel as supporting material that acts as a glue to cohere these building blocks. Usually, bio-ink is either spherical or cylindrical in shape and is prepared on the basis of sorting and tissue liquidity. Maintaining the homogeneity of the cell suspension during the printing process is challenging yet critical to achieve reproducible bioprinting results (Guillotin and Guillemot, 2011). Proposed schemes for the organization of microvascular networks from either single cells or multicellular aggregates by self-assembling principles are shown in Fig. 4.3. To obtain the desired

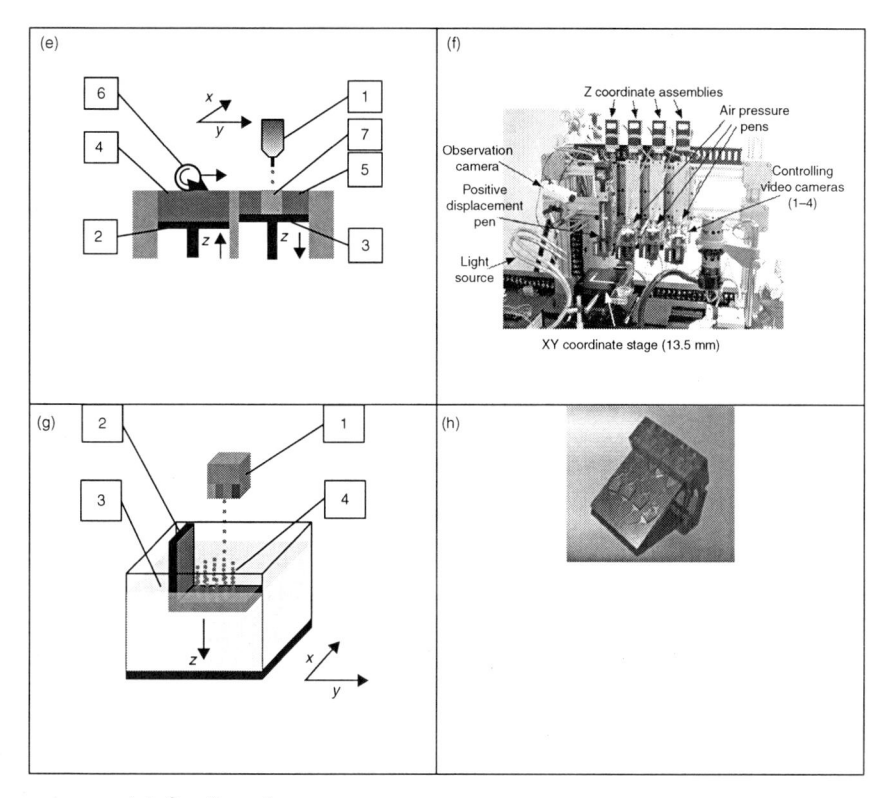

4.4 Continued.

intricate geometry, research priority must be given to the fabrication of tissue spheroids, thick sheets, and straight or branched tubes.

More than any current approach, tissue printing strongly depends on the development of rapid prototyping techniques. Any cellular systems, cell encapsulate hydrogels, or cell aggregates share an indispensable need for stringent biological fabricating conditions to prevent any costly contamination. Ideally, these conditions should be provided in a sterile enclosure and controlled environment by an automated process. Like hydrogel deposition systems, control of the gelation state is necessary to achieve successful printing. Speed and timing are of tremendous importance to enhance cell survival, especially when there are thousands of cells in each aggregate and millions of aggregates in each tissue and organ. Precise deposition of bio-ink and bio-paper is achievable only within RP technologies.

Many types of RP technologies have been developed and may be used for tissue regeneration applications. The major types are summarized in Fig. 4.4.

4.3 Laser-assisted techniques

The common feature of this group of fabrication technology is the use of optical energy to bond loose particles or to assist phase transfer. Typically, laser-assisted devices produce porous scaffolds with well-controlled macroporosity, microporosity, pore distribution, and interconnectivity.

4.3.1 Photopolymerization

Photopolymerization or photoinitiated polymerization is a material synthesis technique in which photochemical reactions occur in mild temperatures and neutral pH conditions (Fisher *et al.*, 2001). Photopolymerization employs optical energy to irradiate a thin layer of photopolymer resin, which is a mixture of low molecular weight monomers capable of chain-reacting to form long-chain polymers under a radiation source of appropriate wavelength (Yang *et al.*, 2002). Ultraviolet (UV), 100–400 nm, is the most commonly used radiation, with the next most commonly used being visible light, 400–800 nm, and sometimes infrared (IR) at 800–2500 nm.

Stereolithography (SLA)

SLA, the first rapid prototyping technique, was introduced in the late 1980s by 3D systems™ and is also the most developed by far. The standard industrial data format for rapid prototyping systems, STL, has been derived from this technique. The SLA apparatus utilizes an ultraviolet laser to solidify a resin, often an epoxy resin in commercial SLA (Fig. 4.4a). A UV laser illuminates the pattern of a single cross-section on the photocurable resin. The model cross-section is solidified leaving the surrounding area in its initial liquid form. The solidified polymer is then dropped down a predefined penetrating depth (D_p) by an elevator and so recoated with the liquid resin for curing the next layer. D_p is the penetration depth of the optical energy into the resin and represents the depth at which the irradiance becomes $1/e$ times that at the surface (Arcaute *et al.*, 2006). The thickness of each solidified layer, its cure depth (C_d μm), is governed by the irradiation dose (E mJ/cm^2) and the critical energy (E_C mJ/cm^2)

$$C_d = D_p \ln(E/E_C) \quad [4.1]$$

(Jacobs, 1992). Experimentally, C_d is also affected by the concentration of the photopolymer while D_P varies depending on the photoinitiator type and concentration and the wavelength of the laser (Fig. 4.5b). The critical energy represents the minimum energy to transform the liquid resin to the solid phase. After sintering, excess resin is removed through draining, washing, or drying, depending on the resin material.

Resin is the most important element of SLA and yet it accounts for its major drawback as well. In SLA applications, the resin should be a liquid that quickly solidifies upon irradiation. Resins which have low molecular weight multi-functional monomers which form highly crosslinked networks often generate glassy, rigid, and brittle materials. Resin viscosity also ought to be engineered since a highly viscous resin will make processing more difficult, particularly during the movement of the elevator. The highest resin viscosities suitable for SLA applications are estimated around 5 Pa/s (Melchels *et al.*, 2010). Furthermore, employing more than one resin in a single part is not yet possible, mostly due to complicated polymerization and removal of uncured resin. As a consequence, structures combining multiple materials have been difficult to build. In addition to resin, a photoinitiator is another key element which, together with the irradiation source, serves

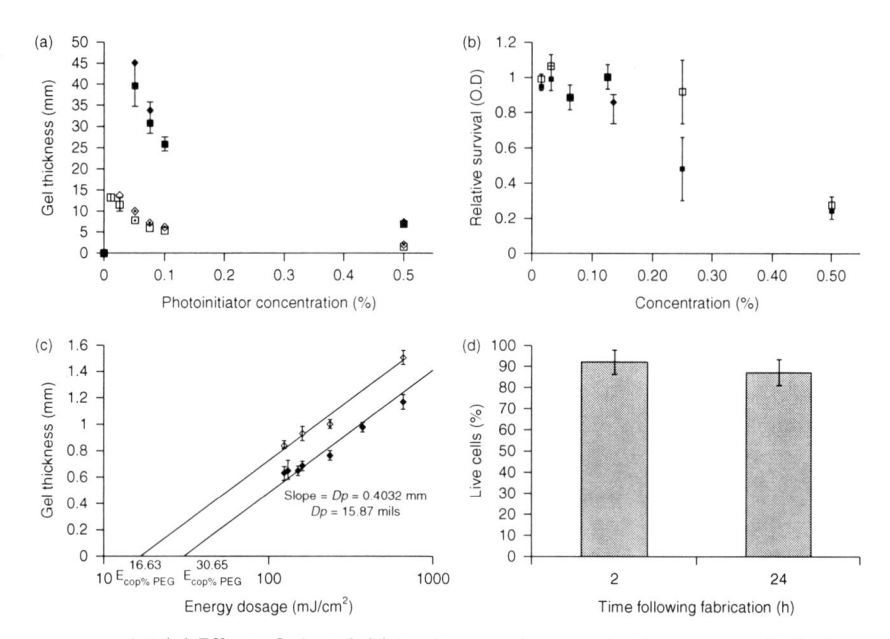

4.5 (a) Effect of photoinitiator type and concentration on crosslinked depth or cure depth. Solid markers correspond to HMPP and hollow markers correspond to I-2959: • 20% PEG-DMA; ♦30% PEG-DMA.
(b) Effect of energy dosage on the crosslinked depth for a PEG-DMA solution with 0.5% (w/v) of I-2959. Solid markers correspond to 20% (w/v) PED-DMA, and hollow markers correspond to 30% (w/v).
(c) Cytotoxicity of photoinitiators as a function of photoinitiator concentration for an exposure time of 48 h. Solid bars and markers correspond to HMPP, hollow bars and markers correspond to I-2959.
(d) Viability of human dermal fibroblasts encapsulated in PEG hydrogels photocrosslinked, crosslinking time 5 s (Arcaute *et al.*, 2006).

to crosslink polymers. Commonly used photoinitiators are Irgacure® 2959, Irgacure® 369, and Luricin® TPO-L.

For its original purposes, SLA is used to produce implants or a negative replica from ceramic slurries; the resin acts as a binder to hold ceramic particles together and is burnt away afterwards. More recent developments have advanced SLA to a higher degree of material types (resin), resulting in applications directed toward tissue regeneration. Poly(propylene fumarate) (PPF), poly-D-L-lactide (PDLLA) and PCL are among several biodegradable macromers and resins that have been studied for SLA uses (Elomaa et al., 2011). For example, PPF is a photocrosslinkable, biodegradable polymer with acidic degradation products containing primarily propylene glycol and fumaric acid (Fisher et al., 2002). The high molecular weight nature of PPF resin requires a dilutent to reduce its high viscosity. Low viscosity dilutent diethyl fumarate (DEF) can be combined with PPF for rapid prototyping applications. Lan et al. (2009) fabricated PPF-DEF (70:30) scaffolds with 65% porosity and highly interconnected channels. The scaffold surface was modified with accelerated biomimetic apatite and arginine-glycine-aspartic acid (RGD) to enhance cell attachment. Histologic staining at 2 weeks showed that MC3T2-E1 cells survived, proliferated, and migrated in between channels. In the same way, resin of PDLLA and its dilutent ethyl lactate were used to manufacture a 73% porous gyroid architecture on which mouse pre-osteoblasts readily adhered, spread, and proliferated (Fig. 4.6). The gyroid had an average pore size of 200 μm and 73% porosity (Melchels et al., 2009). Additionally, resin can also be mixed with specific binding protein molecules (Farsari et al., 2007) for promoting cell response or labeled with fluorescent dyes for imaging purposes (Arcaute et al., 2010).

Bioceramic powers were also added to resins to improve the bioactivity of bone scaffolds. Ceramics such as hydroxyapatite (HA) or tricalcium phosphate (TCP) have been added to polymer composites either to enhance biological response, to promote differentiation, or to improve mechanical properties. For example, the addition of HA to PCL composites enhanced the compressive modulus of PCL (Azevedo et al., 2003; Eosoly et al., 2010). Poly(propylene fumarate) and 7% (w/v) HA particles were mixed to form a resin and used with the DEF dilutent (Lee et al., 2009a). Significantly, more MC3T3-E1 cells adhered and proliferated on these scaffolds compared to PPF-DEF alone. The addition of HA content reduced hydrophobicity and increased wettability of scaffold surface; however, this may increase the viscosity of the photocurable resins. In addition to synthetic resins, some natural polymers have been modified for use with SLA. Photo-curing engineered chitosan was found to evoke good endochondral ossification upon implantation (Qiu et al., 2009). Other modified natural polymers have included methacrylate-functionalized gelatin (Schuster et al., 2009), methacrylated

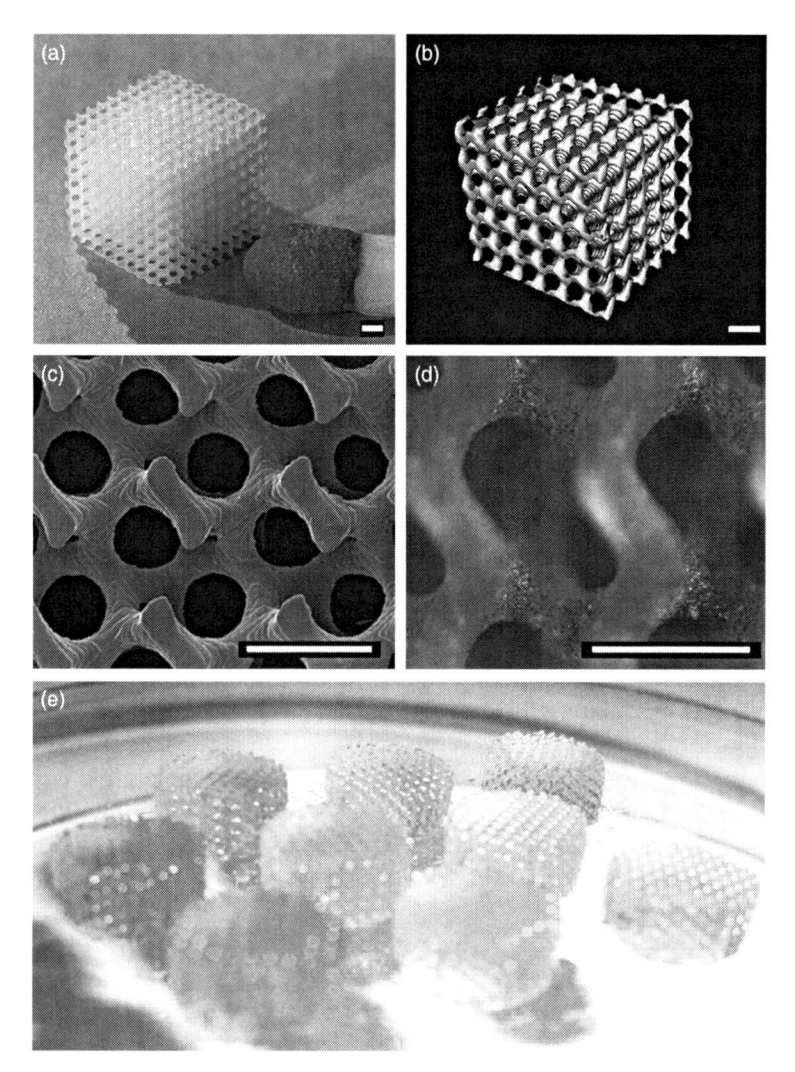

4.6 SLA fabricated scaffolds (a–d) PDLLA gyroid solid structure (Melchels *et al.*, 2009). Scale bar 500 μm. (e) PEG/PDLLA hydrogels of gyroid architecture with high repeatability (Seck *et al.*, 2010).

oligopeptides (Zimmermann *et al.*, 2002), and methacrylated hyaluronic acid (Smeds *et al.*, 1999).

Recently, cell encapsulation was attempted with the SLA technologies using compatible hydrogels. Poly(ethylene glycol) (PEG) is the most heavily studied biocompatible hydrogel material so far. Although the polymer has a stable hydrophilic property, it can be excreted by the kidneys if its molecular weight is less than 30 kg/mol (Yamaoka *et al.*, 1994). Higher molecular weight PEG can be rendered by including biodegradable segments, such as poly(lactic acid) (PLA) or proteolytically degradable peptide sequences (Arcaute *et al.*, 2006; Burdick and Anseth, 2002). PEG can be photopolymerized with the attachment of photoreactive and crosslinkable groups, such as acrylates or methacrylates to its backbone, and so under UV it is crosslinked to form hydrogels. PEG hydrogels have been shown to demonstrate good permeability to oxygen, nutrients, as well as water-soluble metabolites while maintaining soft tissue-like characteristics. Arcaute *et al.* (2006) utilized SLA to photocrosslink poly(ethylene glycol) dimethacrylate (PEG-DA). The acryloyl-PEG was conjugated with bioactive tetrapeptide Arg-Gly-Asp-Ser (RGDS) creating a peptide PEG. When hydrogel solution was mixed with human dermal fibroblasts under a He-Cd laser (325 nm wavelength), cell loaded hydrogels with accurate 3D organization were obtained (Arcaute *et al.*, 2006). In the absence of bioactive molecules, human mesenchymal stem cells (MSCs) appeared to adhere, spread, and proliferate well on the 3D structure. Williams *et al.* (2003) achieved chondrogenesis through a culture of photoencapsulated MSCs in PEG-DA hydrogel *in vitro* for 6 weeks. One common issue with hydrogel-based 3D structures is their low elastic modulus, which is typically lower than 1 kPa. Much development still needs to take place before these hydrogels can be used even for engineering soft tissues. When encapsulating hydrogels with cells in SLA, care also must be taken to maintain cell viability due to the potential cytotoxicity of the photoinitiator, even if the photoinitiator is considered cytocompatible under other circumstances. Cell viability can be enhanced if both exposure time and concentration of the photoinitiator are reduced (Fig. 4.5c). For gelling time of 5 s, cell survival was 90% 2 h following the fabrication in a poly(ethylene glycol) dimethacrylate (PEG-DMA) resin containing 0.5% (w/v) of the least cytotoxic of known photoinitiators, Irgacure 2959 (Arcaute *et al.*, 2006) (Fig. 4.5d). The success of using SLA in tissue regeneration greatly relies on the development of photocurable biomaterials. Given the capacity for SLA to produce structures with accuracy up to several micrometers, the key limitations in resolution and size are the types of resin used, device set-up, penetrating depth, and other processing parameters (Kim *et al.*, 2010).

Two photon polymerization (2PP)

2PP is a modified version of SLA with improved resolution. Its major advance from conventional SLA is its ability to fabricate complex 3D micro/nanodevices where spatial resolution is at the 100–200 nm level (Lee *et al.*, 2008). The superiority of 2PP over the original SLA is the simultaneous absorption of two photons with relatively low density when resin is focused by an ultrashort-pulsed laser beam. The absorption produces enough energy to excite the photoinitiator and initiate chain-reaction polymerization. As a result of the absorption of the threshold energy for photopolymerization, polymerization occurs in a highly localized area around the center of the focused beam; hence, the higher resolution. The polymerization rate of this process is proportional to the square of the laser intensity as compared to the linear relationship in SLA (Melchels *et al.*, 2010). Therefore, 2PP is often referred to as 3D laser nonlinear lithography in some literature. Ovsianikov *et al.* (2011) reported the use of 2PP in fabricating a Polyethylene Glycol Diacrylate (PEGDA) structure with minimum lateral feature size as small as 200 nm and hollow cylinders of 25 μm radius. In the work of Claeyssens and colleagues (2009), a PCL-based triblock copolymer was synthesized to build a 3D structure with 4 μm resolution. One should note that the higher the resolution of a structure, the more time it needs to be built. Accordingly, 2PP will face quite a challenge scaling up scaffolds while maintaining its desired resolution.

4.3.2 Selective laser sintering (SLS)

SLS is another technique that uses optical energy. But unlike SLA, it covers a much broader range of materials. SLS uses focused infrared radiation to fuse loose powder particles together by selectively heating them just beyond their melting or sintering temperature, and so theoretically it can sinter any powdered material that will melt but not decompose under the intense energy (Williams *et al.*, 2005). Another promising advantage of SLS is the elimination of solvents, resulting in easier manufacturing and post-manufacturing process. However, the nature of SLS can allow the fabrication of only porous scaffolds or microspheres.

SLS was initially developed by the University of Texas at Austin and subsequently commercialized by DTM Corporation. SLS belongs to a complex heat process in which powder transfers from solid to liquid and then back to the solid phase. These changes involve both absorption and release of thermal energy (Gibson and Shi, 1997). In general, SLS fusing mechanisms can be classified into (1) liquid phase sintering partial melting (LPS), (2) chemically induced binding, (3) solid state sintering (SSS), and (4) full melting (Kruth *et al.*, 2005). The primary working principles of SLS observed in tissue regeneration applications are LPS and SSS. The former may or may not involve the use of a binder whereas the latter offers a binder-free mechanism. Figure 4.7a shows the LPS mechanism (Kruth *et al.*, 2005) in which the powder material is a combination of a structural material remaining solid throughout the process and a binder being liquefied. Usually, binder particles are much smaller than the structure and are burned away in thermal post-treatment. In ceramic scaffolds, the polymer has been used as a binder due to its lower melting temperature. When there is no binder, there will be partial melting of the structural phase, which exhibits molten and non-molten material areas. The molten areas are formed as the grain boundary absorbs heat while insufficient energy reaches the particle center, creating an intact or non-molten core. SSS, on the other hand, takes place at temperatures between $T_{melt}/2$ and T_{melt}, where T_{melt} is the melting point of the powder. The thermal process forms neck regions between adjacent particles (Fig. 4.7b). This diffusion phenomenon is driven by a vacancy concentration gradient due to the lower free energy when particles grow together. The diffusion results in a slow-binding process; thus, to be effective, powders are preheated before sinter to accelerate the diffusion rate.

An SLS apparatus works in the same way as any other layer-by-layer rapid prototyping device (Fig. 4.4b) that comprises a laser, a scanner and a moving stage, a powder supply and powder deliverer (roller or scraper), and a sintering station. Commonly used lasers are CO_2 laser, diode pumped Nd:YAG laser, or a disk or fiber laser. In most of the available SLS devices, temperature control, often by an infrared sensor, has not yet reached optimal performance. Not only is SLS a comprehensive thermal process, but it is also affected by many other factors: chemical, mechanical, physical, laser, and control theories. These factors can be grouped into either SLS process or SLS powder variables. SLS process variables include laser power, scan spacing, scan speed, and beam spatial distribution (Gibson and Shi, 1997). Laser power directly correlates to the sintering temperature by the Stefan-Boltzmann equation. Sintering temperature is the key for process stability and the quality of fabricated scaffolds. For a polymer, if the temperature reaches T_{melt} the material behaves as a highly viscous liquid, but once above T_{melt} it changes into liquid. The flow consequently reduces the porosity and affects the mechanical properties of sintered parts. Undoubtedly, scan spacing should not exceed laser beam diameter as it might cause incomplete sintering. Scan speed is quantified by the exposure time for a focused spot and influences energy density absorbed by the powder,

$$E = \frac{fP}{\text{Speed} \times \text{Spacing}},\qquad [4.2]$$

where f is a conversion factor. Most commercial laser generators provide Gaussian beam spatial distribution, resulting in a stable penetrating depth, or layer thickness. This factor is a function of both process and powder characteristics, including laser power, energy density, particle size, powder density, specific heat, and thermal conductivity. The layer thickness affects fabrication time and surface roughness. A smaller layer generates a smoother surface but requires more build time as a trade-off.

Powder variables, as already briefly mentioned, include particle size, molecular weight, heat and thermal conductivity, and powder density. In the case of composites, powder composition is a decisive factor. Keeping powder variables constant or using the same batch of material throughout

(a)

0. Loose powder
Binding element
Structural element (solid phase)
Porosity
1. Rearrangement
2. Solution reprecipitation
3. Solid state sintering
Post-heat treatment of green product
Freezing of initial binding
Production of green part
Amount of sinter reaction
Time

(b)

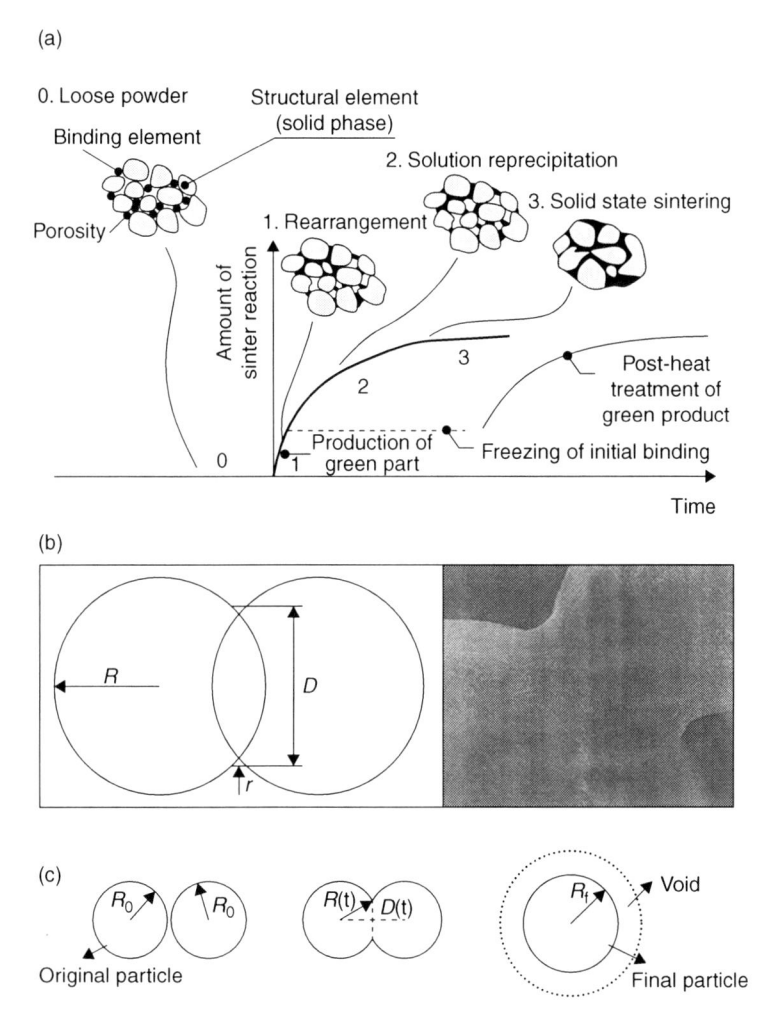

(c) Original particle — Final particle — Void

4.7 (a) Liquid phase sintering partial melting. (b) Neck form between two adjacent particles in solid state sintering (Kruth et al., 2005). (c) Sintering of two particles to eventually form a single larger particle (Pokluda's model): $R_f = 1.2599\, R_o$ (Lu, 2001).

experiments is extremely critical to control scaffold quality (porosity, pore size, density) as well as scaffold comparability (Eosoly et al., 2010; Leong et al., 2001; Savalani et al., 2007). Particle size may play an important role in pore size as well as porosity, thus affecting scaffold design. Experimentally, large particles regulate the granularity of edges and layers, and often result in a rougher surface. Particles as small as 10 μm exhibit poor liquid flow, yielding poor spreading properties and entrapment of particles inside pores (Williams et al., 2005). The recommended range for particles thus lies between 10 and 100 μm. Low molecular weight polymers typically form smoother, flatter, and more dense layers than high molecular weight ones (Das, 2008). Glass transition temperature, T_{glass}, like T_{melt}, is another thermal property of a polymer that has a great impact on the SLS process. The transition from a glass state to a rubber-like state is associated with volume changes, or molecular motion along the polymer chain (Gibson and Shi, 1997). Since SLS is a thermal process, powder density along with thermal conductivity is also of particular importance in tuning the SLS process and improving product quality.

Furthermore, SLS process factors coupled with powder factors determine the accuracy and mechanical properties of fabricated scaffolds, such as strength, hardness, ductility, and stiffness (Gibson and Shi, 1997). In many instances, tensile strength is proportional to the scaffold density which substantially depends on scan spacing (Dewidar et al., 2003). Fabrication position and orientation, especially in manufacturing anatomical parts, also shows a great influence on mechanical properties. The weakest parts, that is those with lowest elastic modulus, are usually in the direction in which scan lines are parallel to the loading direction, whereas the highest elastic modulus values are in the direction where scan lines are perpendicular to loading (Eosoly et al., 2010). SLS processed scaffolds usually inherit rough surfaces, which could help enhance cell adhesion but often may be needed for post-processing with coatings. Resolution of SLS is interactively governed by the powder particle size, focused laser beam diameter, and powder heat conductivity (Williams et al., 2005).

Of the many attempts at exploiting SLS fabricated scaffolds, that by Williams et al. on fabricating a PCL structure-based on an actual pig condyle (Fig. 4.8a) to repair temporo-mandibular joints has been one of the most cited with respect to SLS for skeletal tissue applications (Williams et al., 2005). Designed cylindrical constructs of 1.5 mm pore size, 68% porosity, 5 mm diameter, and 4.5 mm height were seeded with cells before implantation in old immunocompromised mice. Four weeks after implantation, newly formed bone grew onto and within pores as well as on the scaffold exterior. The actual porosity of fabricated scaffolds, 50%, was achieved with 4.5 W power, 1.257 m/s scan speed, and 100 μm layer depth as SLS processing parameters and 10–100 μm particle size, and 50 000 Da molecular weight as PCL parameters. Further, the compressive modulus and strength of the scaffold, ranging from 52 to 67 MPa and from 2 to 3.2 MPa, respectively, fell within the lower range of human trabecular bone. Another example of utilizing SLS fabricated PCL scaffolds for engineering cardiac tissue was demonstrated by Yeong et al (2010). The group was able to achieve a compressive stiffness of 345 kPa and tensile stiffness in the range of 0.43 ± 0.15 MPa. A disk-shaped structure, $1 \times 3 \times 22$ mm^3, consisting of square pyramid cellular units, was designed to 85% porosity with 40–100 μm micropores from 100 μm particle PCL powder. The scanning electron microscopy (SEM) images showed necking regions between sintered particles, with 48% porosity and a surface roughness of 33.7 μm (Fig. 4.8b). Notably, tensile stiffness of PCL scaffolds as a linear function of porosity lower than 80% and a logarithmic function of porosity greater than 80% was established. The construct seemed to sustain a high density of C2C12 myoblast cells at day 4 of culture.

Other SLS manufactured scaffolds are primarily used in hard tissue regeneration. Simpson *et al.* (2008) studied porous 95/5 poly(L-lactide-co-glycolide) (PLGA)/HA and TCP composite scaffolds and also evaluated the effect of fabrication bed temperature, scan speed, and laser power on the degree of sintering, polymer thermal degradation, and material compression strength. Zhou *et al.* (2008) successfully produced rectangular-shaped scaffolds from poly (L-lactic acid) (PLLA)-carbonated HA (CHA) microspheres prepared by emulsion techniques. The use of PLLA-CHA microspheres holds promise in the annihilation of excess powder removal post processing due to their relative small size, 5–30 μm. Recently, Niino *et al.* (2011) reconstructed a highly porous 89% PCL scaffold using a porogen (sodium chloride) leaching SLS-based system. The intricate structure with an embedded flow channel network could be useful for highly oxygen-dependent tissues and organs. The hydrophilicity of PCL was improved by oxygen plasma etching and sodium hydroxide hydrolysis, yet only alkaline hydrolysis showed an improvement in cell adhesion.

4.8 PCL scaffolds fabricated using SLS. (a) Pig condyle and the design of the scaffolds for condyle regeneration (Williams *et al.*, 2005). (b) Pyramid building unit with neck regions (Yeong *et al.*, 2010).

Despite the variety of materials used with SLS systems – ceramics, polymers, and metals – the high temperature procedure limits it to processing only of thermally stable polymer binders. The thermally sensitive nature of SLS makes it an ideal technique for manufacturing strong scaffolds for skeletal tissue regeneration. Because powder stock comes with a wide range of particle sizes, the control of pore dimension is highly restricted. As a result, SLS has a relatively lower dimensional accuracy and porosity compared to SLA.

4.3.3 Other laser-based technologies

In addition to photopolymerization and SLS, there are a few other laser-based techniques. In an ongoing project, our group developed the use of an excimer laser-based device for cutting high resolution patterns on polymer scaffolds, as shown in Fig. 4.9. Another technique, laser-guidance, has been developed based on optical forces for generating laser guidance to manipulate cells or biological particles with high speed in three dimensions (Odde and Renn, 1999). Cell deposition microscope makes use of the technique to micropattern single cells on a substrate according to a predefined geometry with high spatial resolution (Guillotin et al., 2010; Ma et al., 2011). Micropatterning enables studies of heterotypic cell arrangements, cell–cell and cell–ECM interactions, and cellular electrophysiological properties in a single-cell co-culturing environment. Some techniques utilizing laser light to induce deposition of various biomaterials (proteins, DNA, scaffolding materials, and cells), such as matrix-assisted pulsed laser evaporation direct write (MAPLE DW) and biological laser printing (BioLP™), both operate under the laser induced forward transfer (LIFT) principle. LIFT enables precise placement of cells in a biological support material onto a transfer substrate that will absorb UV light, causing local heating and partial vaporization of the support material, and result in the deposition of the remaining material onto a collector substrate. It is the only nozzle- and orifice-free technique capable of printing cells (Wüst et al., 2011), and the risk of nozzle clogging is therefore eliminated. Resolution of LIFT techniques can reach to 10–100 μm with high cell viability (95–100%) (Guillotin and Guillemot, 2011; Wüst et al., 2011). The major difficulty with LIFT is 3D structure formation. One way to avoid the difficulty is to deposit a hydrogel layer after each patterned cell layer; however, cell–cell contact between layers might be disrupted. BioLP™ represents an advance from MAPLE DW by modifying an indirect contact of laser with the biological material through the use of a laser absorption interlayer between the laser and biomaterials to be printed (Wu and Ringeisen, 2010). Pirlo et al. (2012) implemented BioLP™ to place human umbilical vein endothelial cells (HUVEC) with highly viable rates onto PLGA stacked bio-papers prepared by the solvent casting/particulate leaching method. Built on the principle of LIFT, laser-assisted

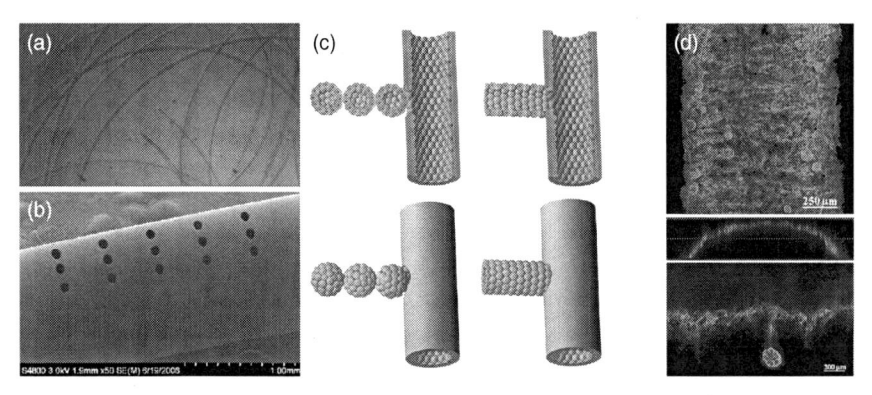

4.9 Using an excimer laser-based device for adding features on polymer scaffolds. (a) Hollow fiber scaffolds fabricated using phase inversion technology (light microscope image). (b) Small holes were cut with excimer laser. (c) Hollow fiber scaffolds with laser-cut holes can be used to study multicellular spheroid fusion and tubular structure sprouting. (d) Kidney epithelial cells occupied the entire hollow fiber lumen and some cells sprouted out from the laser-cut holes.

bioprinting (LAB) has demonstrated a high resolution and high throughput of printing cells individually aligned next to each other at a speed of 5 kHz without causing DNA or cell function damage (Guillotin and Guillemot, 2011). LAB offers printing of solutions of various viscosity (1–300 mPa/s) and cell concentrations up to 1×10^8 cells/mL. However, metallic residues in the printed construct owing to the use of a laser-absorbing metal layer and random location of printed cells compromise LAB performance. It is important to note, however, that all of these methods are more effective in studying the cell biology in a 2D manner due to the limited throughput and hurdles in extending them into 3D constructs and, consequently, tissue construction.

4.4 Extrusion-based techniques

This group, including dispensing-based techniques, emphasizes the extrusion of strands of material through an orifice onto a platform where the material cools, solidifies and attaches to a previously deposited layer. Depending on the set-up, either the orifice or the substrate moves across the plane of each layer. Porosity is defined as the spacing between adjacent filaments/strands.

4.4.1 Fused deposition modeling (FDM)

The origin of FDM refers to extruding thin thermoplastic filaments on an x-y-z stage for building 3D structures (Fig. 4.4c). The raw materials are fed and melted inside a heated liquefier, usually a stainless steel syringe, before extrusion through a nozzle. Once the filaments leave the hot nozzle, they adhere to the previous layer and harden immediately. The platform environment, therefore, is controlled at just below the solidification temperature of the thermoplastics to ensure sufficient fusion energy between layers. Porous scaffolds fabricated by FDM obtain a mesh-like structure, which can be formed from either hollow or solid filaments. Filaments of one layer are usually deposited at some angle (0, 60, 90, 120°, or irregular) in different layers, generating more interconnected lay-down patterns. Scaffold architecture, therefore, is governed by nozzle diameter, deposition speed, spacing between fibers in the same layer, layer thickness, and deposition angle. Essentially, the resolution of FDM is determined by polymer viscosity and the nozzle dimension which can be relatively small, provided there is no extrusion clogging. For tissue regeneration applications, some efforts have focused on structure formation under room temperature by using thermosensitive polymers, such as poloxamer 407 (Pluronic® F127), which has a broad range of viscosities at room temperature (Ruel-Gariépy and Leroux, 2004).

Many porous scaffolds have been fabricated using FDM from different materials, such as PCL, PCL-HA, PCL-TCP, and polypropylene (PP)-TCP with resolutions of 250 μm or higher. PCL is one of the most tested polymers in FDM for tissue regeneration, since PCL is degradable, semicrystalline with a low T_g of −60°C, a melting point of 60°C, and a high decomposition temperature of 350°C (Nair and Laurencin, 2007). It offers a wider range of temperatures for extrusion. Hutmacher *et al.* (2001) characterized PCL honeycomb scaffolds fabricated at different lay-down patterns for fibroblasts and osteoblast-like cell responses to a wide range of porosity, pore sizes, and compressive stiffness and yield strength. Zein *et al.* (2002) constructed similar PCL honeycombs with channel sizes ranging from 160–770 μm and porosity in the range 48–77% from 260–370 μm filaments. PCL scaffolds manufactured by FDM have been used in a pilot study for cranioplasty with a positive clinical outcome after 12 months (Peltola *et al.*, 2008). BurrPlug™, or PCL burr whole plugs, used to patch holes in the skull, had been clinically studied for over 5 years before they gained FDA approval in 2006 and consequently, became commercially available through Osteopore International (Tan, 2004).

Like HA, calcium phosphate (CaP) has been combined with a number of biocompatible polymers, including PCL, especially for bone reconstruction due to the osteoconductivity of the CaP-based bioceramics, higher rates of cell attachment and proliferation, and the differentiation of cells onto the mineralization pathway. Additionally, the presence of CaP buffers the degradation by-products from degradable polyesters (Ignatius *et al.*, 2001). On the basis of the same concept, Schantz *et al.* (2005) studied the osteogenic potential of PCL-CaP honeycomb scaffolds produced by an FDM

3D modeler using PCL-CaP pellets made by solvent casting. Human MSCs were mixed with a fibrin gel and seeded into the scaffolds (Schantz *et al.*, 2005). Biphasic scaffolds produced by FDM also have been investigated by Hutmacher and his group aiming to repair osteochondral defects in the articular joints (Hutmacher *et al.*, 2001). Fibrin-PCL was used as the structure to support cartilage regeneration and porous PCL-TCP was used to support bone regeneration (Swieszkowski *et al.*, 2007).

One of the major deficiencies of FDM is the use of filaments which narrows the range of possible materials, as they must be made into thin fibers and must sustain the heat effect when melted into a semi-liquid phase before extrusion. Natural polymers therefore cannot be readily processed. The heat effect also circumvents the incorporation of biomolecules into scaffolds. Furthermore, FDM produces scaffolds with smooth surfaces that may not optimize cell adhesion and may further necessitate surface modification or coating (Peltola *et al.*, 2008). Moreover, microporosity, a promoting factor for neovascularization and cell attachment, cannot be formed due to the solidification of dense filaments. Chen *et al.* (2011) in their recent bone tissue engineering study embedded hyaluronic acid-methylated collagen terpolymer into PCL honeycombs to increase cellular seeding efficiency, proliferation, distribution and penetration, and osteogenic differentiation through the interaction of integrin ligands and, especially, the additional secondary microporous matrix of the collagen complex to the large pores of the PCL scaffolds.

Since FDM uses microfibers during the fabrication process, it is highly compatible with the nanofiber fabrication process. Electrospinning was combined with FDM to fabricate scaffolds with both microfibers which provide structural stability and nanofibers which support better cell growth (Centola *et al.*, 2010).

The imperfections of FDM have led to advances *via* several modified techniques: 3D fiber-deposition, precision extruding deposition (PED), and low-temperature deposition manufacturing (LDM). FDM and its variations are operated under three primary extrusion mechanisms: pneumatic, mechanical, and pneumatic–mechanical hybrid-powered dispensing (Chang and Sun, 2008; Smith *et al.*, 2007, 2004). While compact mechanical dispensers work better with low viscosity materials, pneumatic systems occupy more room and are suitable for high viscosity materials (Chang *et al.*, 2011). The modification made for the 3D fiber-deposition (3DF) technique is the replacement of pellet or granule as the feedstock material, eliminating precursor filaments as well as the accessories associated with their preparation. Nitrogen pressure is applied to the syringe through a pressurized cap to adjust the flow of polymers for extruding fibers onto the x-y-z stage.

Since many thermoplastic polymers, such as PCL, PLA, polyglycolic acid (PGA), poly[poly(ethylene oxide) terephthalate-co-(butylene) teraphtalate]

(PEOT-PBT), starch-based polymers and co-polymers, can be used in 3DF. It allows the combination of different raw materials during the fabrication process. For example, PCL scaffolds support bone and cartilage regeneration (Schantz et al., 2005; Shao et al., 2006), PLGA enhances microvasculature ingrowth (Laschke et al., 2008a, 2008b; Rücker et al., 2006), PEOT-PBT favors articular cartilage regeneration (Moroni et al., 2007), and many polymers may have applications in neural and vascular regeneration (Moroni et al., 2006). With the use of biphasic polymeric blends of different viscosity, such as PEOT and PBT, more complicated structures can be developed. For example, a shell-core structure or hollow fiber with controllable cavity and shell thickness can be made. A shell-core fiber structure is generated when the less viscous component tends to shift to the wall of a capillary under the shear stresses (Moroni et al., 2007). Hollow fibers can be made using a selective leaching out technique to exclusively dissolve the core polymer by soaking in appropriate solvent (Zhang et al., 2005).

Another modification of 3DF is its capacity to produce hydrogel scaffolds using a modified CNC-milling machine capable of plotting hydrogel material around 60–80°C (Landers et al., 2002). A syringe is wrapped with a temperature-regulating jacket, allowing hydrogel deposition at physiological temperature in either a crosslinked polymer solution or by crosslinking after extrusion (Fedorovich et al., 2010). Among many hydrogels processed with 3DF, alginate, agarose, gelatin, chitosan, collagen gels and gelatinous protein mixtures are often used. These hydrogel matrices are typically dispensed at pressures of 1.5–4 bar to attenuate the shear-thinning effect (Fedorovich et al., 2010). The shear-thinning behavior is known to lower the viscosity of the hydrogels, thus altering their softness, a parameter that determines the fiber thickness in addition to the nozzle size. As a consequence, resolution of the 3DF printed hydrogels is restricted within 100–200 μm. Because of the layer stacking, it is also more difficult to form transversal pores than vertical pores (Fedorovich et al., 2010). Similar to SLA, major development efforts are under way in relation to inverse thermosensitive, photopolymerizable, and viscous ion-sensitive hydrogels. For example, Pluronic®F127 (Fedorovich et al., 2009) has been printed into 3D structures resulting in enhanced mechanical strength, tunable degradation profiles, and prolonged stability.

Efforts are also being made in the integration of multiple printing heads into 3DF to print different cell-laden hydrogels. Multiple cell types can be printed and positioned to mimic normal tissue and organ structures (Fedorovich et al., 2010). One such approach is to fabricate 3D structures closely mimicking the cell population and ECM composition of functional tissues and organs. However, mimicking a native tissue or organ with appropriate spatial organization and anatomical geometry has not been achieved by 3DF or any other printing technology yet. Most of the 3D structures produced are restricted to smaller sizes and simple shapes, as well as simple organization patterns. The perspectives of 3DF lie in the multi-dispensing systems, or multiple head deposition system (MHDS), which may be able to generate more complex structures loaded within multiple bio-factors and cells. Shim *et al.* (2011) used a four-head printer with individual temperature control on each jet to fabricate a hybrid scaffold consisting of blended PCL and PLGA polymers and hydrogel (HA, gelatin, and atelocollagen). MC3T3-E1 cells were embedded in atelocollagen with high viability. Bartolo *et al.* (2011b) expanded a MHDS into a closed sterilization automated system, called BioCell Printing, to reduce the risk of contamination while working with cell assemblies. Like 3DF, PED, which eliminates the need for filament preparation, is implemented with a built-in heater for the extrusion of scaffold materials in granulated form, as shown in Fig. 4.4d. The technique has mostly been applied for producing scaffolds in bone tissue engineering; a pore size of 350 and 250 μm for PCL scaffolds has been fabricated (Shor *et al.*, 2009; Wang *et al.*, 2004).

LDM, the so-called rapid freeze prototyping technique, on the other hand, eliminates the need for heating raw materials. LDM and multinozzle deposition manufacturing (MDM) operate under similar principles with MDM, an advance from LDM by having more than one jetting nozzle. LDM employs both extrusion and a thermally induced phase separation process (Bártolo *et al.*, 2011). In LDM, polymer solutions are deposited on the substrates forming strands that are frozen and lyophilized to remove solvent. LDM allows loading of biomolecules directly into the scaffold during the printing process. Zhu *et al.* (2011) made PLLA strands with premixed chitosan microspheres. Chitosan microspheres could effectively encapsulate and release a variety of biomolecules. Embedding biomolecule-loaded microspheres or nanoparticles into scaffolds potentially prolonged the release of biomolecules.

4.4.2 Bioplotter

In recent years, many rapid prototyping techniques have been introduced, and the terminologies can cause some confusion. Bioplotter, sometimes referred to as the direct write system, is defined as extrusion of continuous filaments (Wüst *et al.*, 2011). It is reviewed herein as an analog to the FDM concepts. Fab@Home, BioAssembly Tool (BAT) and nScript™ are a few examples of commercially developed bioplotters. Fab@Home, a plug-and-play type device with linear-actuator driven syringe originally built by Cornell researchers, has a universal platform with interchangeable syringe tips offering a cheap and accessible solution for various applications. Its alginate hydrogel encapsulating chondrocytes showed good cell viability (Cohen *et al.*, 2006).

Both BAT and nScrypt are pressure operated mechanical extruders capable of working with biomolecules and cells (Jakab et al., 2008). To eject plotting material, two common mechanisms, air pressure and pump-based, are implemented (Nakamura, 2010; Nishiyama et al., 2009; Smith et al., 2004). Thus far, these devices, together with inkjet printing techniques (discussed in the next section), have built the foundation for 3D live tissue printing. BAT, introduced by Sciperio, Inc., is able to print biomolecules, cells, and biomaterials over a wide range of viscosities. The BAT configuration is composed of four extrusion heads, called microdispense pens, with modular pen tips. There are two types of pens, positive displacement and air pressure (pneumatic) pens. Operating temperatures range from −10 to 80°C. Using this set-up, cell survival and resolution appear to be a function of tip diameter (Smith et al., 2004).

Jakab and Forgacs used a similar dispensing tool developed by nScrypt to examine the capacity of cells and tissues to self-assemble into functional living structures of defined shapes for 3D live tissue printing purposes. Building blocks of either cylinder or spheroid shape are fed into the printers. 3D microtissues can be prepared on nonadhesive micromolds or by mechanically cutting a sturdy cylindrical slurry of cell suspension (Jakab et al., 2008) into equal-sized units. These units spontaneously round up to become spherical due to tissue liquidity. Such spheroids can be used as bio-inks or building blocks for 3D tissue and organ fabrication. In multicellular aggregates, composed of multiple cell types, sorting and rounding take place in parallel. However, spheroids can be made only of adhering cells. Non-adhesive cells might need to be embedded into a population of adhering cells or temporarily genetically manipulated with adhesion molecules (Duguay et al., 2003; Steinberg and Takeichi, 1994).

In terms of bio-paper, hydrogels are most frequently used. Of extreme importance is the gelation process that should be fast and homogeneous in order to achieve stable and spatially accurate supporting structures. The ECM-based hydrogels allow for remodeling and removal by cells. Photosensitive and thermoreversible are other types of hydrogels to be considered for bio-papers.

3D tissues including tissue toroids, thick sheets, and straight and branched tubes can be robotically fabricated or formed from the basic elements or building blocks, such as spheroids. When placed the spherical aggregates into designed patterns, toroidal 3D structures or branched structures can be formed through fusion mechanism. Close dispensing of multicellular spheroids on collagen substrate formed a toroid after 60 h in culture. The spheroids are fully fused at 168 h, as shown in Fig. 4.10a. There was evidence of cell–matrix interaction, as the cells pulled on the collagen substrate, inducing contraction of the 3D structure. Additionally, a 6 × 6 array of chick cardiac cell aggregates fused into a thick graft over 70 h, as shown in Fig. 4.10b, and a tube of 12 layers of human umbilical smooth muscle cells (SMCs) was also constructed. All spherical units were prepared at approximately 300–500 μm diameter (Fig. 4.10c). Tubes and branching patterns can be customized using agarose rods and spheroids (Fig. 4.10d). Agarose rods of 300–500 μm are made by gelling liquid agarose in cold phosphate buffered saline (PBS) (Norotte *et al.*, 2009). A small (~1 mm) branching conduit composed of spheroids was fused after 5–7 days. Despite promising results, the fabrication technology is seriously challenged by the difficulty of achieving larger tubular structures in a 3D configuration owing to the requirement for large quantities of spheroids (>1000) and the time for fusion to take place. Spheroid arrays have been used with agarose rods to construct 3D tubular structures, as shown in Fig. 4.11. Pig smooth muscle tubes of 2.5 mm OD and 1.5 mm OD were formed 3 days after patterning. Double-layered tubes, similar to native blood vessel media and adventitia layers, were formed with SMC spheroids and fibroblast spheroids in the inner and the outer walls, respectively. These spheroids fused and form distinct layers.

As organ printing is in its infancy, it is at present unclear how vascular networks will be effectively built from micro-vessels. Viability remains problematic without complex hierarchical vascular scaffolding. This complication arises from the resolution limitations of the device that must harmonize many different elements: micropipette and syringe, the size and properties of building blocks, bio-inks, and bio-papers. As already mentioned, trade-offs of time, resolution, and viability are currently inevitable. Micro-vessel templates require the seeding of three cell types – EC, SMC, and fibroblasts – at precise locations within a single scaffold vascular patterning, and that remains a challenge.

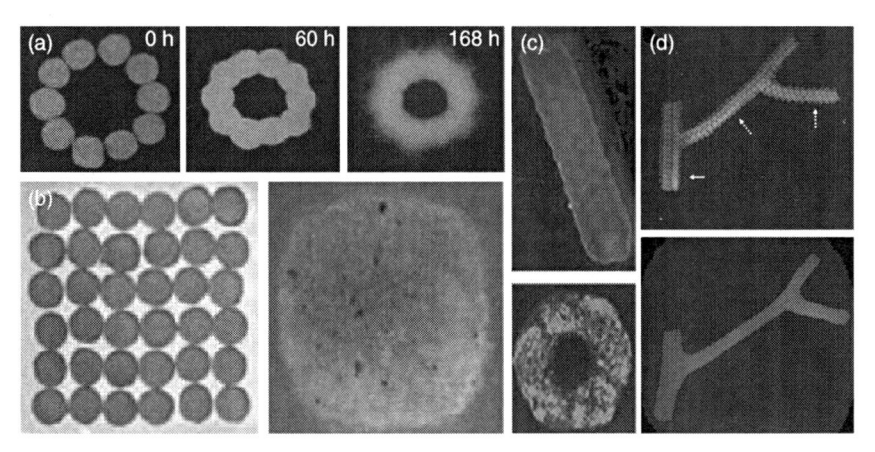

4.10 Fusion of multicellular spheroids. (a) Toriod formation (Jakab *et al.*, 2004). (b) Flat sheet formation (Jakab *et al.*, 2008). (c) Tube formation (Jakab *et al.*, 2010). (d) Branched tube formation (Norotte *et al.*, 2009).

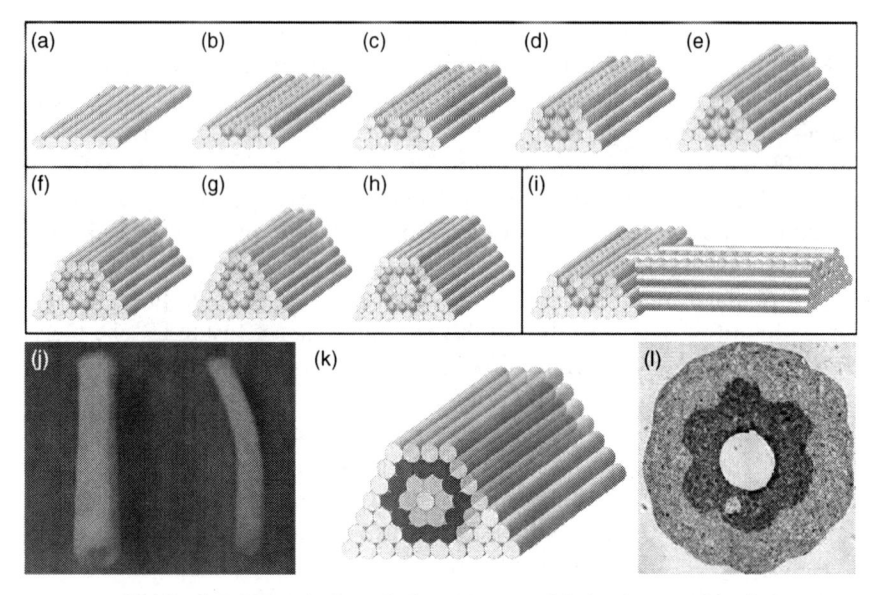

4.11 (a–i) Schematic for tubular structure fabrication; multicellular spheroids are shown in dark grey and agarose rods shown in light grey. (j) Engineered smooth muscle tubes of 2.5 mm OD (right) and 1.5 mm OD (left). (k) Double-layered vascular wall composed of smooth muscle cells (inner wall) and fibroblasts (outer wall). (l) Showed 3 days of post-printed fusion. (Norotte *et al.*, 2009).

4.4.3 Three-dimensional printing (3DP)

3DP™, invented by Sachs *et al.* (1989) and operated on a powder-based system, is a rapid prototyping process that functions in a fashion similar to SLS and yet is much simpler and cheaper (Fig. 4.4e). Each layer is formed by spreading stock powder and selectively joining the powder by microparticles of a liquid material dropped through an ink-jet nozzle or sprayed by a nebulizer. The liquid can serve as a binder, a solvent for dissolving a polymer, or an agent for a reaction such as crystallization to bind the particles together. The bonding is formed by local hardening of the binder and the material particles (Butscher *et al.*, 2011) as the solvent evaporates. Usually, porogens are added to the powder bed (polymers) to generate higher porous constructs. Both powder characteristics and binder droplet size play critical roles in determining 3DP resolution. Depending on the binder, either filtration, cure, pyrolysis, blow, washing, or drying at post-processing is necessary to complete a 3DP scaffold. 3DP works at room temperature on a broad range of materials as long as a suitable binder exists.

In light of the essential requirements for 3DP as well as SLS powdered materials, flowability is necessary to build up a thin and homogeneous powder layer. A powder is able to flow if gravitational forces dominate interparticle forces, the two types of forces acting on particles in bulk powders. Flowability is driven by particle size, shape, chemical composition, and operating temperature (Irsen *et al.*, 2006; Schulze, 1995). Moisture is also important because humidity can easily cause particles to form agglomerations, resulting in caking. Flowability of a bulk solid (ff_c) is dependent on the consolidation stress, σ_1, and the unconfined yield strength, σ_c, quantitatively defined as

$$ff_c = \sigma_1/\sigma_c \qquad [4.3]$$

and can be practically measured using shear testers like Jenike shear and Ring shear testers (Schulze, 1995). Flowability and resolution are mutually dependent since fine powder will give higher resolution but render poorer flowability due to the suppression of interparticle forces over gravitational ones (Zimmermann *et al.*, 2004). Flowability can be tailored up to a factor of 2 by plasma-enhanced chemical vapor deposition treatment (Spillmann *et al.*, 2007), hence potentially leading to a higher resolution for fine powder. Apart from flowability, wettability of the powder by the binder droplet and interaction of the powder with the binder are other critical properties to the powder/binder spreading, printing accuracy, and mechanical strength of the printed scaffolds (Fig. 4.12). Moreover, stability of the powder bed (or powder packing density) can impact binder deposition during spraying, dropping, and recoating, thus influencing the integrity and accuracy of a product. Polymeric additives have been sprayed onto the powder bed to improve particle bonding and flowability. Printed HA scaffolds were sintered for 2 h at 1300°C in a furnace to remove the organic binder compound, containing a water soluble polymer binder, Schelofix, and spray-dried HA-granulate (V5) (Leukers *et al.*, 2005). MC3T3-E1 cells, seeded onto the scaffolds and cultivated under dynamic cell culture, grew in between cavities of the granules. The cells also proliferated deep into the inter-channels, forming close contact with the HA granules. A comprehensive review of 3DP techniques was published by Butscher *et al.* (2011).

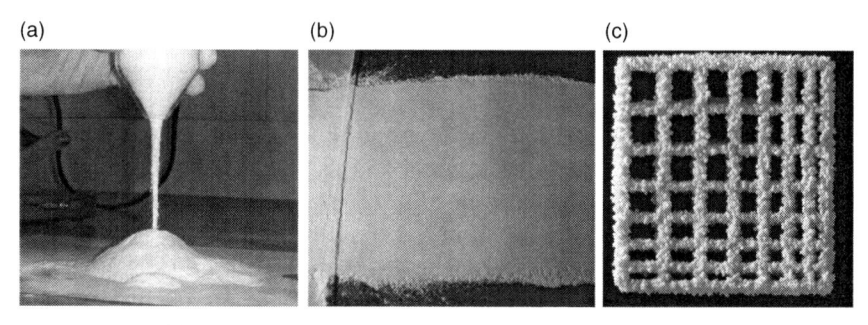

4.12 (a) High flowability of plasma treated powders and (b) a homogeneous bed of the same powder (right) (Butscher *et al.*, 2011). (c) Grid of 7.8 × 7.8 mm² overall dimension with wall thickness of 33 μm (Seitz *et al.*, 2005).

TheriForm™, adapted from 3DP for tissue regeneration, can fabricate drug delivery devices (Bártolo et al., 2011) and porous scaffolds (Castilho et al., 2011; Zeltinger et al., 2001). The common binder used in TheriForm is chloroform, which can be extracted using CO_2. Liquid carbon dioxide extraction has been found to be an effective means of removing residual solvent (Koegler et al., 2002). Additional pores are created after the removal of residual chloroform and particulate leaching. Leaching can also cause shrinkage of the overall dimensions (Sherwood et al., 2002). Zeltinger et al. (2001) confirmed the dependence of different cell type behavior on pore size by TheriForm-built scaffolds. Vascular SMCs favored 90 and 107 µm while dermal fibroblasts showed no preference for particular pore sizes. Also using TheriForm™, Sherwood and co-workers (2002) managed to produce a 90% porous D,L-PLGA/L-PLA cartilage region and a 55% porous bone region from composite scaffolds with a porosity gradient transition portion that formed cartilage after 6 weeks of *in vitro* culture. In the 3DP-based study by Seitz et al. (2005), a ceramic powder, HA, coated with polymer adhesives V5.2 and V12, was bonded by the polymeric binder Schelofix. Shrinkage was observed between 18% and 20% and the finest wall thickness was measured at 330 µm. Internal channels and pore size were reached within a good range for osteointegration, 450–570 µm and 10–30 µm, respectively.

Naturally, the use of solvent potentially poses a constraint to the 3DP approach. It demands a clean removal of solvent residues and solvent-resistant print-heads. Most solvent compatible systems, on the other hand, use low-resolution nozzles that lower the overall resolution of the fabricated scaffolds (Lee et al., 2005). Binder droplets are preferably small to maximize feature fidelity and resolution, yet they must also penetrate layer thickness. Layer thickness, however, must be in an appropriate range with the porogen size, usually ~200–400 µm, which ultimately requires larger binder dimensions (Lee et al., 2005). To overcome these restrictions, an indirect 3DP approach in which the technique is used to create molds for making 3D scaffolds has been investigated. Large pore sizes and finer resolution might be achieved at a larger scale (Lee et al., 2005; Sachlos et al. 2003).

Another extrusion-based process, though not yet as prevalent as others, employs pressure to extrude a material solution through a syringe, and is known as the pressure-assisted microsyringe (PAM) system. This technique involves the deposition of polymer dissolved in solvent through a syringe fitted with a 10–20 µm glass capillary needle (Tsang and Bhatia, 2004). Pressure and solution viscosity must work in harmony to ensure the expulsion of material from a small syringe tip. The PAM method produces structures with greater resolution than FDM but cannot incorporate micropores into structures due to the syringe dimensions. Yan et al. (2005) established a custom-made 3D syringe-based device that printed hepatocytes encapsulated in gelatin/chitosan constructs. After three months, a culture of the constructs displayed a stable 3D gel structure with some morphology changes in hepatocyte aggregates, either spheres or vortexes, with spindle shapes. The device then was upgraded into a double-nozzle PAM system allowing for plotting adipose-derived stromal cells (ADSCs) in gelatin/alginate/fibrinogen precursor to form a vascular-like network surrounding the hepatocytes in gelatin/alginate/chitosan hydrogel. ADSCs were effectually induced to differentiate into endothelial-like cells merely in spindle shapes at the periphery of the strands, probably due to the short culture time (Li et al., 2009). This method yields remarkably high lateral resolution, on a cellular scale: 5–20 μm (Vozzi et al., 2002; Yan et al., 2005). Wonhye et al (2009) mimicked human skin in a culture consisting of fibroblasts and keratinocytes in multi-layered, nebulized, crosslinked collagen scaffolds. Distinct layers of cells showed normal proliferation with experimental resolution differing according to inter-dispensing distance.

4.5 Inkjet printing (IP)

Inkjet printing refers to a non-contact reprographic technique that reproduces digital data in image form onto a substrate *via* ink drops. Whereas in the past it played a classic means of graphical printing, today inkjet printing is experiencing an extensive expansion in biomedical research owing to the feasibility of using off-the-shelf printers, a chief technical advantage of IP over other technologies (Boland et al., 2006; Cai et al., 2009; Okamoto et al., 2000; Sumerel et al., 2006). Inexpensive modification of HP inkjet printers, costing only several hundred US dollars, has enabled many laboratories to investigate IP in a variety of research contexts, primarily the printing of biomolecules onto substrates with almost no bioactivity loss; examples include DNA chips, protein arrays, and cell patterns. It is this technology that gives rise to the live tissue printing concept, and many scientists regard it as the ideal tool for creating living tissue substitutes (Binder et al., 2011; Jakab et al., 2008).

4.5.1 IP fundamentals

Most printers available on the open market operate thermally and piezoelectrically. An inkjet printer has a cartridge filled with ink, which is forced through a microfluidic chamber to an output orifice, forming ink drops (Binder et al., 2011). A thermal inkjet contains a heating element that creates locally small air bubbles in the ink. These air bubbles expand and quickly collapse, generating the necessary pressure pulse, which propagates and ejects ink drops out of the orifice. The energy supplied to the ink is dissipated into kinetic energy and heating of the drop inside the orifice. As the drop is expelled, the energy goes into viscous flow, surface tension of the drop, and kinetic energy (Calvert, 2001). Drop size varies depending on the applied temperature gradient, frequency of the current pulse, and ink viscosity. Although the local heat can actually exceed 300°C, a very small fraction of the ink is vaporized in the chamber because the heating occurs so quickly that it does not diffuse into the bulk liquid of the ink. The heating time during the drop ejection in an HP Deskjet 500 printer is in the order of 3.6 kHz frequency (Cui *et al.*, 2010). The bulk liquid can rise above ambient temperature by 10°C (Xu *et al.*, 2005) or up to 24°C theoretically (Cui *et al.*, 2010) depending on the liquid thermophysical properties of the ink. The temperature can be estimated from

$$T = E/C_p \times V_{\text{drop}},\qquad [4.4]$$

where C_p is the heat capacity of the ink, V_{drop} is the volume of the drop, and E is the energy supplied by the heating unit.

The ink in use for tissue regeneration, bio-ink, is made up of cells in suspension. Even though cells are susceptible to heat and stress, evidence shows a low apoptotic ratio of 3.5% (Cui *et al.*, 2010) or fewer than 10% damaged cells (Xu *et al.*, 2005), indicating that the influence of heat and shear stress applied to cells inside each drop might be negligible (Boland *et al.*, 2006; Cui *et al.*, 2010; Xu *et al.*, 2005). These phenomena, however, develop in transient pores of cell membranes and close within several hours (Bartolo *et al.*, 2011a; Cui *et al.*, 2010; Xu *et al.*, 2009). These temporary microdisruptions in the cell membrane have a positive effect by opening up an opportunity to introduce plasmids into cells for transient transfection purposes (Fig. 4.13). Initial transfection was achieved at more than 30% with Chinese hamster ovary (CHO) cells (Cui *et al.*, 2010). The cell viability was significantly higher while the transfection efficiency was significantly lower than that of electroporation with porcine aortic endothelial cells (EC) (Xu *et al.*, 2009). More studies need to be conducted to understand the mechanism of inkjet-mediated transfection: whether it is direct transfer of DNA into the nucleus or indirect transfer through cell division, and the stability and efficiency of the transfection, as well as its effects on cell viability (Binder *et al.*, 2011). Nevertheless, it is speculated that this gene transfection process could offer high-throughput, hazard-free protocols at low cost for efficient, targeted transgene integration, gene correction, and macroparticle delivery applications in tissue regeneration and gene therapy (Moehle *et al.*, 2007; Urnov *et al.*, 2005).

Piezoelectric inkjet printers have a cylindrical actuator made of a polycrystalline piezoelectric ceramic that surrounds a glass capillary nozzle. The material of the actuator exhibits the reverse piezoelectric effect, thus allowing it to change its shape according to the applied voltage. Under this voltage, the piezoelectric element induces in the microfluidic channel behind an orifice a sequence of pressure waves that propagate and overcome the surface tension to eject a drop. Drop formation is therefore adjusted by the electrical signals driving the piezoelectric actuator, including voltage amplitude, pulse

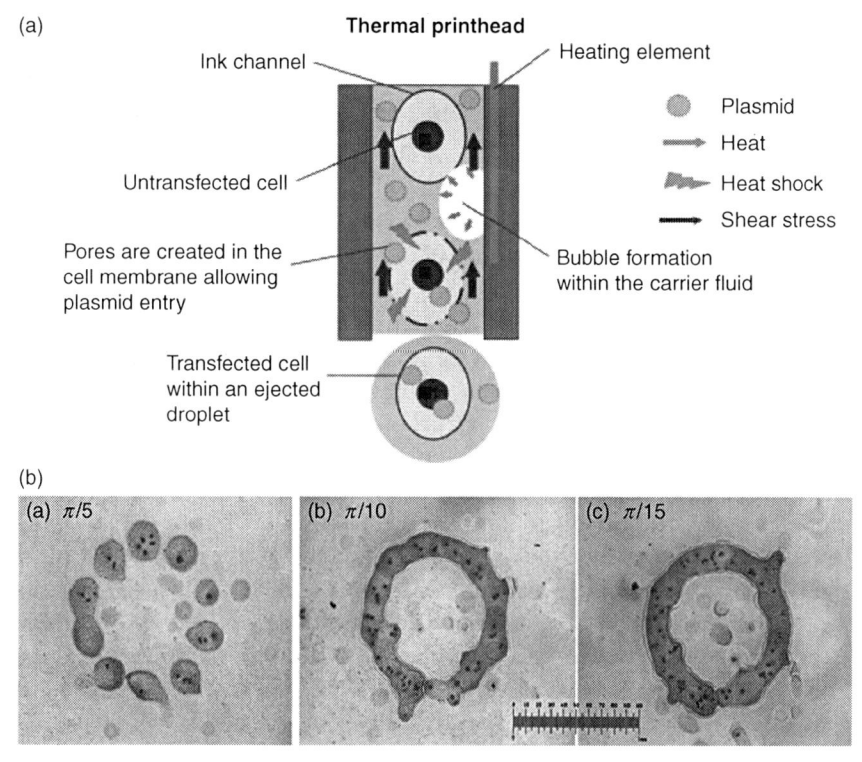

4.13 (a) Possible mechanism for inkjet-induced gene transfection of plasmids into cells through the temporary open pores on the cell membranes (Xu *et al.*, 2009). (b) Formation of satellite drops (Saunders *et al.*, 2010).

duration, and frequency (Reis *et al.*, 2005). Among these, voltage amplitude might alter stresses experienced by cells in suspension through the velocity of the ejected drop. This impact is indirectly associated with a lower rate of cell survival post-printing from 98% to 94% if the actuation voltage is increased from 40 to 80 V (Saunders *et al.*, 2008). The minimum magnitude to produce stable inkjet printing conditions is 40 V; however, the allowable voltage varies to a great extent according to the characteristics of both nozzle geometry and piezoelectric unit, especially the piezoelectric impedance (Parsa *et al.*, 2010). Pulse length seems to have no negative effect on cell survival (Saunders *et al.*, 2008) but affects the size of drops; drop volume increases with increasing pulse duration (Parsa *et al.*, 2010).

Ink used in piezoelectric printers is more viscous than in other inkjets, partly to eliminate ink leakage and reduce mist. Yet it uses more power and higher vibration frequencies to generate ink drops. Frequencies up to 30 kHz and power sources ranging from 12 to 100 W in commercial printers might damage cell walls and cause cell lysis as frequencies from 15 to 25 kHz and power sources from 10 to 375 W are known to disrupt cell membranes (Cui *et al.*, 2010; Xu *et al.*, 2005).

Both thermal and piezoelectric printers exhibit non-uniformity among individual drops. One reason for this variability might be the need to print at lower constraints (heat and voltage) to ensure cell viability. A second cause of the variability might be the non-spherical shape of initial ink drops (Saunders *et al.*, 2008, 2010; Sun *et al.*, 2004). The initial ejected drop exhibits a teardrop shape with a long tail. Depending on the drop distance from the nozzle to the substrate and the size and the fly of the drop, the tail might be pulled into the head of the drop because of the surface tension, forming a sphere before landing on the substrate; alternatively, it may detach from the drop, generating a small satellite drop (Fig. 4.13b). Piezoelectric printers appear to attain more uniform drops at higher voltages than thermal printers (Saunders *et al.*, 2010).

Parsa *et al.* (2010) have added surfactant to the ink to minimize the formation of satellite drops. By doing so, surface tension is lowered, allowing for easier transport of metabolites to cells and extra protection against shear-induced mechanical damage (Mizrahi, 1975). The lowering of surface tension also helps to reduce bubble breaks during printing. The bubble effect may be introduced into the bulk liquid through the invasion of gas bubbles. Rupturing of bubbles at the ink surface causes cell damage and cell lysis when cells are attached to air bubbles (Ma *et al.*, 2004). A surfactant of thermally reversible triblock copolymer PEO-poly(propylene oxide)-PEO (Pluronic™ or Poloxamer) did not reduce cell viability in short-term culture; nevertheless, its impact needs further investigation for long-term culture, as it lowered cell proliferation after 13 days (Parsa *et al.*, 2010).

A common issue with inkjet printing of cells is the inhomogeneous distribution of cells within repeated printing (Binder *et al.*, 2011). The distribution systematically differs according to cell types, concentration, and bio-ink types. Possible reasons for this lack of homogeneity might be the sedimentation of cell suspensions after a period of time, the viscosity of bio-inks, and the drop mass. Fluctuating cell populations in drops challenge the analysis of cell viability and cell functionality, especially in long-term cultures. Overall, owing to stimulation by many of the aforementioned printing factors, mechanical forces, process-induced stresses during flow and injection, drop velocity, dehydration, and substrate properties, cell survival is low and a recovery period is required (Cui *et al.*, 2010; Nair *et al.*, 2009; Xu *et al.*, 2006).

4.5.2 Bioprinting resolution

The resolution of inkjet printers is not the ultimate resolution of the bioprinting system. Their resolution is defined as the minimum drop size, whereas the final resolution of the printed structure is dependent on the way in which the drop spreads on the substrate; the difference between these resolutions is termed the resolution error (Binder et al., 2011). Minimum drop size on the other hand is a function of multiple variables: printer type, printhead type (air pressure or pump-based microvalves), and ink viscosity and density (Reis et al., 2005). Smaller nozzles permit smaller drops and higher resolution. Nozzle sizes range from 20 μm in thermal printers to 100 μm in piezoelectrics; drops, therefore, are in the range of 10 pL to 100 μL (Parsa et al., 2010; Reis et al., 2005).

Ink viscosity is the key to printing. Low viscous bio-inks of living cells are preferable to prevent nozzle clogging, often caused by drying of the ink; low-viscosity inks also allow for quick refills in situations where repeated printing is necessary. Soluble low-viscosity surfactants – for instance, Pluronic – can be added to a medium to reduce nozzle clogging (Calvert, 2001). Natural and synthetic hydrogels can also be used with cells (Boland et al., 2003). Collagen gels have a high degree of viscosity but are widely used due to their quick gelation. Low-viscosity calcium chloride is also used to reduce clogging as a crosslinking agent which diffuses into alginic acid as a substrate. Nonetheless, this diffusion lowers the drop placement resolution (Moon et al., 2010) and also influences cell viability through pH changes (Nakamura et al., 2008). Additionally, cellular debris and contaminants can also cause nozzle clogging. Thus, customized printers are desirable. A nozzle-free strategy has been attempted to prevent clogging during drop generation (Demirci and Montesano, 2007b). Further, a pressure approach to overcome the surface tension of high viscosity liquids by using a mechanical valve ejector was attempted (Demirci and Montesano, 2007a; Lee et al., 2009b). Moon et al. (2010) developed air-pressured mechanical valves to print SMC loaded collagen gels. Cell seeding uniformity was achieved at 26 ± 2 cells/mm^2 at 10^6 cells/mL, 122 ± 20 cells/mm^2 at 5×10^6 cells/mL, and 216 ± 38 cells/mm^2 at 10×10^6 cells/mL.

Furthermore, nozzle fouling increases with increased cell density since commercial printers are designed for liquid printing rather than suspensions of living cells (Arai et al., 2011). The maximum linear cell printing resolution for a 10 μm diameter cell is 1×10^6 cells/cm, which equals a cell density of 1×10^9 cells/mL. To prevent nozzle obstruction from cell sedimentation and aggregations, lower cell concentrations of 5×10^6 cells/mL (Xu et al., 2005) to 6×10^6 cells/mL (Arai et al., 2011) are favorable. It is critical to note that, for the maximum density 1×10^9 cells/mL, 27 h would be necessary to complete a tissue construct of 1 cm^3 at single-cell resolution and a 10 kHz deposition rate (Guillotin and Guillemot, 2011). Resolution and speed are therefore of mutual concession.

The resolution error relies on drop distance and velocity, and contact surface properties. Typically, printheads are located a few millimeters above the substrate – the further the distance, the bigger the error. Likewise, drop distance influences the precision of spatial distributions of the printed structure (Binder et al., 2011). Drop velocity, mainly decided by the pressure generated inside the orifice, might impact the interactions with the substrate, as well as the mechanical stress applied on the drop at landing. The influence of surface stiffness on printing resolution is inevitable as a drop can splash or disintegrate upon collision with the substrate (Rein, 1993). Soft and viscous surfaces can reduce the effects of stress impact on the drop, but very soft drops can consolidate, a phenomenon called bleeding. When cells are printed on stiff substrates, strain energy is higher and cells are more rounded but cell viability is low. If substrates are solid and strong, cell morphology is roundest but the cell viability is the lowest. To achieve high resolution, a relevant substrate should, therefore, be soft enough to absorb the kinetic energy of drops for cell survival yet stiff enough to preserve spatial formations without drop coalescence (Tirella et al., 2011). In other words, a trade-off between high cell viability and spatial resolution is unavoidable. Yet the substrate must also provide an appropriate environment for cell fusion and cell movement at a desired rate (Jakab et al., 2008). In essence, interfacial tension between substrate and bio-ink should be tunable for controlling the desired structure.

In light of the ideal characteristics of substrates or bio-papers used in IP, hydrogels are of interest for printing with live cells. Biomolecules can easily be mixed onto a hydrogel substrate before deposition of cells to enhance cell–gel interactions. Another reason to use hydrogels is that bio-inks, when ejected into drops of picoliter sized (droplets), can dry out very quickly. Hydrogel can keep the moisture for a much longer time. Various hydrogels have been used for inkjet printing. For instance, soy agar hydrogels are used for anchorage-independent CHO cells to promote cell fusion into a ring pattern (Norotte et al., 2009). Collagen gel has been used as a substrate for neuronal anchorage-dependent printing to support neuron differentiation (Xu et al., 2005) or for constructing cardiac structures (Jakab et al., 2008). In a gelation approach to overcome dehydration and bleeding, cells can be suspended in a gel precursor as a bio-ink and deposited onto a gel reactant as bio-paper. Nakamura (2010) used 0.8–1% sodium alginate and a 2% calcium chloride solution as a gel precursor and gel reactant, respectively, to fabricate double-walled tube structures with vascular endothelial cells (VEC) inside and SMCs outside, along with a long gel tube of 18 cm length and less than 1 mm diameter. Although biocompatible, alginate hydrogel, formed by the ioned bonding, may prevent cell adhesion and tissue growth, causing apoptosis (Frisch and Francis, 1994; Giancotti and Ruoslahti, 1999), and may not degrade.

As opposed to alginate hydrogel, fibrin – a natural or therapeutically derived biopolymer – has adhesion and enzymatic degradation capabilities (Rosso et al., 2005). A fibrin structure is formed by the fast reaction of fibrinogen with thrombin. Cell behavior, attachment, expansion, and proliferation, was observed in and on the fibrin gel during a week of incubation (Nakamura et al., 2010). The results support the properties of fibrin hydrogels for promoting cell migration, proliferation, and matrix synthesis through the release of growth factors, such as platelet-derived growth factor and the transforming growth factor beta (TGF-β) (Schense and Hubbell, 1998). However, the rapid degradation of the gel, with subsequent loss of shape and volume, renders a soft, fragile, and unstable 3D structure (Ahmed et al., 2008). Fibrin gel also suffers from shrinkage and low mechanical stiffness when compared to alginate hydrogels. These limits can be improved by optimizing the pH, the concentrations of fibrinogen and calcium ions, and crosslinking or combining fibrin with other materials (Ahmed et al., 2008). A biodegradable water-soluble polyurethane with good mechanical properties, used as bio-paper by Zhang et al. (2008), might find useful applications in vascular tissue engineering as it also possesses good elasticity and anti-platelet adhesion. A suitable material to fabricate cell constructs while also yielding a high resolution of printing is yet to be developed. To date, only resolutions ranging from 85 to 300 μm have been achieved with nozzle sizes of 30 to 200 μm (Wüst et al., 2011; Xu et al., 2006).

4.5.3 Bioprinted structures

So far, the types of structure fabricated by inkjet printing consist of fibers, 2D cell sheets, and multilayered sheets (Nishiyama et al., 2009). Many of these structures are printed by dropping sodium alginate as the hydrogel precursor into a bath of calcium chloride solution as the crosslinker. Due to the printhead size, it is difficult to print cells in the range 10–30 μm even though the finest features of non-living material can reach 50 μm (Wüst et al., 2011). Precise spatial control of growth factors deposited onto an ECM pattern can direct muscle-derived stem cells into differentiation fates such as osteoblasts, tenocytes, and myocytes; therefore, this approach might have potential in the regeneration of multi-tissue units (Hannachi et al., 2009; Ker et al., 2011). A 3D porous sheet of 25 mm × 5 mm × 1 mm dimensions, deposited by the alternate printing of fibrinogen, thrombin, and neural cells (embryonic hippocampal and cortical neurons), demonstrated the maintenance of their phenotypes as well as their basic electrophysiological functions.

Cell sheets produced by this approach can be rolled into a tubular shape to form blood vessels, yet these methods cannot produce 3D tissues or organs in a fully robotic way (Cui and Boland, 2009; Hannachi et al., 2009; McAllister et al.; Shimizu et al., 2002). In the report by Cui et al. a tubular structure mimicking human microvasculature of 500 μm diameter was fabricated with microvascular ECs and thrombin with Ca^{2+} printed on a fibrinogen substrate (Cui and Boland, 2009). After 21 days of culture, ECs formed a confluent lining along with the fibrin fibers. A hierarchical design of 3D pseudo-cardiac scaffolds of 1 cm inner diameter with two connected ventricles and a rectangular sheet of

$3 \times 0.8 \times 0.5$ cm^3 was constructed using cardiomyocytes micro-entrapped with alginate/gelatin gels and crosslinked with ejected calcium chloride drops (Fig. 4.14a-b). The droplets had hollow shells with diameters of about 25 μm; some were surrounded by satellite dots and some were collapsed into micro-shells. Although the scaffold wall was 300–800 μm, it appeared to support oxygen diffusion, evidenced by the beating of the whole construct under electrical stimulations and without stimulation. Whether inkjet printing can provide the promise of a potential approach that harnesses the capability of printing human vasculature remains to be seen. Limited attempts to use inkjet printers to prove the 3D printing concept for live tissues have been made. In the early works of Boland *et al.* (2003) sequential layers of EC aggregates were

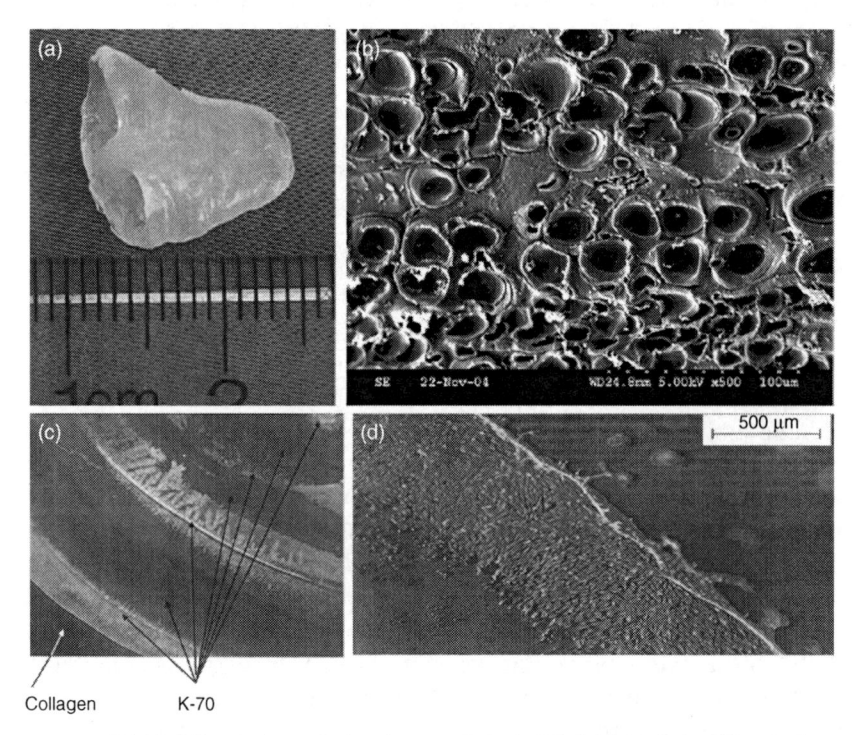

4.14 (a) Heart-shaped structures printed with inkjet printer (Xu *et al.*, 2009). (b) The bubble structures formed inside the printed structure (Xu *et al.*, 2009). (c) Top view of seven alternate layers of clear and dyed gels printed using inkjet printer. The edges between the layers reveal the amount of mixing that occurs when a new drop liquefies some of the existing structure before it is gelled itself. Optimizing gelling kinetics, drop size, and deposition rate may minimize this effect. (d) Cell aggregate fusion on a printed collagen gel ring. The image obtained on an epifluorescent microscope shows fused aggregates and cells that migrated across the gel. Cells that migrated toward the center of the ring proved nonviable, possibly because they became dry outside the gel (Boland *et al.*, 2003).

closely printed on thermosensitive gels to fuse into assemblies according to a 3D gel pattern (Fig. 4.14c and 4.14d).

Although inkjet printing is cheap and fast, the technique inherits a potential lack of capability for printing the high cell densities and large 3D structures required for tissue and organ function. High printing speed might also cause considerable cell damage. Because nozzle diameters are less than 300 μm, allowing maximum drop volumes around 0.015 μL per nozzle, this limits the inks which can be printed and only individual cells or small cell aggregates (sand) can be used as live ink (Wilson and Boland, 2003). Small orifices are not suitable for dispensing large cell aggregates (bricks). The gentle conditions provided by pressure-operated extruders appear to offer a more promising candidate technique (Jakab et al., 2004). A typical inkjet cartridge can print only 400 000 cells before failure due to clogging and deposition of solid salts in the microfluidic chamber. Adding chelating agents, such as ethylenediamine tetraacetic acid (EDTA), to bio-ink as an anti-aggregant helps to increase inkjet cartridge longevity (Parzel et al., 2009) and also to reduce nozzle clogging. Taken together, inkjet printing is a cost-effective tool but is hampered by low throughput; thus, whether it becomes the preferred tool for tissue printing is far from clear.

4.6 Conclusion

Each rapid prototyping system has its own pros and cons, but together they comprise the means to design, fabricate, model, and control constructed structures. Rapid prototyping technologies have not yet led to the construction of hierarchically organized tissues but have nevertheless revolutionized tissue engineering for translation into clinical application. Tissue regeneration therapies with live tissue printing are estimated to become a market of $10–30 billion within the next 20–30 years (Bártolo et al., 2011). For rapid prototyping systems to thrive in tissue regeneration, current state-of-the-art RP is dependent on its technical capabilities integrating with the development of biomaterials and obtaining sufficient cells for printing human-sized tissues and organs.

The development of advanced materials to mimic tissue and organ properties appears to be only a matter of time and continued investigation. Existing biocompatible materials have been mechanically and physiologically measured for their potential roles in biomedical applications (Table 4.1). Polymers, ceramics, bioglass, and their biocomposites have been tested in engineering hard tissues. Hydrogels play a significant role in soft tissue regeneration and live structure printing, yet they lack the mechanical strength necessary for the reconstruction of 3D soft tissues. Whether developing new biomaterials or chemically/physically modified existing biomaterials, none has yet been developed that can degrade appropriately without eliciting foreign body reaction and can match the remodeling rates of host tissues. To date, no particular single material can mimic human tissue elasticity, strength, and fracture toughness. Moreover, the number of biomaterials processable with rapid prototyping devices is still low. Each RP technique requires specific materials in certain forms, such as filaments, powders, solid pellets, or solutions. Practically, marketed rapid prototyping devices have not been specifically developed for use with biomaterials. Instead, the selected biomaterials have to be made compatible with rapid prototyping processes, particularly thermal ones, for preserving their functionality (Table 4.2). SLS and 3DP are most flexible in terms of materials. In porous scaffold fabrication, the surface roughness and aggregation or viscosity of support materials impact the efficiency of removal of trapped materials within the pore network (Yang *et al.*, 2002). In addition, there is a general lack of systematic knowledge about the optimal size and geometry of particles for powdered rapid prototyping systems. Experiments have been conducted mostly on a trial-and-error basis.

There has unfortunately always been a trade-off between the biological requirements of tissue regeneration and the technical feasibility of RP systems. Some systems have limited accuracy and poor raw material availability. Those commercial apparatuses, like SLA or SLS, involve a serious financial investment that does not allow easy access for all tissue engineering laboratories. Low-cost tools, like some bioplotters or inkjet printers, permit the practical construction of 3D structures, but those materials used have been exploited because they met some requirements, such as the ability to mix with cells or cell biocompatibility, without fulfilling other crucial parameters, such as degradation or mechanical strength. This has primarily led to proof-of-concept results in live tissue printing rather the production of actual live tissues or organs. Furthermore, the selection of RP techniques strongly depends on the application. In cardiovascular tissue engineering, for instance, a scaffold-based approach is not favored since it may reduce cell–cell interaction in the regenerated heart tissues as well as cause incorrect depositions of ECM alignment, thus influencing the force-generating ability of myocardial constructs (Norotte *et al.*, 2009). Whether or not new rapid prototyping tools arise to meet the needs of specific tissue regeneration applications, RP systems are likely to become more affordable and common for biological, clinical, and industrial research laboratories.

One of the benefits of using rapid prototyping technologies for tissue regeneration is their reproducibility and the ability to scale-up. Therefore, to accelerate the progress of tissue regeneration, development is needed in the quantitative measurements of living and non-living scaffolds and systematic analyses of cell and tissue behaviors. Similar to these requirements for advances in materials, there is a need for systematic investigation of process parameters, optimization measures, and standardized evaluation methods for the 3D constructs produced by rapid prototyping systems. Process parameters

Table 4.1 Properties of common biomaterials used in tissue regeneration by rapid prototyping systems

Biomaterial	Biocompatibility	Biodegradability: mechanism and timeline	Applications
Synthetic			
PCL	Minimal inflammation	Hydrolysis. Bulk. More than 3 years	Skin. Cartilage. Bone. Ligaments. Tendons. Vessels. Nerves.
PEOT/PBT	No inflammation. Mild foreign body reaction.	Hydrolysis. Bulk. 1 month–5 years.	Skin. Cartilage. Bone. Muscle.
PEG-DA	Minimal foreign body reaction.	Hydrolysis. Bulk. 1–3 weeks.	Skin. Vessels. Nerves. Smooth muscle.
PLA	Accumulative degradation product, lactic acid, causes delayed inflammatory response.	Hydrolysis. Bulk. 5 months–5 years.	Skin. Cartilage. Bone. Ligaments. Tendons. Vessels. Nerves. Bladder. Liver.
PGA	Accumulative degradation product, glycolic acid, causes delayed inflammatory response.	Hydrolysis. Bulk. 1–12 months.	
PLGA	Delayed inflammatory response.	Hydrolysis. Bulk. Adjustable.	
PPE	Minimal foreign body reaction. Minimal inflammation.	Hydrolysis. Surface. 1–3 years.	Cartilage. Bone. Nerves. Liver.
PPF-DA	Minimal foreign body reaction. Mild inflammation.	Hydrolysis. Bulk. 6–3 years.	Bone. Vessels.
Natural derived			
Alginate	No foreign body reaction. No inflammation.	Dissolution or hydrolysis depending on crosslinker. Bulk. 1 day–3 months.	Skin. Cartilage. Bone. Nerves. Muscle. Pancreas. Liver.
Chitosan	No foreign body reaction. No inflammation.	Enzymatic cleavage (lysozyme). Bulk. 3 days–6 months.	Skin. Cartilage. Bone. Nerves. Vessels. Liver. Pancreas.
Collagen	No foreign body reaction. Minimal inflammation.	Enzymatic cleavage (metalloproteases). Bulk. 2–24 weeks (depending on crosslinker).	Skin. Cartilage. Bone. Ligaments. Tendons. Vessels. Nerves. Bladder. Adipose tissues.

(Continued)

Table 4.1 Continued

Biomaterial	Biocompatibility	Biodegradability: mechanism and timeline	Applications
Hyaluronate	No foreign body reaction. No inflammation.	Enzymatic cleavage. (hylauronidase) Bulk. Hours–1 month.	Skin. Cartilage. Bone. Ligaments. Nerves. Vessels. Liver.
Ceramics			
HA		Dissolution, resorbed by osteoclasts. Poor degradation.	Bone.
TCP		Dissolution, resorbed by osteoclasts. Bulk. 8–24 weeks.	Bone.

Sources: Drury and Mooney, 2003; Leong et al., 2008; Moroni et al., 2008; Weinand et al., 2006; Yang et al., 2001.

Table 4.2 Comparison of major rapid prototyping techniques

RP system	Resolution (μm)	Material	Cell-embedded capability	Advantages	Disadvantages
Photopolymerization SLA (Hutmacher et al., 2004; Yang et al., 2002); (Kim et al., 2010)	100	PEG, PEGDA, PEO, PPF, modified PDLLA and PCL.	Low survival.	High resolution. Fine features. Easy to remove support materials.	Limited range of photosensitive polymers and resin.
2PP (Ovsianikov et al., 2011)	0.1–0.2				Difficult to scale up.
SLS (Tan et al., 2003; Varghese et al., 2005; Yang et al., 2002)	500	Polymers, ceramics (PCL, HA, TCP).	None.	Wide range of materials. Good mechanical strength. High accuracy. High porosity.	Materials in powder form. Powdery surface finish. Chance of degrading materials due to high processing temperature. Difficult to remove trapped materials.
FDM (Tellis et al., 2008)	250	Thermoplastic polymers and their composites (PCL, PP-TCP, PCL-HA, PCL-TCP).	None.	Easy set-up. Low costs. No trapped material.	Materials in filament form. High temperature. Smooth surface.
3DF (Boland et al., 2003)	100–500	Thermoplastic polymers and their composites. Hydrogels (alginate, agarose, gelatin, chitosan, collagen).	Survival rate varies.	Easy set-up. Low costs. Material in pellet. Incorporate bioactive molecules.	Thermal process.

(Continued)

Table 4.2 Continued

RP system	Resolution (μm)	Material	Cell-embedded capability	Advantages	Disadvantages
Bioplotter (Lam et al., 2002; Landers and Mulhaupt 2000; Wang et al., 2006; Yan et al., 2005)	10	Hydrogels.	High survival.	Incorporate bioactive molecules.	Low mechanical strength. Unstable final structure. Low accuracy. Slow processing.
3DP (Lam et al., 2002)	200–300	Ceramics, polymers.	None.	Low costs. Fast processing. Wide range of materials. Microporosity induced after removal of binders.	Required binders. Weak bonding between powder particles. Powdery surface. Post-processing.
Inkjet (Xu et al., 2006)	85–300	Hydrogels.	High survival.	Cheapest.	High shear forces. Clogging. Thin final structure. Hard to scale up. Unstable final structure. Slow processing. Dependence on self-assembly

like deposition velocity and pressure have been probed for the quality of printed alginate hydrogels by Tirella *et al.* (2009). Likewise, dispensing pressure and nozzle diameter have been scrutinized for cell viability and recovery (Chang and Sun, 2008). However, optimization cannot be completed without extensive mathematical modeling and computer simulation, for example, *via* MatLab and CAD modeling and Monte Carlo simulation. Gravimetry, mercury intrusion, liquid displacement, SEM, and computed tomography are typically used to quantify scaffold geometry, including pore parameters (Ballyns *et al.*, 2010; Butscher *et al.*, 2011; Karageorgiou and Kaplan, 2005). Quantitative imaging techniques and an adequate statistical assessment will be crucial for evaluating precision, quality, and reproducibility of rapid prototyping products. Complicated imaging techniques will also help in detecting the mechanical strength of living/non-living scaffolds based on macro- and micro-features. However, the gain in biological performance of rapid prototyping fabricated constructs as compared to conventional methods has not been thoroughly investigated. The comprehensive analyses of cell viability and proliferation in various matrices using different delivery approaches have yet to be performed for a cross reference comparison (Binder *et al.*, 2011).

Numerous rapid prototyping systems have been demonstrated to achieve features at the microscale (Table 4.2). Fine features help cells to better communicate, adhere, infiltrate, and grow. However, at increasingly high resolutions, the architecture might acquire lower mechanical strength. An optimal structure design demands a harmonized work of modeling, resolution, materials, and mechanical properties. In bioprinting, the highest resolution is defined as the continuous deposition of one cell or one cell aggregate adjacent to another, or one cell or cell aggregate in a single droplet. With this resolution, printing of massive amounts of cells or cell aggregates and large-scale constructs will require significant time. An ever-limiting trade off will be that increasing deposition speed is likely to affect cell viability and functionality. Compared to current methods, the flexibility that rapid prototyping technologies offers is outstanding in many technical aspects, including resolution, speed, scalability, time effectiveness, and most importantly, complex scaffolds and bioprinting. A largely unresolved matter remains, however — to what degree can rapid prototyping recapitulate the nature of tissues and organs and how far can engineered tissues and organs go in imitating nature? Since the underlying biology is not completely understood, the patterns that will be produced by self-assembly of single cells or multiple cell types are not fully predictable. Moreover, scaling up a structure by packing building blocks to generate a tissue construct at the macroscopic level is still questionable and challenging owing to the inherent trade-off between resolution and manufacturing scale. For the construction of organ subunits of a few centimeters in dimension, neither the size nor mechanical supports from an existing material is yet feasible. It is foreseeable, though it is far from reality, that bioprinting will lead to tissue regeneration therapies with improved clinical outcomes.

4.7 References

Ahmed TAE, Dare EV and Hincke M. (2008). Fibrin: A versatile scaffold for tissue engineering applications. *Tissue Engineering Part B-Reviews* **14**(2):199–215.

Arai K, Iwanaga S, Toda H, Genci C, Nishiyama Y and Nakamura M. (2011). Three-dimensional inkjet biofabrication based on designed images. *Biofabrication* **3**(3): 034113. Availabel online at: http://iopscience.iop.org/1758-5090/3/3/034113/.

Arcaute K, Mann B and Wicker R. (2010). Stereolithography of spatially controlled multi-material bioactive poly(ethylene glycol) scaffolds. *Acta Biomaterialia* **6**(3):1047–1054.

Arcaute K, Mann BK and Wicker RB. (2006). Stereolithography of three-dimensional bioactive poly(ethylene glycol) constructs with encapsulated cells. *Annals of Biomedical Engineering* **34**(9):1429–1441.

Azevedo MC, Reis RL, Claase MB, Grijpma DW and Feijen J. (2003). Development and properties of polycaprolactone/hydroxyapatite composite biomaterials. *Journal of Materials Science: Materials in Medicine* **14**(2):103–107.

Ballyns JJ, Cohen DL, Malone E, Maher SA, Potter HG, Wright T, Lipson H and Bonassar LJ. (2010). An optical method for evaluation of geometric fidelity for anatomically shaped tissue-engineered constructs. *Tissue Engineering Part C-Methods* **16**(4):693–703.

Bartolo P, Domingos M, Gloria A and Ciurana J. (2011a). BioCell printing: Integrated automated assembly system for tissue engineering constructs. *Cirp Annals-Manufacturing Technology* **60**(1):271–274.

Bartolo P, Domingos M, Gloria A and Ciurana J. (2011b). BioCell printing: Integrated automated assembly system for tissue engineering constructs. *CIRP Annals-Manufacturing Technology* **60**(1):271–274.

Bártolo PJ, Domingos M, Patrício T, Cometa S and Mironov V. (2011). *Biofabrication Strategies for Tissue Engineering Advances on Modeling in Tissue Engineering*. In: Fernandes PR, Bártolo PJ, editors: Springer, The Netherlands. pp. 137–176.

Binder K, Allen A, Yoo J and Atala A. (2011). Drop-on-demand inkjet bioprinting: A primer. *Gene Therapy and Regulation (GTR)* **6**(1):17.

Boland T, Mironov V, Gutowska A, Roth EA and Markwald RR. (2003). Cell and organ printing 2: Fusion of cell aggregates in three-dimensional gels. *Anatomical Record Part A-Discoveries in Molecular Cellular and Evolutionary Biology* **272A**(2):497–502.

Boland T, Xu T, Damon B and Cui X. (2006). Application of inkjet printing to tissue engineering. *Biotechnology Journal* **1**(9):910–917.

Burdick JA and Anseth KS. (2002). Photoencapsulation of osteoblasts in injectable RGD-modified PEG hydrogels for bone tissue engineering. *Biomaterials* **23**(22):4315–4323.

Butscher A, Bohner M, Hofmann S, Gauckler L and Muller R. (2011). Structural and material approaches to bone tissue engineering in powder-based three-dimensional printing. *Acta Biomaterialia* **7**(3):907–920.

Cai K, Dong H, Chen C, Yang L, Jandt KD and Deng L. (2009). Inkjet printing of laminin gradient to investigate endothelial cellular alignment. *Colloids and Surfaces B: Biointerfaces* **72**(2):230–235.

Calvert P. (2001). Inkjet printing for materials and devices. *Chemistry of Materials* **13**(10):3299–3305.

Castilho M, Pires I, Gouveia B and Rodrigues J. (2011). Structural evaluation of scaffolds prototypes produced by three-dimensional printing. *International Journal of Advanced Manufacturing Technology* **56**(5–8):561–569.

Centola M, Rainer A, Spadaccio C, De Porcellinis S, Genovese JA and Trombetta M. (2010). Combining electrospinning and fused deposition modeling for the fabrication of a hybrid vascular graft. *Biofabrication* **2**(1):014102. Avialable online at: http://iopscience.iop.org/1758-5090/2/1/014102/.

Chang CC, Boland ED, Williams SK and Hoying JB. (2011). Direct-write bioprinting three-dimensional biohybrid systems for future regenerative therapies. *Journal of Biomedical Materials Research Part B-Applied Biomaterials* **98B**(1):160–170.

Chang R and Sun W. (2008). Effects of dispensing pressure and nozzle diameter on cell survival from solid freeform fabrication-based direct cell writing. *Tissue Engineering Part A* **14**(1):41–48.

Chen YH, Zhou SW and Li Q. (2011). Microstructure design of biodegradable scaffold and its effect on tissue regeneration. *Biomaterials* **32**(22):5003–5014.

Claeyssens F, Hasan EA, Gaidukeviciute A, Achilleos DS, Ranella A, Reinhardt C, Ovsianikov A, Xiao S, Fotakis C, Vamvakaki M, Chichkov BN and Farsari M. (2009). Three-dimensional biodegradable structures fabricated by two-photon polymerization. *Langmuir* **25**(5):3219–3223.

Cohen DL, Malone E, Lipson H and Bonassar LJ. (2006). Direct freeform fabrication of seeded hydrogels in arbitrary geometries. *Tissue Engineering* **12**(5):1325–1335.

Cui X, Dean D, Ruggeri ZM and Boland T. (2010). Cell damage evaluation of thermal inkjet printed Chinese hamster ovary cells. *Biotechnology and Bioengineering* **106**(6):963–969.

Cui XF and Boland T. (2009). Human microvasculature fabrication using thermal inkjet printing technology. *Biomaterials* **30**(31):6221–6227.

Das S. (2008). Selective laser sintering of polymers and polymer-ceramic composites. In: Bopaya Bidanda PJB, editor. *Virtual Prototyping & Bio Manufacturing in Medical Applications*: Springer Science+Business Media, LLC, New York, NY 10013, USA. pp. 229–260.

David LR, Proffer P, Hurst WJ, Glazier S and Argenta LC. (2004). Spring-mediated cranial reshaping for craniosynostosis. *Journal of Craniofacial Surgery* **15**(5):810–816.

Demirci U and Montesano G. (2007a). Cell encapsulating droplet vitrification. *Lab on a Chip* **7**(11):1428–1433.

Demirci U and Montesano G. (2007b). Single cell epitaxy by acoustic picolitre droplets. *Lab on a Chip* **7**(9):1139–1145.

Dewidar MM, Dalgarno KW and Wright CS. (2003). Processing conditions and mechanical properties of high-speed steel parts fabricated using direct selective laser sintering. *Proceedings of the Institution of Mechanical Engineers Part B-Journal of Engineering Manufacture* **217**(12):1651–1663.

Drury JL and Mooney DJ. (2003). Hydrogels for tissue engineering: scaffold design variables and applications. *Biomaterials* **24**(24):4337–4351.

Duguay D, Foty RA and Steinberg MS. (2003). Cadherin-mediated cell adhesion and tissue segregation: Qualitative and quantitative determinants. *Developmental Biology* **253**(2):309–323.

Elisseeff J. (2004). Injectable cartilage tissue engineering. *Expert Opinion on Biological Therapy* **4**(12):1849–1859.

Elomaa L, Teixeira S, Hakala R, Korhonen H, Grijpma DW and Seppala JV. (2011). Preparation of poly(epsilon-caprolactone)-based tissue engineering scaffolds by stereolithography. *Acta Biomaterialia* **7**(11):3850–3856.

Eosoly S, Brabazon D, Lohfeld S and Looney L. (2010). Selective laser sintering of hydroxyapatite/poly-epsilon-caprolactone scaffolds. *Acta Biomaterialia* **6**(7):2511–2517.

Farsari M, Filippidis G, Drakakis TS, Sambani K, Georgiou S, Papadakis G, Gizeli E and Fotakis C. (2007). Three-dimensional biomolecule patterning. *Applied Surface Science* **253**(19):8115–8118.

Fedorovich NE, Dewijn JR, Verbout AJ, Alblas J and Dhert WJA. (2008). Three-dimensional fiber deposition of cell-laden, viable, patterned constructs for bone tissue printing. *Tissue Engineering Part A* **14**(1):127–133.

Fedorovich NE, Moroni L, Malda J, Alblas J, Blitterswijk CA and Dhert WJA. (2010). *3D-Fiber Deposition for Tissue Engineering and Organ Printing Applications Cell and Organ Printing*. In: Ringeisen BR, Spargo BJ, Wu PK, editors: Springer, The Netherlands. pp. 225–239.

Fedorovich NE, Swennen I, Girones J, Moroni L, van Blitterswijk CA, Schacht E, Alblas J and Dhert WJA. (2009). Evaluation of photocrosslinked lutrol hydrogel for tissue printing applications. *Biomacromolecules* **10**(7):1689–1696.

Fisher JP, Dean D, Engel PS and Mikos AG. (2001). Photoinitiated polymerization of biomaterials. *Annual Review of Materials Research* **31**:171–181.

Fisher JP, Vehof JWM, Dean D, van der Waerden J, Holland TA, Mikos AG and Jansen JA. (2002). Soft and hard tissue response to photocrosslinked poly(propylene fumarate) scaffolds in a rabbit model. *Journal of Biomedical Materials Research* **59**(3):547–556.

Frisch S and Francis H. (1994). Disruption of epithelial cell-matrix interactions induces apoptosis. *The Journal of Cell Biology* **124**(4):619–626.

Giancotti FG and Ruoslahti E. (1999). Integrin signaling. *Science* **285**(5430):1028–1033.

Gibson I and Shi DP. (1997). Material properties and fabrication parameters in selective laser sintering process. *Rapid Prototyping Journal* **3**(4):129–136.

Goiato MC, Santos MR, Pesqueira AA, Moreno A, dos Santos DM and Haddad MF. (2011). Prototyping for surgical and prosthetic treatment. *Journal of Craniofacial Surgery* **22**(3):914–917.

Griffith LG and Swartz MA. (2006). Capturing complex 3D tissue physiology in vitro. *Nature Reviews Molecular Cell Biology* **7**(3):211–224.

Guillotin B and Guillemot F. (2011). Cell patterning technologies for organotypic tissue fabrication. *Trends in Biotechnology* **29**(4):183–190.

Guillotin B, Souquet A, Catros S, Duocastella M, Pippenger B, Bellance S, Bareille R, Remy M, Bordenave L, Amedee J and Guillemot F. (2010). Laser assisted bioprinting of engineered tissue with high cell density and microscale organization. *Biomaterials* **31**(28):7250–7256.

Hannachi IE, Yamato M and Okano T. (2009). Cell sheet technology and cell patterning for biofabrication. *Biofabrication* **1**(2):022002.

Hutmacher DW, Loessner D, Rizzi S, Kaplan DL, Mooney DJ and Clements JA. (2010). Can tissue engineering concepts advance tumor biology research? *Trends in Biotechnology* **28**(3):125–133.

Hutmacher DW, Schantz JT, Lam CXF, Tan KC and Lim TC. (2007). State of the art and future directions of scaffold-based bone engineering from a biomaterials perspective. *Journal of Tissue Engineering and Regenerative Medicine* **1**(4):245–260.

Hutmacher DW, Schantz T, Zein I, Ng KW, Teoh SH and Tan KC. (2001). Mechanical properties and cell cultural response of polycaprolactone scaffolds designed and fabricated via fused deposition modeling. *Journal of Biomedical Materials Research* **55**(2):203–216.

Hutmacher DW, Sittinger M, Risbud MV. (2004). Scaffold-based tissue engineering: rationale for computer-aided design and solid free-form fabrication systems. *Trends in Biotechnology* **22**(7):354–362.

Ignatius AA, Ohnmacht M, Claes LE, Kreidler J and Palm F. (2001). A composite polymer/tricalcium phosphate membrane for guided bone regeneration in maxillofacial surgery. *Journal of Biomedical Materials Research* **58**(5):564–569.

Irsen SH, Leukers B, Hockling C, Tille C and Seitz H. (2006). Bioceramic granulates for use in 3D printing: Process engineering aspects. *Materialwissenschaft Und Werkstofftechnik* **37**(6):533–537.

Ishaug SL, Crane GM, Miller MJ, Yasko AW, Yaszemski MJ and Mikos AG. (1997). Bone formation by three-dimensional stromal osteoblast culture in biodegradable polymer scaffolds. *Journal of Biomedical Materials Research* **36**(1):17–28.

Jacobs PF. (1992). *Rapid Prototyping and Manufacturing: Fundamentals of Stereolithography*. Dearborn, Mich: Society of Manufacturing Engineers:16.

Jakab K, Neagu A, Mironov V, Markwald RR and Forgacs G. (2004). Engineering biological structures of prescribed shape using self-assembling multicellular systems. *Proceedings of the National Academy of Sciences of the United States of America* **101**(9):2864–2869. PNAS 2004 March 2.

Jakab K, Norotte C, Damon B, Marga F, Neagu A, Besch-Williford CL, Kachurin A, Church KH, Park H, Mironov V, Markwald R, Vunjak-Novakovic G and Forgacs G. (2008). Tissue engineering by self-assembly of cells printed into topologically defined structures. *Tissue Engineering Part A* **14**(3):413–421.

Jakab K, Norotte C, Marga F, Murphy K, Vunjak-Novakovic G and Forgacs G. (2010). Tissue engineering by self-assembly and bio-printing of living cells. *Biofabrication* **2**(2):022001. Available online at http://iopscience.iop.org/1758-5090/2/2/022001/

Karageorgiou V and Kaplan D. (2005). Porosity of 3D biomaterial scaffolds and osteogenesis. *Biomaterials* **26**(27):5474–5491.

Ker EDF, Chu B, Phillippi JA, Gharaibeh B, Huard J, Weiss LE and Campbell PG. (2011). Engineering spatial control of multiple differentiation fates within a stem cell population. *Biomaterials* **32**(13):3413–3422.

Kim K, Yeatts A, Dean D and Fisher JP. (2010). Stereolithographic bone scaffold design parameters: Osteogenic differentiation and signal expression. *Tissue Engineering Part B-Reviews* **16**(5):523–539.

Koegler WS, Patrick C, Cima MJ and Griffith LG. (2002). Carbon dioxide extraction of residual chloroform from biodegradable polymers. *Journal of Biomedical Materials Research* **63**(5):567–576.

Kruth JP, Mercelis P, Van Vaerenbergh J, Froyen L and Rombouts M. (2005). Binding mechanisms in selective laser sintering and selective laser melting. *Rapid Prototyping Journal* **11**(1):26–36.

Kyle W Binder A., Yoo JJ, Atala A. (2011). Drop-on-demand inkjet bioprinting: A primer. *Gene Therapy and Regulation (GTR)* **6**(1):17.

Lalan BAS, Pomerantseva MDI and Vacanti MDJP. (2001). Tissue engineering and its potential impact on surgery. *World Journal of Surgery* **25**(11):1458–1466.

Lam CXF, Mo XM, Teoh SH, Hutmacher DW. (2002). Scaffold development using 3D printing with a starch-based polymer. *Materials Science & Engineering C-Biomimetic and Supramolecular Systems* **20**(1–2):49–56.

Lan PX, Lee JW, Seol YJ and Cho DW. (2009). Development of 3D PPF/DEF scaffolds using micro-stereolithography and surface modification. *Journal of Materials Science-Materials in Medicine* **20**(1):271–279.

Landers R, Mulhaupt R. (2000). Desktop manufacturing of complex objects, prototypes and biomedical scaffolds by means of computer-assisted design combined with computer-guided 3D plotting of polymers and reactive oligomers. *Macromolecular Materials and Engineering* **282**(9):17–21.

Landers R, Pfister A, Hübner U, John H, Schmelzeisen R and Mülhaupt R. (2002). Fabrication of soft tissue engineering scaffolds by means of rapid prototyping techniques. *Journal of Materials Science* **37**(15):3107–3116.

Langer R and Vacanti JP. (1993). Tissue engineering. *Science* **260**(5110):920–926.

Laschke MW, Rücker M, Jensen G, Carvalho C, Mülhaupt R, Gellrich N-C and Menger MD. (2008a). Improvement of vascularization of PLGA scaffolds by inosculation of in situ-preformed functional blood vessels with the host microvasculature. *Annals of Surgery* **248**(6):939–948 doi: 10.1097/SLA.0b013e31818fa52f.

Laschke MW, Rücker M, Jensen G, Carvalho C, Mülhaupt R, Gellrich NC and Menger MD. (2008b). Incorporation of growth factor containing Matrigel promotes vascularization of porous PLGA scaffolds. *Journal of Biomedical Materials Research Part A* **85A**(2):397–407.

Lee JW, Ahn G, Kim DS and Cho DW. (2009a). Development of nano- and microscale composite 3D scaffolds using PPF/DEF-HA and micro-stereolithography. *Microelectronic Engineering* **86**(4–6):1465–1467.

Lee KS, Kim RH, Yang DY and Park SH. (2008). Advances in 3D nano/microfabrication using two-photon initiated polymerization. *Progress in Polymer Science* **33**(6):631–681.

Lee KY, Alsberg E and Mooney DJ. (2001). Degradable and injectable poly(aldehyde guluronate) hydrogels for bone tissue engineering. *Journal of Biomedical Materials Research* **56**(2):228–233.

Lee M, Dunn JCY and Wu BM. (2005). Scaffold fabrication by indirect three-dimensional printing. *Biomaterials* **26**(20):4281–4289.

Lee W, Debasitis JC, Lee VK, Lee J-H, Fischer K, Edminster K, Park J-K and Yoo S-S. (2009b). Multi-layered culture of human skin fibroblasts and keratinocytes through three-dimensional freeform fabrication. *Biomaterials* **30**(8):1587–1595.

Leong KF, Cheah CM and Chua CK. (2003). Solid freeform fabrication of three-dimensional scaffolds for engineering replacement tissues and organs. *Biomaterials* **24**(13):2363–2378.

Leong KF, Chua CK, Sudarmadji N and Yeong WY. (2008). Engineering functionally graded tissue engineering scaffolds. *Journal of the Mechanical Behavior of Biomedical Materials* **1**(2):140–152.

Leong KF, Phua KKS, Chua CK, Du ZH and Teo KOM. (2001). Fabrication of porous polymeric matrix drug delivery devices using the selective laser sintering technique. *Proceedings of the Institution of Mechanical Engineers Part H-Journal of Engineering in Medicine* **215**(H2):191–201.

Leukers B, Gulkan H, Irsen SH, Milz S, Tille C, Schieker M and Seitz H. (2005). Hydroxyapatite scaffolds for bone tissue engineering made by 3D printing. *Journal of Materials Science-Materials in Medicine* **16**(12):1121–1124.

Li S, Xiong Z, Wang X, Yan Y, Liu H and Zhang R. (2009). Direct fabrication of a hybrid cell/hydrogel construct by a double-nozzle assembling technology. *Journal of Bioactive and Compatible Polymers* **24**(3):249–265.

Lisignoli G, Fini M, Giavaresi G, Nicoli Aldini N, Toneguzzi S and Facchini A. (2002). Osteogenesis of large segmental radius defects enhanced by basic fibroblast growth factor activated bone marrow stromal cells grown on non-woven hyaluronic acid-based polymer scaffold. *Biomaterials* **23**(4):1043–1051.

Ma NN, Chalmers JJ, Aunins JG, Zhou WC and Xie LZ. (2004). Quantitative studies of cell-bubble interactions and cell damage at different pluronic F-68 and cell concentrations. *Biotechnology Progress* **20**(4):1183–1191.

Ma Z, Pirlo RK, Wan Q, Yun JX, Yuan XC, Xiang P, Borg TK and Gao BZ. (2011). Laser-guidance-based cell deposition microscope for heterotypic single-cell micropatterning. *Biofabrication* **3**(3):034107. Available online at: http://iopscience.iop.org/1758-5090/3/3/034107/.

McAllister TN, Maruszewski M, Garrido SA, Wystrychowski W, Dusserre N, Marini A, Zagalski K, Fiorillo A, Avila H, Manglano X, Antonelli J, Kocher A, Zembala M, Cierpka L, de la Fuente LM and L'heureux N. Effectiveness of haemodialysis access with an autologous tissue-engineered vascular graft: A multicentre cohort study. *The Lancet* **373**(9673):1440–1446.

Melchels FPW, Feijen J and Grijpma DW. (2009). A poly(D,L-lactide) resin for the preparation of tissue engineering scaffolds by stereolithography. *Biomaterials* **30**(23–24):3801–3809.

Melchels FPW, Feijen J and Grijpma DW. (2010). A review on stereolithography and its applications in biomedical engineering. *Biomaterials* **31**(24):6121–6130.

Mironov V, Boland T, Trusk T, Forgacs G and Markwald RR. (2003). Organ printing: Computer-aided jet-based 3D tissue engineering. *Trends in Biotechnology* **21**(4):157–161.

Mizrahi A. (1975). Pluronic polyols in human lymphocyte cell line cultures. *Journal of Clinical Microbiology* **2**(1):11–13.

Moehle EA, Rock JM, Lee YL, Jouvenot Y, DeKelver RC, Gregory PD, Urnov FD and Holmes MC. (2007). Targeted gene addition into a specified location in the human genome using designed zinc finger nucleases. *Proceedings of the National Academy of Sciences of the United States of America* **104**(9):3055–3060.

Moon S, Hasan SK, Song YS, Xu F, Keles HO, Manzur F, Mikkilineni S, Hong JW, Nagatomi J, Haeggstrom E, Khademhosseini A and Demirci U. (2010). Layer by layer three-dimensional tissue epitaxy by cell-laden hydrogel droplets. *Tissue Engineering Part C-Methods* **16**(1):157–166.

Moroni L, De Wijn JR, Van Blitterswijk CA. (2008). Integrating novel technologies to fabricate smart scaffolds. *Journal of Biomaterials Science-Polymer Edition* **19**(5):543–572.

Moroni L, Hendriks JAA, Schotel R, De Wijn JR and Van Blitterswijk CA. (2007). Design of biphasic polymeric 3-dimensional fiber deposited scaffolds for cartilage tissue engineering applications. *Tissue Engineering* **13**(2):361–371.

Moroni L, Schotel R, Sohier J, de Wijn JR and van Blitterswijk CA. (2006). Polymer hollow fiber three-dimensional matrices with controllable cavity and shell thickness. *Biomaterials* **27**(35):5918–5926.

Nair K, Gandhi M, Khalil S, Yan KC, Marcolongo M, Barbee K and Sun W. (2009). Characterization of cell viability during bioprinting processes. *Biotechnology Journal* **4**(8):1168–1177.

Nair LS and Laurencin CT. (2007). Biodegradable polymers as biomaterials. *Progress in Polymer Science* **32**(8–9):762–798.

Nakamura M. (2010). *Reconstruction of Biological Three-Dimensional Tissues: Bioprinting and Biofabrication Using Inkjet Technology Cell and Organ Printing.*

In: Ringeisen BR, Spargo BJ, Wu PK, editors: Springer, The Netherlands. pp. 23–33.

Nakamura M, Iwanaga S, Henmi C, Arai K and Nishiyama Y. (2010). Biomatrices and biomaterials for future developments of bioprinting and biofabrication. *Biofabrication* **2**(1): 014110. Available online at: http://iopscience.iop.org/1758-5090/2/1/014110?fromSearchPage=true.

Nakamura M, Nishiyama Y, Henmi C, Iwanaga S, Nakagawa H, Yamaguchi K, Akita K, Mochizuki S and Takiura K. (2008). Ink jet three-dimensional digital fabrication for biological tissue manufacturing: Analysis of alginate microgel beads produced by ink jet droplets for three dimensional tissue fabrication. *Journal of Imaging Science and Technology* **52**(6):060201.

Nicodemus GD and Bryant SJ. (2008). Cell encapsulation in biodegradable hydrogels for tissue engineering applications. *Tissue Engineering Part B-Reviews* **14**(2):149–165.

Niino T, Hamajima D, Montagne K, Oizumi S, Naruke H, Huang H, Sakai Y, Kinoshita H and Fujii T. (2011). Laser sintering fabrication of three-dimensional tissue engineering scaffolds with a flow channel network. *Biofabrication* **3**(3):034104. Available online at: http://iopscience.iop.org/1758-5090/3/3/034104.

Nishiyama Y, Nakamura M, Henmi C, Yamaguchi K, Mochizuki S, Nakagawa H and Takiura K. (2009). Development of a three-dimensional bioprinter: Construction of cell supporting structures using hydrogel and state-of-the-art inkjet technology. *Journal of Biomechanical Engineering-Transactions of the ASME* **131**(3): 035001.

Norotte C, Marga FS, Niklason LE and Forgacs G. (2009). Scaffold-free vascular tissue engineering using bioprinting. *Biomaterials* **30**(30):5910–5917.

Odde DJ and Renn MJ. (1999). Laser-guided direct writing for applications in biotechnology. *Trends in Biotechnology* **17**(10):385–389.

Okamoto T, Suzuki T and Yamamoto N. (2000). Microarray fabrication with covalent attachment of DNA using Bubble Jet technology. *Nature Biotechnology* **18**(4):438–441.

Ovsianikov A, Malinauskas M, Schlie S, Chichkov B, Gittard S, Narayan R, Lobler M, Sternberg K, Schmitz KP and Haverich A. (2011). Three-dimensional laser micro- and nano-structuring of acrylated poly(ethylene glycol) materials and evaluation of their cytoxicity for tissue engineering applications. *Acta Biomaterialia* **7**(3):967–974.

Parsa S, Gupta M, Loizeau F and Cheung KC. (2010). Effects of surfactant and gentle agitation on inkjet dispensing of living cells. *Biofabrication* **2**(2):025003. Available online at: http://iopscience.iop.org/1758-5090/2/2/025003/.

Parzel CA, Pepper ME, Burg T, Groff RE and Burg KJL. (2009). EDTA enhances high-throughput two-dimensional bioprinting by inhibiting salt scaling and cell aggregation at the nozzle surface. *Journal of Tissue Engineering and Regenerative Medicine* **3**(4):260–268.

Peltola SM, Melchels FPW, Grijpma DW and Kellomaki M. (2008). A review of rapid prototyping techniques for tissue engineering purposes. *Annals of Medicine* **40**(4):268–280.

Pérez-Pomares JM and Foty RA. (2006). Tissue fusion and cell sorting in embryonic development and disease: Biomedical implications. *BioEssays* **28**(8):809–821.

Pirlo RK, Wu P, Liu J and Ringeisen B. (2012). PLGA/hydrogel biopapers as a stackable substrate for printing HUVEC networks via BioLP (TM). *Biotechnology and Bioengineering* **109**(1):262–273.

Qiu Y, Zhang N, Kang Q, An Y and Wen X. (2009). Chemically modified light-curable chitosans with enhanced potential for bone tissue repair. *Journal of Biomedical Materials Research A* **89**(3):772–9.

Rein M. (1993). Phenomena of liquid-drop impact on solid and liquid surfaces. *Fluid Dynamics Research* **12**(2):61–93.

Reis N, Ainsley C and Derby B. (2005). Ink-jet delivery of particle suspensions by piezoelectric droplet ejectors. *Journal of Applied Physics* **97**(9) 094903.

Rezwan K, Chen QZ, Blaker JJ and Boccaccini AR. (2006). Biodegradable and bioactive porous polymer/inorganic composite scaffolds for bone tissue engineering. *Biomaterials* **27**(18):3413–3431.

Rosso F, Marino G, Giordano A, Barbarisi M, Parmeggiani D and Barbarisi A. (2005). Smart materials as scaffolds for tissue engineering. *Journal of Cellular Physiology* **203**(3):465–470.

Rowley JA and Mooney DJ. (2002). Alginate type and RGD density control myoblast phenotype. *Journal of Biomedical Materials Research* **60**(2):217–223.

Rücker M, Laschke MW, Junker D, Carvalho C, Schramm A, Mülhaupt R, Gellrich N-C and Menger MD. (2006). Angiogenic and inflammatory response to biodegradable scaffolds in dorsal skinfold chambers of mice. *Biomaterials* **27**(29):5027–5038.

Ruel-Gariépy E and Leroux J-C. (2004). In situ-forming hydrogels – review of temperature-sensitive systems. *European Journal of Pharmaceutics and Biopharmaceutics* **58**(2):409–426.

Sachlos E, Reis N, Ainsley C, Derby B and Czernuszka JT. (2003). Novel collagen scaffolds with predefined internal morphology made by solid freeform fabrication. *Biomaterials* **24**(8):1487–1497.

Sachs EM HJ, Cima MS and Williams PA. (1989). Three-dimensional printing techniques US patent 5204055.

Salem AK, Stevens R, Pearson RG, Davies MC, Tendler SJ, Roberts CJ, Williams PM and Shakesheff KM. (2002). Interactions of 3T3 fibroblasts and endothelial cells with defined pore features. *Journal of Biomedical Material Research* **61**(2):212–217.

Saunders RJG and Derby B. (2010). Piezoelectric inkjet printing of cells and biomaterials. In: Ringeisen BR, editor. *Cell and Organ Printing*: Springer Science+Business Media. Netherlands p. 16.

Saunders R, Gough J and Derby B. (2010). *Piezoelectric Inkjet Printing of Cells and Biomaterials Cell and Organ Printing*. In: Ringeisen BR, Spargo BJ, Wu PK, editors: Springer, The Netherlands. pp. 35–50.

Saunders RE, Gough JE and Derby B. (2008). Delivery of human fibroblast cells by piezoelectric drop-on-demand inkjet printing. *Biomaterials* **29**(2):193–203.

Savalani MM, Hao L, Zhang Y, Tanner KE and Harris RA. (2007). Fabrication of porous bioactive structures using the selective laser sintering technique. *Proceedings of the Institution of Mechanical Engineers Part H-Journal of Engineering in Medicine* **221**(H8):873–886.

Schantz JT, Brandwood A, Hutmacher DW, Khor HL and Bittner K. (2005). Osteogenic differentiation of mesenchymal progenitor cells in computer designed fibrin-polymer-ceramic scaffolds manufactured by fused deposition modeling. *Journal of Materials Science-Materials in Medicine* **16**(9):807–819.

Schense JC and Hubbell JA. (1998). Cross-linking exogenous bifunctional peptides into fibrin gels with Factor XIIIa. *Bioconjugate Chemistry* **10**(1):75–81.

Schulze D. (1995). Flowability of bulk solids – definition and measuring principles. *Chemie Ingenieur Technik* **67**(1):60–68.

Schuster M, Turecek C, Weigel G, Saf R, Stampfl J, Varga F and Liska R. (2009). Gelatin-based photopolymers for bone replacement materials. *Journal of Polymer Science Part A: Polymer Chemistry* **47**(24):7078–7089.

Seitz H, Rieder W, Irsen S, Leukers B, Tille C. (2005). Three-dimensional printing of porous ceramic scaffolds for bone tissue engineering. *Journal of Biomedical Material Research Part B Applied Biomaterials* **74**(2):782–788.

Shao XX, Hutmacher DW, Ho ST, Goh JCH and Lee EH. (2006). Evaluation of a hybrid scaffold/cell construct in repair of high-load-bearing osteochondral defects in rabbits. *Biomaterials* **27**(7):1071–1080.

Sherwood JK, Riley SL, Palazzolo R, Brown SC, Monkhouse DC, Coates M, Griffith LG, Landeen LK and Ratcliffe A. (2002). A three-dimensional osteochondral composite scaffold for articular cartilage repair. *Biomaterials* **23**(24):4739–4751.

Shim JH, Kim JY, Park M, Park J and Cho DW. (2011). Development of a hybrid scaffold with synthetic biomaterials and hydrogel using solid freeform fabrication technology. *Biofabrication* **3**(3):034102. Available online at: http://iopscience.iop.org/1758-5090/3/3/034102/

Shimizu T, Yamato M, Akutsu T, Shibata T, Isoi Y, Kikuchi A, Umezu M and Okano T. (2002). Electrically communicating three-dimensional cardiac tissue mimic fabricated by layered cultured cardiomyocyte sheets. *Journal of Biomedical Materials Research* **60**(1):110–117.

Shor L, Guceri S, Chang R, Gordon J, Kang Q, Hartsock L, An YH and Sun W. (2009). Precision extruding deposition (PED) fabrication of polycaprolactone (PCL) scaffolds for bone tissue engineering. *Biofabrication* **1**(1): 015003. Available online at: http://iopscience.iop.org/1758-5090/1/1/015003/.

Simpson RL, Wiria FE, Amis AA, Chua CK, Leong KF, Hansen UN, Chandrasekaran M and Lee MW. (2008). Development of a 95/5 poly(L-lactide-co-glycolide)/hydroxylapatite and beta-tricalcium phosphate scaffold as bone replacement material via selective laser sintering. *Journal of Biomedical Materials Research Part B-Applied Biomaterials* **84B**(1):17–25.

Smeds KA, Pfister-Serres A, Hatchell DL and Grinstaff MW. (1999). Synthesis of a novel polysaccharide hydrogel. *Journal of Macromolecular Science, Part A* **36**(7–8):981–989.

Smith CM, Christian JJ, Warren WL and Williams SK. (2007). Characterizing environmental factors that impact the viability of tissue-engineered constructs fabricated by a direct-write bioassembly tool. *Tissue Engineering* **13**(2):373–383.

Smith CM, Stone AL, Parkhill RL, Stewart RL, Simpkins MW, Kachurin AM, Warren WL and Williams SK. (2004). Three-dimensional bioassembly tool for generating viable tissue-engineered constructs. *Tissue Engineering* **10**(9–10):1566–1576.

Sokolsky-Papkov M, Agashi K, Olaye A, Shakesheff K and Domb AJ. (2007). Polymer carriers for drug delivery in tissue engineering. *Advanced Drug Delivery Reviews* **59**(4–5):187–206.

Spillmann A, Sonnenfeld A and von Rohr PR. (2007). Flowability modification of lactose powder by plasma enhanced chemical vapor deposition. *Plasma Processes and Polymers* **4**:S16–S20.

Steinberg MS and Takeichi M. (1994). Experimental specification of cell sorting, tissue spreading, and specific spatial patterning by quantitative differences in cadherin expression. *Proceedings of the National Academy of Sciences* **91**(1):206–209.

Sturm S, Zhou SW, Mai YW and Li Q. (2010). On stiffness of scaffolds for bone tissue engineering-a numerical study. *Journal of Biomechanics* **43**(9):1738–1744.

Sumerel J, Lewis J, Doraiswamy A, Deravi LF, Sewell SL, Gerdon AE, Wright DW and Narayan RJ. (2006). Piezoelectric ink jet processing of materials for medical and biological applications. *Biotechnology Journal* **1**(9):976–987.
Sun W, Darling A, Starly B and Nam J. (2004). Computer-aided tissue engineering: Overview, scope and challenges. *Biotechnology and Applied Biochemistry* **39**:29–47.
Suwińska A, Czołowska R, Oźdźeński W and Tarkowski AK. (2008). Blastomeres of the mouse embryo lose totipotency after the fifth cleavage division: Expression of Cdx2 and Oct4 and developmental potential of inner and outer blastomeres of 16- and 32-cell embryos. *Developmental Biology* **322**(1):133–144.
Swieszkowski W, Tuan BHS, Kurzydlowski KJ and Hutmacher DW. (2007). Repair and regeneration of osteochondral defects in the articular joints. *Biomolecular Engineering* **24**(5):489–495
Tan KH, Chua CK, Leong KF, Cheah CM, Cheang P, Abu Bakar MS, Cha SW. (2003). Scaffold development using selective laser sintering of polyetheretherketone-hydroxyapatite biocomposite blends. *Biomaterials* **24**(18):3115–3123.
Tan LL. (2004). Plugging bone the painless way, *Innovation, The Magazine of Research and Technology* **4**(3):60–61. http://www.innovationmagazine.com/innovation/volumes/v4n3/features1.shtml.
Tellis BC, Szivek JA, Bliss CL, Margolis DS, Vaidyanathan RK, Calvert P. (2008). Trabecular scaffolds created using micro CT guided fused deposition modeling. *Materials Science & Engineering C-Biomimetic and Supramolecular Systems* **28**(1):171–178.
Temenoff JS and Mikos AG. (2000). Review: Tissue engineering for regeneration of articular cartilage. *Biomaterials* **21**(5):431–440.
Tirella A, Orsini A, Vozzi G and Ahluwalia A. (2009). A phase diagram for microfabrication of geometrically controlled hydrogel scaffolds. *Biofabrication* **1**(4):045002. Availabel online at: http://iopscience.iop.org/1758-5090/1/4/045002/
Tirella A, Vozzi F, De Maria C, Vozzi G, Sandri T, Sassano D, Cognolato L and Ahluwalia A. (2011). Substrate stiffness influences high resolution printing of living cells with an ink-jet system. *Journal of Bioscience and Bioengineering* **112**(1):79–85.
Tsang VL and Bhatia SN. (2004). Three-dimensional tissue fabrication. *Advanced Drug Delivery Reviews* **56**(11):1635–1647.
Upcraft S and Fletcher R. (2003). The rapid prototyping technologies. *Assembly Automation* **23**(4):318–330.
Urnov FD, Miller JC, Lee YL, Beausejour CM, Rock JM, Augustus S, Jamieson AC, Porteus MH, Gregory PD and Holmes MC. (2005). Highly efficient endogenous human gene correction using designed zinc-finger nucleases. *Nature* **435**(7042):646–651.
von Heimburg D, Zachariah S, Kühling H, Heschel I, Schoof H, Hafemann B and Pallua N. (2001). Human preadipocytes seeded on freeze-dried collagen scaffolds investigated in vitro and in vivo. *Biomaterials* **22**(5):429–438.
Vozzi G, Previti A, De Rossi D and Ahluwalia A. (2002). Microsyringe-based deposition of two-dimensional and three-dimensional polymer scaffolds with a well-defined geometry for application to tissue engineering. *Tissue Engineering* **8**(6):1089–1098.
Wang F, Shor L, Darling A, Khalil S, Sun W, Guceri S and Lau A. (2004). Precision extruding deposition and characterization of cellular poly-epsilon-caprolactone tissue scaffolds. *Rapid Prototyping Journal* **10**(1):42–49.

Wang N, Naruse K, Stamenovic D, Fredberg JJ, Mijailovich SM, Toric-Norrelykke IM, Polte T, Mannix R and Ingber DE. (2001). Mechanical behavior in living cells consistent with the tensegrity model. *Proceedings of the National Academy of Sciences of the United States of America* **98**(14):7765–7770.

Wang XH, Yan YN, Pan YQ, Xiong Z, Liu HX, Cheng B, Liu F, Lin F, Wu RD, Zhang RJ and others. (2006). Generation of three-dimensional hepatocyte/gelatin structures with rapid prototyping system. *Tissue Engineering* **12**(1):83–90.

Webb PA. (2000). A review of rapid prototyping (RP) techniques in the medical and biomedical sector. *Journal of Medical Engineering and Technology* **24**(4):149–153.

Weinand C, Pomerantseva I, Neville CM, Gupta R, Weinberg E, Madisch I, Shapiro F, Abukawa H, Troulis MJ and Vacanti JP. (2006). Hydrogel-beta-TCP scaffolds and stem cells for tissue engineering bone. *Bone* **38**(4):555–563.

Whitesides GM and Grzybowski B. (2002). Self-assembly at all scales. *Science* **295**(5564):2418–2421.

Williams CG, Kim TK, Taboas A, Malik A, Manson P and Elisseeff J. (2003). In vitro chondrogenesis of bone marrow-derived mesenchymal stem cells in a photopolymerizing hydrogel. *Tissue Engineering* **9**(4):679–688.

Williams JM, Adewunmi A, Schek RM, Flanagan CL, Krebsbach PH, Feinberg SE, Hollister SJ and Das S. (2005). Bone tissue engineering using polycaprolactone scaffolds fabricated via selective laser sintering. *Biomaterials* **26**(23):4817–4827.

Wilson WC and Boland T. (2003). Cell and organ printing 1: Protein and cell printers. *Anatomical Record Part A-Discoveries in Molecular Cellular and Evolutionary Biology* **272A**(2):491–496.

Wonhye L, Debasitis JC, Lee VK, Jong-Hwan L, Fischer K, Edminster K, Je-Kyun P and Seung-Schik Y. (2009). Multi-layered culture of human skin fibroblasts and keratinocytes through three-dimensional freeform fabrication. *Biomaterials* **30**(8): 1587–95

Wu PK and Ringeisen BR. (2010). Development of human umbilical vein endothelial cell (HUVEC) and human umbilical vein smooth muscle cell (HUVSMC) branch/stem structures on hydrogel layers via biological laser printing (BioLP). *Biofabrication* **2**(1):014111. Available online at: http://iopscience.iop.org/1758-5090/2/1/014111/

Wüst S, Müller R and Hofmann S. (2011). Controlled positioning of cells in biomaterials – approaches towards 3D tissue printing. *Journal of Functional Biomaterials* **2**(3):119–154.

Xu JW, Zaporojan V, Peretti GM, Roses RE, Morse KB, Roy AK, Mesa JM, Randolph MA, Bonassar LJ and Yaremchuk MJ. (2004). Injectable tissue-engineered cartilage with different chondrocyte sources. *Plastic and Reconstructive Surgery* **113**(5):1361–1371.

Xu T, Gregory CA, Molnar P, Cui X, Jalota S, Bhaduri SB and Boland T. (2006). Viability and electrophysiology of neural cell structures generated by the inkjet printing method. *Biomaterials* **27**(19):3580–3588.

Xu T, Jin J, Gregory C, Hickman JJ and Boland T. (2005). Inkjet printing of viable mammalian cells. *Biomaterials* **26**(1):93–99.

Xu T, Rohozinski J, Zhao W, Moorefield EC, Atala A and Yoo JJ. (2009). Inkjet-mediated gene transfection into living cells combined with targeted delivery. *Tissue Engineering Part A* **15**(1):95–101.

Yamaoka T, Tabata Y and Ikada Y. (1994). Distribution and tissue uptake of poly(ethylene glycol) with different molecular weights after intravenous administration to mice. *Journal of Pharmaceutical Sciences* **83**(4):601–606.

Yan Y, Wang X, Pan Y, Liu H, Cheng J, Xiong Z, Lin F, Wu R, Zhang R and Lu Q. (2005). Fabrication of viable tissue-engineered constructs with 3D cell-assembly technique. *Biomaterials* **26**(29):5864–5871.

Yang SF, Leong KF, Du ZH and Chua CK. (2001). The design of scaffolds for use in tissue engineering. Part 1. Traditional factors. *Tissue Engineering* **7**(6):679–689.

Yang SF, Leong KF, Du ZH and Chua CK. (2002). The design of scaffolds for use in tissue engineering. Part II. Rapid prototyping techniques. *Tissue Engineering* **8**(1):1–11.

Yeong WY, Chua CK, Leong KF and Chandrasekaran M. (2004). Rapid prototyping in tissue engineering: Challenges and potential. *Trends in Biotechnology* **22**(12):643–652.

Yeong WY, Sudarmadji N, Yu HY, Chua CK, Leong KF, Venkatraman SS, Boey YCF and Tan LP. (2010). Porous polycaprolactone scaffold for cardiac tissue engineering fabricated by selective laser sintering. *Acta Biomaterialia* **6**(6):2028–2034.

Zein I, Hutmacher DW, Tan KC and Teoh SH. (2002). Fused deposition modeling of novel scaffold architectures for tissue engineering applications. *Biomaterials* **23**(4):1169–1185.

Zeltinger J, Sherwood JK, Graham DA, Mueller R and Griffith LG. (2001). Effect of pore size and void fraction on cellular adhesion, proliferation, and matrix deposition. *Tissue Engineering* **7**(5):557–572.

Zhang CH, Wen XJ, Vyavahare NR and Boland T. (2008). Synthesis and characterization of biodegradable elastomeric polyurethane scaffolds fabricated by the inkjet technique. *Biomaterials* **29**(28):3781–3791.

Zhang N, Zhang C and Wen X. (2005). Fabrication of semipermeable hollow fiber membranes with highly aligned texture for nerve guidance. *Journal of Biomedical Materials Research Part A* **75A**(4):941–949.

Zhang ZY, Teoh SH, Chong WS, Foo TT, Chng YC, Choolani M and Chan J. (2009). A biaxial rotating bioreactor for the culture of fetal mesenchymal stem cells for bone tissue engineering. *Biomaterials* **30**(14):2694–2704.

Zhong J, Chan A, Morad L, Kornblum HI, Fan GP and Carmichael ST. (2010). Hydrogel matrix to support stem cell survival after brain transplantation in stroke. *Neurorehabilitation and Neural Repair* **24**(7):636–644.

Zhou W, Lee S, Wang M, Cheung W and Ip W. (2008). Selective laser sintering of porous tissue engineering scaffolds from poly(L-lactide)/carbonated hydroxyapatite nanocomposite microspheres. *Journal of Materials Science: Materials in Medicine* **19**(7):2535–2540.

Zhu N, Li MG, Cooper D and Chen XB. (2011). Development of novel hybrid poly(L-lactide)/chitosan scaffolds using the rapid freeze prototyping technique. *Biofabrication* **3**(3).

Zimmermann I, Eber M and Meyer K. (2004). Nanomaterials as flow regulators in dry powders. Zeitschrift Fur Physikalische Chemie-International *Journal of Research in Physical Chemistry and Chemical Physics* **218**(1):51–102.

Zimmermann J, Bittner K, Stark B and Mülhaupt R. (2002). Novel hydrogels as supports for in vitro cell growth: Poly(ethylene glycol)- and gelatine-based (meth) acrylamidopeptide macromonomers. *Biomaterials* **23**(10):2127–2134.

5
Rapid prototyping of complex tissues with laser assisted bioprinting (LAB)

B. GUILLOTIN, S. CATROS, V. KERIQUEL,
A. SOUQUET, A. FONTAINE, M. REMY, J.-C. FRICAIN
and F. GUILLEMOT, INSERM U1026, France and Université
Bordeaux Segalen, France

DOI: 10.1533/9780857097217.156

Abstract: Laser assisted bioprinting (LAB) is an emerging tool for tissue engineering (TE) that allows printing of cells and liquid materials with a cell-level resolution, to reproduce the anisotropy of living tissues. We have integrated LAB in a dedicated workstation to allow rapid prototyping applications. After introducing the rationale of LAB in TE, we present the physical parameters related to laser induced forward transfer technique (LIFT), implemented in LAB. These parameters must be tuned to print viable cell patterns with respect to cell-level histological organization, and to high throughput manufacturing. Finally, we present some typical multi-component printing, 3D printing approaches, *in vitro* as well as *in vivo*.

Key words: laser assisted bioprinting, rapid prototyping, tissue engineering, regenerative medicine.

5.1 Introduction

In parallel with scaffold-based approaches involving cell seeding of porous structures (Hutmacher, 2000), some authors have suggested that three-dimensional (3D) biological structures can be built from the bottom up using the technology of bioprinting, the automated, computer-aided deposition of cells, cell aggregates and biomaterials (Klebe *et al.*, 1994; Klebe, 1988; Mironov, 2003; Mironov *et al.*, 2009; Guillemot *et al.*, 2010a; Jakab *et al.*, 2010). To this end, ink-jet printers have been successfully used to pattern biological assemblies according to a computer-aided design (CAD) template (Nakamura *et al.*, 2005; Boland *et al.*, 2006; Saunders *et al.*, 2008). Pressure-operated mechanical extruders have been also developed to handle living cells and cell aggregates (Jakab *et al.*, 2010). In parallel with these methods, laser-assisted printing technologies have emerged as alternative methods for the assembly and micropatterning of biomaterials and cells. Laser printing of biological material in general, and living cells in particular, is based on the laser-induced forward-transfer (LIFT, see Figs 5.1 and 5.2) technique, in which a pulsed laser is used to induce the transfer of material from a source film to a receiving substrate (Brisbane, 1971; Bohandy *et al.*, 1986; Young *et al.*, 2002). Several variations of LIFT have been implemented to print living cells (Guillemot *et al.*, 2010; Schiele *et al.*, 2010). For convenience, we will use the general term of LAB, although we will mainly discuss results obtained with a set-up that requires an intermediate light-absorbing layer of metal. Thus, under suitable irradiation conditions, and for liquids presenting a wide range of rheologies, the material can be deposited in the form of well-defined circular droplets with a high degree of spatial resolution (Fig. 5.3; Barron *et al.*, 2004, 2005; Guillotin *et al.*, 2010).

In this chapter, we will briefly introduce conceptual challenges in the field of TE and the technical properties that make LAB a suitable tool for TE. Next, we will present the physical parameters that must be tuned to print viable cell patterns with respect to cell-level spatial resolution, in high throughput conditions. Finally, we will present TE applications of LAB for basic research in biology as well as for regenerative medicine.

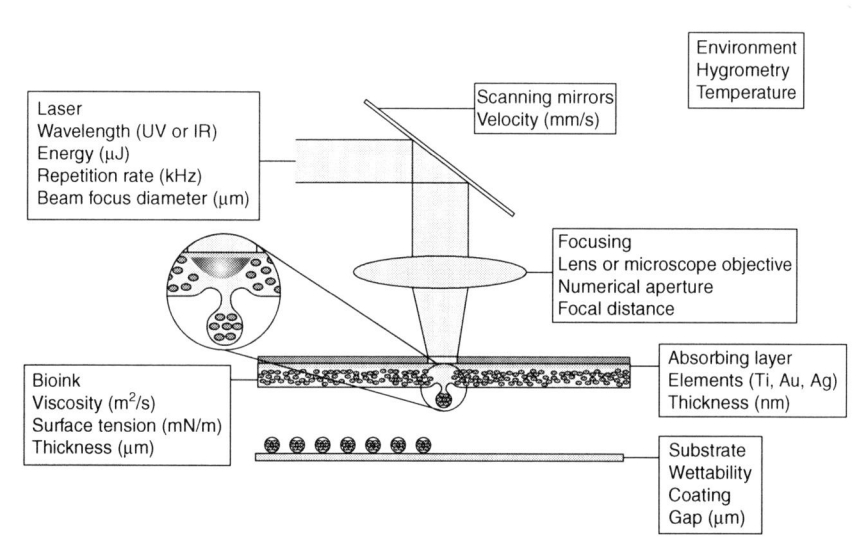

5.1 A typical LIFT experimental set-up is generally composed of three elements: a pulsed laser source, a target or ribbon coated with the material to be transferred, and a receiving substrate. The ribbon is a three layer component: a support – transparent to the laser radiation wavelength – coated with a thin absorbing layer (50 nm), itself coated with a transfer layer (50 μm) – named bioink – that contains the elements to be printed such as biomaterials, cells, or biomolecules (Guillemot *et al.*, 2010).

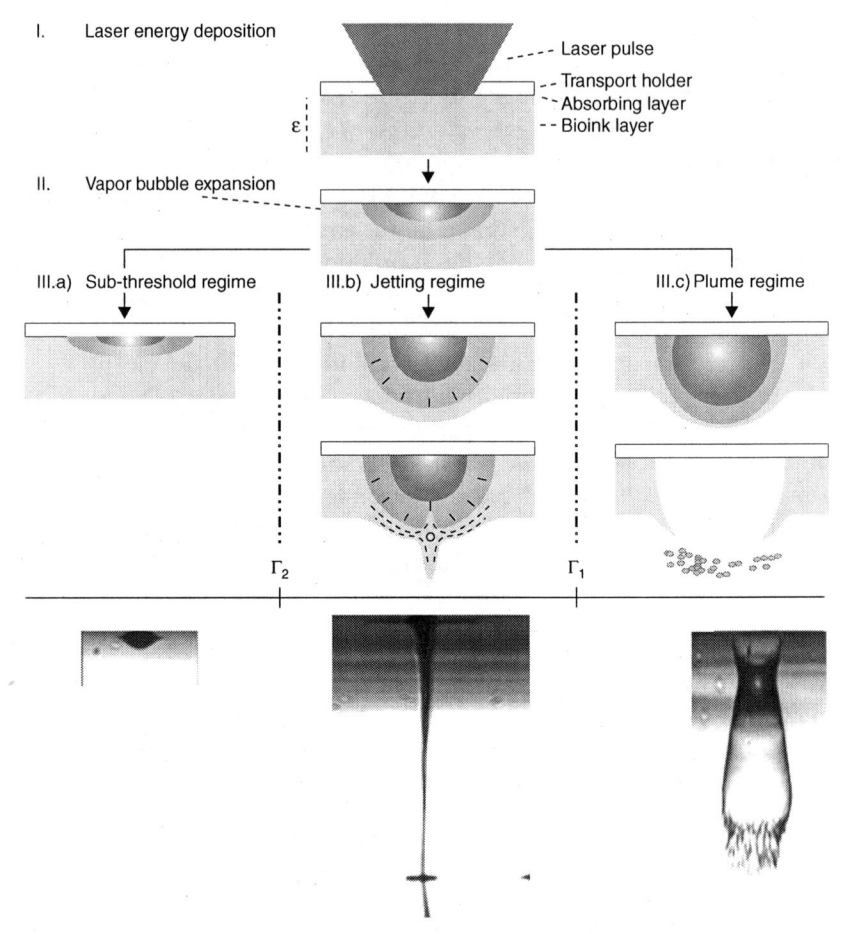

5.2 Mechanism for laser-induced droplet ejection. A vapor bubble is generated by vaporization of the absorbing layer and/or the first molecular layers of the liquid film. At given bioink viscosity and film thickness, jetting is observed for intermediary values of laser fluences: $\Gamma_1 < \Gamma < \Gamma_2$ (see III.b). For a lower fluence $\Gamma < \Gamma_2$, the bubble collapses far from the free surface without generating a jet (see III.a). For a higher fluence $\Gamma < \Gamma_1$, the bubble bursts to the surface, generating sub-micrometer droplets (see III .c). Increasing film thickness or bioink viscosity leads to increased threshold Γ values. The three regimes are revealed by umbroscopy in time resolved imaging (unpublished data).

5.2 Rationale for using laser assisted bioprinting (LAB) in tissue engineering

The architecture of a given tissue is characterized by its cell types, the biochemical and mechanical properties of its extra cellular matrix (ECM), and the geometry of these components. Living tissues and organs are composed

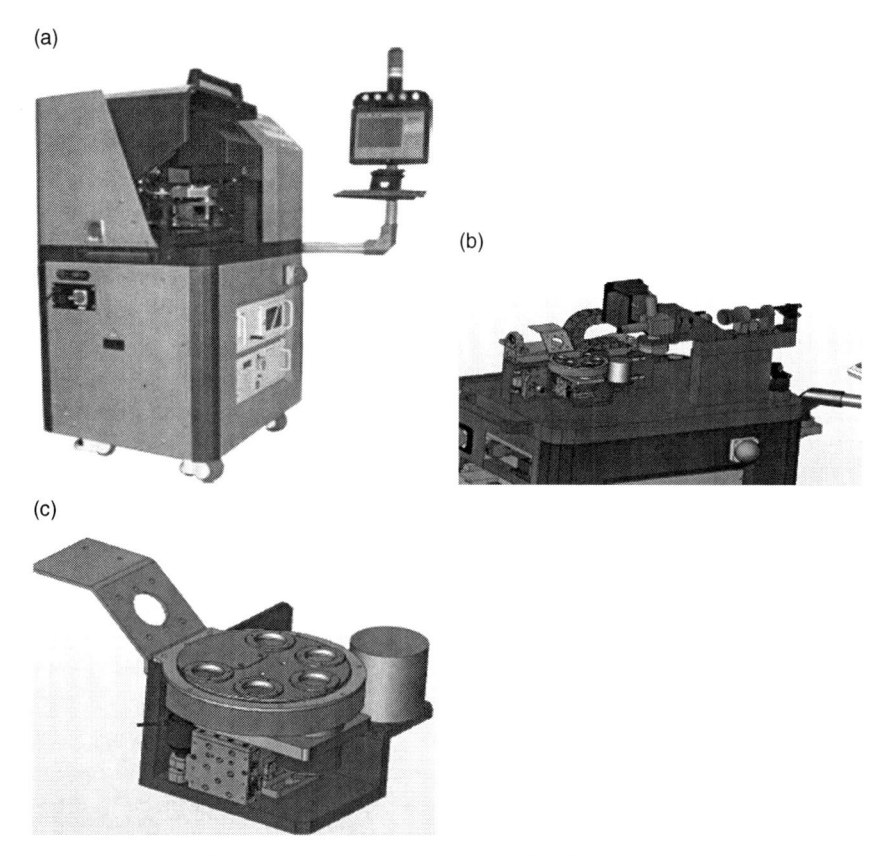

5.3 (a) View of the high-throughput LAB; (b) optomechanical set-up and (c) high resolution positioning system placed below the carousel holder with a loading capacity of five different ribbons.

of multiple cell types, which are assembled and interfaced at the micrometer scale. In the liver, for example, columns of hepatocytes are interfaced with biliary capillaries on the apical side and with sinusoidal blood vessels on the basal side to form lobules. Such high density, compartmentalized and integrated cellular organization has two major functional outcomes: (i) homeostasis, in particular metabolic exchange, is optimized; and (ii) functional units are packed together to form organs with a physiological efficiency that is compatible with living tissues. Consequently, miniaturization of TE processes (i.e., microscale organization of cells) is necessary to fabricate organotypic structures that compare favorably with the physiology of organs.

Tissue engineering approaches can be divided into three strategies based on the scale of spatial organization. First, macroscopic strategy can be likened to traditional TE, in which cells are seeded onto a macroporous scaffold. Cells are expected to colonize the inner volume of the scaffold by cell mobility and proliferation, and fluid flow. However, scaffolds do not present the ability to mimic the functional multicellular anisotropy and density of the host tissue. Second, mesostructures or modular blocks, also termed organoids (McGuigan and Sefton, 2006; McGuigan et al., 2008; Mironov et al., 2009), are based on the ability of the cells to self-assemble and their capacity to maintain viability and function when located within the diffusion limit of nutrient supply. The modular approach enables the production of 3D modules in a variety of shapes (e.g., cylinders with a diameter between 40 and 1000 μm and cell densities of 10^5–10^8 cells/cm^2), and to allow fabrication of multicellular constructs (e.g., a bone-mimicking construct including both osteoblasts, osteoclasts, and endothelial cells). However, neither macroscopic nor mesoscopic approaches have demonstrated the ability to mimic the functional multicellular anisotropy and density of the host tissue. Consequently, it is also conceptually challenging to design an efficient perfusion system that is physiologically interfaced with the engineered tissue and that will branch to the host vasculature at implantation.

Reproducing the local cell microenvironment can be considered the ultimate target for TE and cell patterning (Nelson et al., 2006, 2009). Conceptually, it could be defined as the capacity for positioning a single cell into its most suitable environment. Coordinated interactions between soluble factors, different cell types, and extracellular matrices (i.e., mechanical and biochemical cues) should be taken into account. Such a 'cell niche' manufacturing approach is unique in its purpose of dealing with tissue complexity and engineering a desired tissue from the bottom up. A microscopic scaffold-free, bottom-up approach to TE has been proposed (Voldman, 2006). Accordingly, computer-assisted design/computer-assisted manufacturing (CAD/CAM) LAB workstations have been designed and used to print viable cells and to organize them with cell-level resolution, with high throughput, in 2- and 3-dimensional tissue constructs that mimic the functional histology of living tissues.

5.3 Terms of reference for LAB

In this section, we discuss experimental conditions that matter in LAB process.

5.3.1 Workstation requirements

In designing a LAB workstation, various pulsed lasers were evaluated for their suitability with living cells and biomaterials as well as for rapid prototyping applications, all leading to comparable conclusions in terms of cell viability and printing resolution. Major requirements considered were as follows:

- The wavelength (λ) should not induce alteration of biological materials. Due to the potential denaturation of DNA by ultraviolet (UV) lighting,

and despite short pulse duration and the presence of the light absorbing metal layer, near infrared lasers might be preferable to UV lasers.
- The pulse duration (τ) and the repetition rate (f) must be considered with the purpose of high throughput processes.
- The beam quality, including divergence (q), spatial mode, and pulse-to-pulse (ptp) stability is important to ensure the reproducibility, stability, and high resolution of the system.
- laser-based workstation dedicated to TE applications should be designed with the purpose of executing various tasks rather than solely bioprinting. Consequently, the mean laser power should be high enough to perform additional processes like photolithography, photopolymerization, machining, sintering, and foaming (Duncan *et al.*, 2002; Lazare *et al.*, 2005; Claeyssens *et al.*, 2009).
- Laser repetition rate (in combination with scanning velocity, see below) has to be taken into account to avoid coalescence of vapor bubbles within the thickness of the bioink, and reciprocal perturbation of consecutive jets. For example, successive laser spots should not overlap each other to avoid an alteration of the energy conversed by the absorbing layer. Also, sufficient distance should be considered to prevent one droplet from landing on and displacing the preceding droplet that has been printed onto the substrate. Figure 5.4 is an example of partial overlap of jets due to spatio-temporal pulse proximity. Hence, the spatio-temporal proximity between both laser pulses and ejected droplets has to be taken into account in order to obtain a high printing resolution and reproducible results.
- Galvanometric mirrors are useful to drive the laser beam onto the ribbon. This approach offers a higher manufacturing speed compared to a design where the laser beam is fixed and the sample is translated by a motorized stage (Gaebel *et al.*, 2011). However, a refractive lens is required to correct the incidence of the laser beam onto the ribbon (Guillemot *et al.*, 2010).
- By analogy with inkjet printer cartridges, several ribbons of different materials can be processed by the LAB. We have designed a carousel wheel on which up to five elements can be loaded, and automatically presented under the laser beam at will (Fig. 5.3; Guillemot *et al.*, 2010).
- Dedicated CAD/CAM software interfaces with the workstation (Fig. 5.4).

5.3.2 Droplet ejection mechanism

A typical LIFT experimental set-up is generally composed of three elements: a pulsed laser source, a target coated with the material to be printed, and a receiving substrate. In addition, as numbered in Fig. 5.1, micro droplet ejection depends on numerous parameters which need to be considered to print

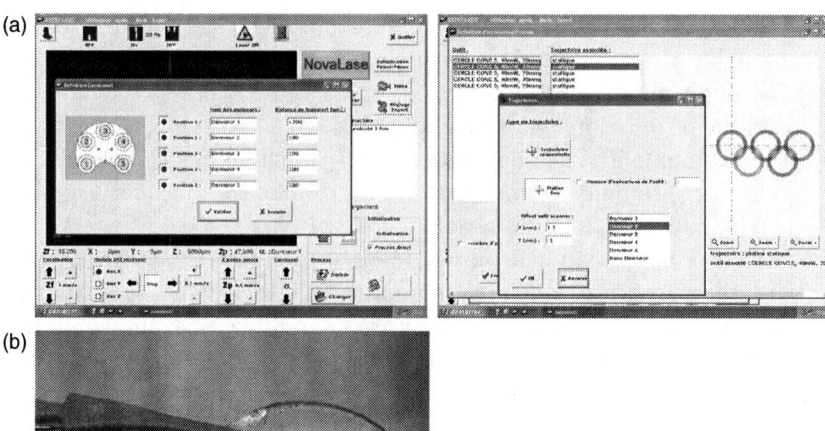

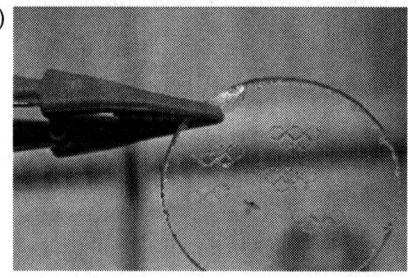

5.4 (a) Example of CAD/CAM software to help design the pattern of manufacturing as well as to pilot the laser beam accurately; (b) printed pattern according to the CAD/CAM design.

viable cells with high printing resolution and high throughput. As described in the literature (Young *et al.*, 2002; Duocastella *et al.*, 2007; Duocastella *et al.*, 2008; Mezel *et al.*, 2009), the generation of microdroplets by LIFT proceeds through six consecutive steps: (i) laser energy is first deposited into the skin depth of the absorbing layer; (ii) the absorbing layer is then heated in its skin depth which (iii) induces heating of a very thin film of the bioink near the absorbing layer; (iv) depending on the laser irradiation intensity, a vapor bubble is generated; (v) the vapor bubble expands which (vi) triggers the bioink–air interface deformation.

The growth phase of water bubbles in the bulk of a liquid has been studied in detail (see (Brennen, 1995). It has been shown that the growing–collapsing process could be described by the Rayleigh–Plesset equation, which states (Xiu-Mei *et al.*, 2008):

$$R\ddot{R} + \frac{3}{2}\dot{R}^2 = \frac{kP_l}{\rho_l}\left(\frac{R_0}{R}\right)^{3\gamma} - \frac{P_l - P_v}{\rho_l} - 4v\frac{\dot{R}}{R} - \frac{2\sigma_l}{\rho_l R} \quad [5.1]$$

where R is the vapor bubble radius; ρ_l the liquid density; P_l the liquid pressure; v the kinematic viscosity; σ_l the surface tension; γ the ratio of specific heats; P_v the saturated pressure in the bubble; and k the ratio of gas pressure in the bubble, P_g, to the liquid pressure P_l. This equation is the expression of the vapor bubble radius evolution versus time. It mainly depends on the liquid kinematic viscosity and surface tension.

The Rayleigh–Plesset equation describes bubble dynamics in an infinite volume. Because the size of the vapor bubble is not negligible compared to the bioink thickness, the interactions of the bubble with the free surface have to be taken into account. To this end, Pearson *et al.*(2004) and Robinson *et al.* (2001) have demonstrated that, (i) when the bubble reaches its maximum diameter, it begins to collapse under the external pressure strengths, and (ii) a jet may be formed according to the dimensionless standoff distance

$$\Gamma = \frac{h}{R_{max}} \qquad [5.2]$$

which is the ratio between the distance h (distance between the initial vapor bubble centroid and the free surface) and the maximum bubble radius, R_{max}. In Equation 5.2, the maximum bubble radius, R_{max}, depends on laser energy, through the k ratio in the Rayleigh–Plesset equation, and on the viscosity of the bioink, while h is related to the thickness of the bioink film.

Consequently, the three regimes generally observed in LIFT experiments (sub-threshold, jetting, and plume regimes (Fig. 5.2; Young *et al.*, 2002) are not solely the result of laser irradiation conditions but also of rheological properties and film thickness of the bioink. In other words, jetting does not simply occur on the basis of an energy threshold mechanism (Duocastella *et al.*, 2007) but rather on the basis of a complex Γ (E, ν, ε) threshold mechanism (Fig. 5.2). Over a given laser energy for which a vapor bubble is formed at the absorbing layer–bioink interface, the three above-mentioned regimes can be distinguished:

- If Γ is higher than a threshold value Γ_1: the droplet ejection cannot occur, unless the substrate is close to the target. These are the conditions to observe the sub-threshold regime (Fig. 5.2a).
- When Γ is lower than a threshold value Γ_2 the bubble expands until it bursts, giving rise to the plume regime (Fig. 5.2b).
- If Γ is ranged between Γ_1 and Γ_2 the bubble expands, then collapses, and finally a jet is formed (Fig. 5.2c).

With regard to TE, the mechanism of LAB still has several drawbacks, the main one being the implementation of a laser-absorbing layer of metal, which is vaporized with the printed material into the final TE product. Recent and promising developments to avoid this limitation include the use of a polyimide membrane as the laser-absorbing layer, which is capable of dissipating shock energy through elastoplastic deformation (Brown *et al.*,

2010), or the generation of the cavitation bubble without the use of a laser-absorbing layer (Duocastella *et al.*, 2010).

5.4 LAB parameters for cell printing

In this section, we introduce the experimental conditions which are necessary to control to achieve cell printing with high resolution.

5.4.1 Bioink composition considerations

LAB requires cells to be suspended in a liquid bioink prior to being printed onto the substrate. Also, in order to print a 3D material containing cells, the bioink should gel after printing onto the substrate. With respect to the layer-by-layer 3D building strategy, the gelling process is necessary (i) to stabilize the printed 2D pattern and (ii) to support the subsequent ink layer for 3D constructs using the layer-by-layer approach (see below). In addition, the gelling should not be harmful to the cells. In addition to these properties, which are specific to LAB applications, the bioink should possess properties reminiscent of the physiological ECM, which is critical for cell homeostasis *in vivo* (Engler *et al.*, 2006, 2009). According to these terms of reference, cells have been successfully printed using various solutions such as culture medium alone (Wu and Ringeisen, 2010), or in combination with sodium alginate (Catros *et al.*, 2011a), thrombin (Guillotin *et al.*, 2010), some combination of hyaluronic acid and fibrinogen (Gruene *et al.*, 2011), or a combination of blood plasma and sodium alginate (Koch *et al.*, 2009; Gruene *et al.*, 2011). Alginate, type I collagen and Matrigel™ can also be used in the liquid form. Using slightly different laser assisted printing techniques, mouse embryonic stem cells have been successfully printed using a gel form of Matrigel™ or 20% gelatin (Schiele *et al.*, 2009; Raof *et al.*, 2011).

Another concern is the evaporation of the bioink film, which is typically spread on the ribbon as a 50 µm thin layer. Accordingly, cell printing should be performed under a controlled atmosphere in terms of hygrometry (such conditions are also desired when biological material is processed). Othon *et al.* (2008) have proposed the use of methyl-cellulose in the bioink to prevent evaporation.

5.4.2 Viscosity of the bioink and laser energy influence on cell viability

Printed cells may not retain viability for any conditions of LAB. It has been previously shown and numerically modeled that a minimum shock-absorbing mattress of hydrogel, like Matrigel™, was required as a receiving substrate, in order to absorb mechanical shock in the printed cells (Ringeisen *et al.*, 2004; Wang *et al.*, 2008). Increasing the viscosity of the bioink, using sodium alginate, could improve cell viability if the mattress thickness is not sufficient to absorb mechanical shock (Catros *et al.*, 2011). Although the viscosity of the bioink is critical, the laser energy deposit should also be reduced to print viable cells on a thin film of shock-absorbing substrate. No LAB-induced alteration of cells (in terms of phenotype and DNA nicks) has been detected so far (Hopp *et al.*, 2005; Othon *et al.*, 2008; Gruene *et al.*, 2010; Raof *et al.*, 2011), which validates *de facto* the use of LAB for engineering cell-containing tissues for basic research. However, further studies should be implemented to rule out any genotoxicity of LAB, in case of use of LAB for cell-based clinical applications.

5.5 High resolution and high throughput needs and limits

In applying LAB to organ printing, cells should be printed with a density and an organization similar to a parenchyma, that is, a tissue essentially made of cells, with virtually no Extracellular Matrix (ECM), high histological organization with rich blood perfusion, and multiple cell types. Assuming that LAB is able to print cells one by one and next to each other, each cell having a diameter of 10 μm, it would require 28 h to print a volume of 1 cm^3 with single cell resolution at a printing frequency of 10 kHz. In other words, to fabricate a given volume, given a constant printing speed, increasing the organization of said volume would be more time consuming. Further, the printing of highly complex cell patterns at a high printing speed is an important objective. We have previously demonstrated that high-throughput printing by LAB was achievable (Guillotin *et al.*, 2010; Guillemot *et al.*, 2010). Concerning cell printing, we have reached a throughput threshold around 5 kHz. Indeed, there are some spatio-temporal limits to jet formation and ejection due to fluid dynamics that limit resolution (Fig. 5.5). The time of ejection of a jet is in the order of several tens of microseconds, which compares to the 5 kHz frequency that we classically use. The LIFT model strongly suggests that the jet produces some draining effect at the surface of the bioink film (Mezel *et al.*, 2009) that most likely accounts for the perturbation of the subsequent jet (Fig. 5.6). Given a constant scanning speed, the pulse frequency and the draining effect imply that two consecutive jets cannot be fired adjacent to each other above a certain firing frequency. This means that the laser beam should be driven fast enough (by galvanometric mirror), or the droplet-receiving substrate holder should be moved fast enough between two consecutive pulses, so that the subsequent jet occurs where the bioink film has not been disturbed by the preceding jet. According to these physical limits, LAB printing speed is theoretically merely limited by the highest laser pulse frequency available, according to the fastest galvanometric mirror available, under the condition that intact bioink film is presented under the laser beam at the same time.

5.5 Spatio-temporal limit of cell printing by LAB. Cells were printed onto glass according to five parallel lines of varying scanning speed (from top to bottom): 100, 200, 400, 800, and 1600 mm/s (laser pulse frequency of 5 kHz). The highest printing resolution of cells is achieved with scanning speed of 400 mm/s. With lower scanning speed, the precision of droplet deposit is impaired, most likely due to the overlap of successive jets (Guillotin *et al.*, 2010).

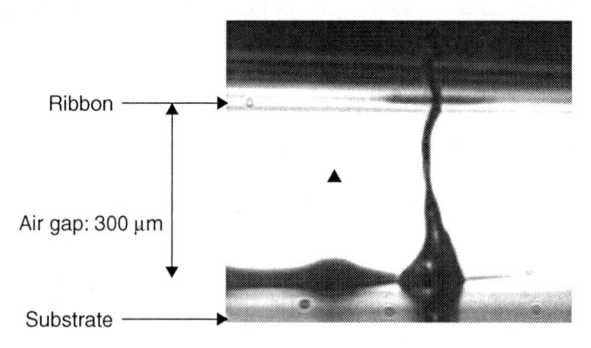

5.6 Spatio-temporal limit of jet formation. The laser beam is moved from left to right. Two consecutive jets are observed by umbroscopy (unpublished data). The first jet (indicated by the arrow head) is retracting into the film of bioink, while the following jet is observed. Contact of the jet with the receiving substrate can be seen. Distance between the ribbon and the substrate is 300 μm. The two consecutive jets overlap partially due to draining/capillary effect.

5.5.1 Viscosity of the bioink and laser energy influence on printing resolution

Printing resolution can be defined as the number of printed droplets per length. This implies that the highest resolution is obtained by printing droplets as small as possible and adjacent to each other. High resolution printing of liquids by LAB has been demonstrated (Fig. 5.7 and (Colina *et al.*, 2005; Dinca *et al.*, 2008; Serra *et al.*, 2009)). The size of the droplet depends on the ink viscosity and the laser energy deposit. Laser energy deposit can be modulated by tuning the energy of the laser source and/or by cutting the laser beam with a diaphragm aperture stop. The higher the viscosity and/or the lower the energy, the smaller the droplet diameter. By adapting the energy to a given viscosity, a wide range of ECMs, characterized by many different viscosities, can be printed at a similar resolution.

5.5.2 High cell density printing with cell-level resolution

To achieve micron-scale cell printing resolution (Fig. 5.5), cells should be printed with a minimal volume of surrounding ECM (or bioink). However, because LAB is a LIFT-based and nozzle free device, the number of cells in each ejected droplet is statistic (Barron et al., 2005). LAB nozzle-free set-up precludes the cell printing process from clogging issues. Thus, bioinks loaded with cell densities comparable to cell confluence observed in living tissue can be used. The presence of a single cell in each printed droplet is challenged by the use of a bioink with a low cell concentration, for example, 5×10^7 cells/mL. If cell density is too low on the ribbon, the ejected droplet of ink may not contain any cell (Guillotin et al., 2010). To overcome this problem, at least two strategies can be proposed: (i) increased laser energy deposit leads to the ejection of bigger droplets. As a result, cells are more likely to be dragged off by draining/capillary effect. (ii) Cell density can be increased up to the point that cells are touching each other at the surface of the ribbon, that is, 1×10^8 cells/mL. If droplet diameter is large enough, two contiguous printed droplets may coalesce, thus drawing a continuous line of cells or material (Fig. 5.7). Although increasing the size of the droplet leads to printing more cells at a time, the trade-off is a decrease in cell printing resolution. If single cell printing resolution is desired, that is, one-by-one, next to each other, then the smallest droplet should be ejected, using the lowest possible laser energy above cell printing threshold, and implying that (i) cell density is high enough so that there is always one cell in the field of the laser beam, or (ii) further development may focus on the implementation of a cell recognition scanning technology onto the ribbon prior to printing, so that the laser beam could exactly aim one single cell per pulse (Schiele et al., 2009). Accordingly, LAB can print cells virtually one by one from a high cell concentration bioink (1×10^8 cells/mL), to fabricate a tissue engineered product with organization and cell density comparable with living tissues, in which multiple cell types are in physical contact with each other.

5.5.3 Three-dimensional printing: the layer-by-layer approach

For TE applications, 3D printing is admitted to be challenging, especially in terms of managing perfusion and histological complexity within the entire

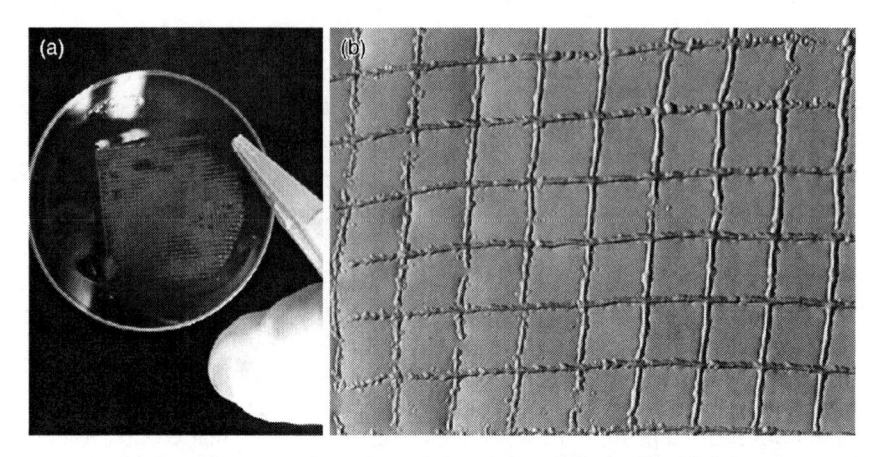

5.7 Rapid prototyping of a soft free form of fibrin. The bioink composed of thrombin (250 UI) and $CaCl_2$ (40 mM) has been printed onto a layer of fibrinogen (90 mg/mL), according to a computer designed matrix of orthogonal lines (2 cm length, 500 μm pitch) at a speed of 5 kHz and 200 mm/s. (a) Photograph of the fibrin pattern, scale 1:1; (b) fibrin pattern observed with a phase contrast optical microscope, magnification 25×.

volume of the construct (Mironov et al., 2009; Guillotin and Guillemot, 2011). However, LAB is not completely suited for building tissue constructs with cm^3 size and handling capacity. Indeed, the characteristic droplet volume is in the order of 1 pL (Guillemot et al., 2010). Some material can then be used in a layer-by-layer approach to provide volume and or biochemical properties that the bioink does not supply, to stabilize the pattern of the printed cells, and support the 3D construct. The layer-by-layer approach is commonly used in the bioprinting approach in general (Moon et al., 2000). For instance, a 3D structure was achieved with adipose-tissue derived stem cells (ADSCs) in a mixture of blood plasma and sodium alginate (Gruene et al., 2011). The final construct was then solidified by spraying the $CaCl_2$ crosslinker. The same team successfully stacked two arrays of spatially organized endothelial colony-forming cells (ECFCs) and ADSCs using a bioink composed of fibrinogen and hyaluronic acid. Each layer of array was wetted in, or sprayed, with thrombin solution resulting in a stable construct made of fibrin (Gruene et al., 2011).

Another approach aims to use thin solid films of biomaterials, coined as biopapers, as stackable layers that support bioprinted material or cells. The behavior of such three-dimensional constructs was addressed using electrospun 100 μm thick scaffolds of polycaprolactone (PCL) and cells printed by LAB (Catros et al., 2012). Also, poly (L-lactide–co-glycolide) (PLGA)/hydrogel (type I collagen or Matrigel™) biopapers have been used to print human umbilical vein endothelial cells (HUVECs) (Lee et al., 2009a, 2009b;

Pirlo *et al.*, 2012). Such biopapers may be patterned to provide additional control on possible cell migration and differentiation. Alternatively, two different cell types have been co-cultured to stabilize each other in the initial printed pattern (Wu and Ringeisen, 2010; Gaebel *et al.*, 2011).

5.6 Applications of LAB

In addition to liquids and all kinds of organic molecules in solution, numerous studies have shown successful laser assisted printing of a fairly broad range of prokaryotic cells and eukaryotic cells (see Schiele *et al.*, 2010 for a review). Considering human primary cells in particular, the following types have been printed by LAB: HUVECs and human umbilical vein smooth muscle cells (HUVSMCs) (Wu and Ringeisen 2010; Gaebel *et al.*, 2011), human mesenchymal stem cells (hMSCs) (Koch *et al.*, 2009; Gruene *et al.*, 2010; Gaebel *et al.*, 2011), ADSCs and ECFCs (Gruene *et al.*, 2011), and human bone-marrow derived osteo-progenitors (HOPs) (Catros *et al.*, 2011).

5.6.1 Multicolor printing at the micro-scale

Since biological tissues are composed of multiple components in close interactions with each other (cells of different types, proteins, and other components of the ECM), not only three-dimensional structures but also 'multicolor' printing, that is, the printing of multiple biological components, should be considered in TE products. Together with the high resolution printing capability of LAB, it is possible to print different cell types in close contact with each other, with a high cell concentration, according to a desired spatial organization (Guillotin *et al.*, 2010). The micrometric printing resolution achievable by LAB for multiple cell types and materials is consistent with the study of cell-to-cell, or cell-to-material interactions as well.

5.6.2 LAB engineered stem cell niche and tissue chips

Printing bioactive factors (e.g., morphogens, growth factors) onto inserted materials in order to regulate cell fate (in terms of migration and differentiation, for example), or to position cells at the desired coordinates, is achievable by LAB. Another, non-exclusive approach, would be to use LAB for printing resolutive patterns of material itself prior to printing cells onto the patterned material (Guillotin *et al.*, 2010). Studies would also deal with generating artificial cell niches by co-depositing a suitable combination of stem cells with ECM components (Lutolf and Blau, 2009). Indeed, embryonic stem cells printed by matrix assisted pulsed laser evaporation direct write (MAPLE-DW) formed embryoid bodies with retained pluripotency (Raof *et al.*, 2011). In relation to these issues, mechanical and topological cues should be studied using bottom-up approaches for engineering tissues. Future studies will focus on organizing multiple elements like cells, ECM-like materials, and growth factors at different scales of histology.

5.6.3 Addressing the perfusion issue in tissue engineering

The main limitation with respect to thick cellularized tissue constructs is the time required for the assembly and maturation of a perfused vascular network throughout the entire tissue construct. In certain cases, the assembly and maturation time might be longer than the cell survival time. Micropattern-guided vasculogenesis might accelerate vascular lumen formation as well as branching between the host and the tissue construct. To this end, endothelial cords have auto-assembled consecutively to endothelial cell alignment with cell-scale accuracy by LAB (Wu and Ringeisen, 2010). Such an approach to model endothelial tube formation might be fruitful in the field of vasculogenesis and angiogenesis related research.

5.6.4 Laser assisted engineering of transplants

To our knowledge, two studies report *in vivo* transplantation of LAB engineered tissue constructs. Gaebel *et al.* (2011) have used the LAB to fabricate a cardiac patch for cardiac regeneration in a rat model of acute myocardial infarction. LAB was used to pattern a co-culture of HUVECs and hMSCs onto a 300 µm thick, 8 mm diameter disk of polyester urethane urea. Healing potential of patterned patch compared similarly to the unpatterned patch, demonstrating the suitability of the LAB procedure for tissue regeneration, and suggesting that patterning may favor faster vasculogenesis and grafting.

The influence of the three-dimensional organization of MG63 cells and electrospun 100 µm thick PCL biopapers was evaluated regarding cell proliferation *in vitro* and *in vivo* (Catros *et al.*, 2011). For this purpose, a layer-by-layer sandwich model of assembly was compared to a control hybrid material made of the same amount of material with an alternative 3D arrangement. These constructs were both evaluated *in vitro* and *in vivo* in mouse calvarial defect reconstruction model. Results underscore the benefit of the layer-by-layer approach to encapsulate cells within a sandwich of PCL biopapers, either *in vitro* or *in vivo*, as far as cell viability and cell proliferation are concerned.

In both reports, cell patterning at cell-level resolution was not studied. Increasing the resolution may help guide faster tissue-specific organization like vasculogenesis, which may support morphogenesis or healing processes.

5.6.5 *In vivo* printing

To the best of our knowledge, all previous studies report on LAB *in vitro* fabrication. Our team has performed some preliminary assays for *in vivo* printing (Keriquel *et al.*, 2010). More precisely, the purpose of our study was to fill a critical sized calvarial bone defect in mouse by printing bioink composed of a nano-sized hydroxyapatite slurry. A specific mouse holder was designed in order to position the surface of mice dura mater instead of the quartz substrate. Then, 30 layers of hydroxyapatite were printed inside one defect. The histological results have shown that the material was present in the test defects of all groups. However, bone repair was inconsistent. As a conclusion, we have shown that *in vivo* bioprinting is possible. Future experiments in this model should improve the mechanical and biological properties of the printed material.

5.7 Conclusion

In this chapter, we have shown that LAB of cells requires an understanding of vapor bubble dynamics which governs the droplet ejection. Consequently, the bioink should be designed accordingly and spatio-temporal proximity of consecutive laser-induced jets should be considered for optimal printing resolution. Several studies taken together demonstrate the capability of LAB to print virtually all cell types although many human cell types remain to be validated. These cells can be printed onto numerous biomaterials, either solid or gel, like biopolymers and nano-sized particles of hydroxyapatite. The potential of LAB to fabricate functional cell-containing transplants for tissue repair has been demonstrated, together with the possibility of shunting the transplantation process by operating LAB directly *in vivo*. Combining LAB with other laser-assisted processes, such as machining and polymerization, should be addressed with specific attention to integrating these different processes in the same workstation to guarantee subcellular resolution. As a direct write method of living cells, LAB can be combined with other TE methods (Guillotin and Guillemot, 2011).

Concerning the layer-by-layer microfabrication of functional tissues that mimic their *in vivo* counterparts, it remains to be determined whether the exact reproduction of the histoarchitecture of living tissue is crucial; in other words, to what extent and resolution cellular self-assembly has to be guided. Future studies involving pattern formation in morphogenesis, specifically the relationship between form and function, should advise this aim. Moreover, the engineering of realistic tissue constructs will help further understanding of tissue physiology and function; this, in turn, will refine TE strategies and optimize blueprints.

5.8 Acknowledgements

We acknowledge the financial support from GIS-AMA (Advanced Materials in Aquitaine), ANR (Agence Nationale pour la Recherche), IFRO (Institut Français de la Recherche Odontologique), FRM (Fondation de la Recherche Médicale) and Région Aquitaine.

5.9 References

Barron, J.A. Wu, P., Ladouceur, H.D. and Ringeisen, B.R. (2004). Biological laser printing: a novel technique for creating heterogeneous 3-dimensional cell patterns. *Biomedical Microdevices*, **6**(2), pp. 139–147.

Barron, J.A., Krizman, D.B. and Ringeisen, B.R. (2005). Laser printing of single cells: statistical analysis, cell viability, and stress. *Annals of Biomedical Engineering*, **33**(2), pp. 121–130.

Bohandy, J., Kim, B.F. and Adrian, F.J. (1986). Metal deposition from a supported metal film using an excimer laser. *Journal of Applied Physics*, **60**(4), p. 1538.

Boland, T. Xu, T., Damon, B and Cui, X. (2006). Application of inkjet printing to tissue engineering. *Biotechnology Journal*, **1**(9), pp. 910–917.

Brennen, C.E. (1995). *Cavitation and Bubble Dynamics*, Available at: http://resolver.caltech.edu/CaltechBOOK:1995.001 (Consulted 1 August, 2009).

Brisbane, C.E. (1971). PATTERN DEPOSIT BY LASER - Google Patents. Available at: http://www.google.com/patents?hl=en&lr=&vid=USPAT3560258&id=x4F UAAAAEBAJ&oi=fnd&dq=Pattern+deposit+by+laser+filed+for+patent+ 1967+%22BRISBANE%22&printsec=abstract#v=onepage&q=Pattern%20 deposit%20by%20laser%20filed%20for%20patent%201967%20 %22BRISBANE%22&f=false (Consulted 26 September, 2011).

Brown, M.S., Kattamis, N.T. and Arnold, C.B. (2010). Time-resolved study of polyimide absorption layers for blister-actuated laser-induced forward transfer. *Journal of Applied Physics*, **107**(8), p. 083103.

Catros, S., Guillotin, B. Bacakova, M., Fricain, J. and Guillemot, F. (2011a). Effect of laser energy, substrate film thickness and bioink viscosity on viability of endothelial cells printed by Laser-Assisted Bioprinting. *Applied Surface Science*, **257**(12), pp. 5142–5147.

Catros, S., Fricain, J.-C. Guillotin, B., Pippenger, B., Bareille, R., Remy, M., Lebraud, E., Desbat, B., Amédée, J. and Guillemot, F. (2011b). Laser-assisted bioprinting for creating on-demand patterns of human osteoprogenitor cells and nano-hydroxyapatite. *Biofabrication*, **3**(2), p. 025001.

Catros, S., Guillemot, F. Nandakumar, A., Ziane, S., Moroni, L., Habibovic, P., van Blitterswijk, C., Rousseau, B., Chassande, O., Amédée, J. and Fricain, J.C. (2012). Layer-by-layer tissue microfabrication supports cell proliferation in vitro and in vivo. *Tissue Engineering Part C: Methods*, **18**(1), pp. 62–70.

Claeyssens, F. Hasan, E.A., Gaidukeviciute, A., Achilleos, D.S., Ranella, A., Reinhardt, C., Ovsianikov, A., Shizhou, X., Fotakis, C., Vamvakaki, M., Chichkov, B.N. and Farsari, M. (2009). Three-dimensional biodegradable structures fabricated by two-photon polymerization. *Langmuir*, **25**, pp. 3219–3223.

Colina, M. Serra, P., Fernández-Pradas, J.M., Sevilla, L. and Morenza, J.L. (2005). DNA deposition through laser induced forward transfer. *Biosensors and Bioelectronics*, **20**(8), pp. 1638–1642.

Dinca, V. Kasotakis, E., Catherine, J., Mourka, A., Ranella, A., Ovsianikov, A., Chichkov, B.N., Farsari, M., Mitraki, A. and Fotakis, C. (2008). Directed three-dimensional patterning of self-assembled peptide fibrils. *Nano Letters*, **8**(2), pp. 538–543.

Duncan, A.C. Weisbuch, F., Rouais, F., Lazare, S. and Baquey, C. (2002). Laser microfabricated model surfaces for controlled cell growth. *Biosensors and Bioelectronics*, **17**(5), pp. 413–426.

Duocastella, M. Fernandez-Pradas, J.M., Serra, P. and Morenza, J.L. (2008). Jet formation in the laser forward transfer of liquids. *Applied Physics A*, **93**(2), pp. 453–456.

Duocastella, M. Fernandez-Pradas, J.M., Morenza, J.L., Zafra, D. and Serra, P. (2010). Novel laser printing technique for miniaturized biosensors preparation. *Sensors and Actuators B: Chemical*, **145**(1), pp. 596–600.

Duocastella, M. Colina, M., Fernandez-Pradas, J.M., Serra, P. and Morenza, J.L. (2007). Study of the laser-induced forward transfer of liquids for laser bioprinting. *Applied Surface Science*, **253**(19), pp. 7855–7859.

Engler, A.J. Sen, S., Sweeney, H.L. and Discher, D.E. (2006). Matrix elasticity directs stem cell lineage specification. *Cell*, **126**(4), pp. 677–689.

Engler, A.J. Humbert, P.O., Wehrle-Haller, B. and Weaver, V.M. (2009). Multiscale modeling of form and function. *Science*, **324**(5924), pp. 208–212.

Fedorovich, N.E. De Wijn, J.R., Verbout, A.J., Alblas, J. and Dhert, W.J. (2008). Three-dimensional fiber deposition of cell-laden, viable, patterned constructs for bone tissue printing. *Tissue Engineering*, **14**(1), pp. 127–133.

Gaebel, R. Ma, N., Liu, J., Guan, J., Koch, L., Klopsch, C., Gruene, M, Toelk, A., Wang, W., Mark, P., Wang, F., Chichkov, B., Li, W. and Steinhoff, G. (2011). Patterning human stem cells and endothelial cells with laser printing for cardiac regeneration. *Biomaterials*, **32**(35), pp. 9218–9230.

Gruene, M. Deiwick, A., Koch, L., Schlie, S., Unger, C., Hofmann, N., Bernemann, I., Glasmacher, B. and Chichkovm, B. (2010). Laser printing of stem cells for biofabrication of scaffold-free autologous grafts. *Tissue Engineering Part C: Methods*, p. 100830145320029.

Gruene, M. Pflaum, M., Deiwick, A., Koch, L., Schlie, S., Unger, C., Wilhelmi, M., Haverich, A. and Chichkov, B.N. (2011a). Adipogenic differentiation of laser-printed 3D tissue grafts consisting of human adipose-derived stem cells. *Biofabrication*, **3**, p. 015005.

Gruene, M. Pflaum, M., Hess, C., Diamantouros, S., Schlie, S., Deiwick, A., Koch, L., Wilhelmi, M., Jockenhoevel, S., Haverich, A. and Chichkov, B. (2011b). Laser printing of three-dimensional multicellular arrays for studies of cell–cell and cell–environment interactions. *Tissue Engineering Part C: Methods*, p. 110629135038006.

Guillemot, F., Souquet, A. Souquet, A., Catros, S., Guillotin, B., Lopez, J., Faucon, M., Pippenger, B., Bareille, R., Rémy, M., Bellance, S., Chabassier, P., Fricain, J.C. and Amédée, J. (2010a). High-throughput laser printing of cells and biomaterials for tissue engineering. *Acta Biomaterialia*, **6**(7), pp. 2494–2500.

Guillemot, F., Souquet, A. *et al.* (2010b). Laser-assisted cell printing: principle, physical parameters versus cell fate and perspectives in tissue engineering. *Nanomedicine*, **5**(3), pp. 507–515.

Guillemot, F., Mironov, V. and Nakamura, M. (2010c). Bioprinting is coming of age: report from the International Conference on Bioprinting and Biofabrication in Bordeaux (3B'09). *Biofabrication*, **2**(1), p. 010201.

Guillotin, B. *et al.* (2010). Laser assisted bioprinting of engineered tissue with high cell density and microscale organization. *Biomaterials*, **31**(28), pp. 7250–7256.

Guillotin, B. and Guillemot, F. (2011). Cell patterning technologies for organotypic tissue fabrication. *Trends in Biotechnology*, **29**(4), pp. 183–190.

Hopp, B. Smausz, T., Kresz, N., Barna, N., Bor, Z., Kolozsvári, L., Chrisey, D.B., Szabó, A. and Nógrádi, A. (2005). Survival and proliferative ability of various living cell types after laser-induced forward transfer. *Tissue Engineering*, **11**(11–12), pp. 1817–1823.

Hutmacher, D.W. (2000). Scaffolds in tissue engineering bone and cartilage. *Biomaterials*, **21**(24), pp. 2529–2543.

Jakab, K. Norotte, C., Marga, F., Murphy, K., Vunjak-Novakovic, G. and Forgacs, G. (2010). Tissue engineering by self-assembly and bio-printing of living cells. *Biofabrication*, **2**(2), p. 022001.

Keriquel, V. Guillemot, F., Arnault, I., Guillotin, B., Miraux, S., Amédée, J., Fricain, J.C. and Catros, S. (2010). In vivo bioprinting for computer- and robotic-assisted medical intervention: preliminary study in mice. *Biofabrication*, **2**(1), p. 014101.

Klebe, R.J. (1988). Cytoscribing: a method for micropositioning cells and the construction of two- and three-dimensional synthetic tissues. *Experimental Cell Research*, **179**(2), pp. 362–373.

Klebe, R.J. Thomas, C.A., Grant, G.M., Grant, A. and Gosh, P. (1994). Cytoscription: computer controlled micropositioning of cell adhesion proteins and cells. *Methods in Cell Science*, **16**(3), pp. 189–192.

Koch, L. *et al.* (2009). Laser printing of skin cells and human stem cells. *Tissue Engineering Part C: Methods*, p. 091221133515000.

Lazare, S. Tokarev, V., Sionkowska, A. and Wisniewski, M. (2005). Surface foaming of collagen, chitosan and other biopolymer films by KrF excimer laser ablation in the photomechanical regime. *Applied Physics A: Materials Science & Processing*, **81**(3), pp. 465–470.

Lee, W., Debasitis, J.C. Debasitis, J.C., Lee, V.K., Lee, J.H., Fischer, K., Edminster, K., Park, J.K. and Yoo, S.S. (2009a). Multi-layered culture of human skin fibroblasts and keratinocytes through three-dimensional freeform fabrication. *Biomaterials*, **30**(8), pp. 1587–1595.

Lee, W., Pinckney, J. Pinckney, J., Lee, V., Lee, J.H., Fischer, K., Polio, S., Park, J.K. and Yoo S.S. (2009b). Three-dimensional bioprinting of rat embryonic neural cells. *NeuroReport*, **20**(8), pp. 798–803.

Lutolf, M.P. and Blau, H.M. (2009). Artificial stem cell niches. *Advanced Materials*, **21**(32–33), pp. 3255–3268.

McGuigan, A.P. Bruzewicz, D.A., Glavan, A., Butte, M.J. and Whitesides, G.M. (2008). Cell encapsulation in sub-mm sized gel modules using replica molding. *PLoS ONE*, **3**(5), p. e2258.

McGuigan, A.P. and Sefton, M.V. (2006). Vascularized organoid engineered by modular assembly enables blood perfusion. *Proceedings of the National Academy of Sciences*, **103**(31), pp. 11461–11466.

Mezel, C. Hallo, L., Souquet, A., Breil, J., Hébert, D. and Guillemot, F. (2009). Self-consistent modeling of jet formation process in the nanosecond laser pulse regime. *Physics of Plasmas*, **16**(12), pp. 123112-123214.

Mironov, V. (2003). Organ printing: computer-aided jet-based 3D tissue engineering. *Trends in Biotechnology*, **21**(4), pp. 157–161.

Mironov, V. Visconti, R.P., Kasyanov, V., Forgacs, G., Drake, C.J. and Markwald, R.R. (2009). Organ printing: tissue spheroids as building blocks. *Biomaterials*, **30**(12), pp. 2164–2174.

Moon, S. *et al.* (2010). Layer by layer three-dimensional tissue epitaxy by cell-laden hydrogel droplets. *Tissue Engineering Part C: Methods*, **16**(1), pp. 157–166.

Nakamura, M. Kobayashi, A., Takagi, F., Watanabe, A., Hiruma, Y., Ohuchi, K., Iwasaki, Y., Horie, M., Morita, I. and Takatani, S. (2005). Biocompatible inkjet printing technique for designed seeding of individual living cells. *Tissue Engineering*, **11**(11–12), pp. 1658–1666.

Nelson, C.M. (2009). Geometric control of tissue morphogenesis. *Biochimica et Biophysica Acta (BBA) – Molecular Cell Research*, **1793**(5), pp. 903–910.

Nelson, C.M. and Tien, J. (2006). Microstructured extracellular matrices in tissue engineering and development. *Current Opinion in Biotechnology*, **17**(5), pp. 518–523.

Othon, C.M. Wu, X., Anders, J.J. and Ringeisen, B.R. (2008). Single-cell printing to form three-dimensional lines of olfactory ensheathing cells. *Biomedical Materials*, **3**(3), p. 034101.

Pearson, A. Cox, E., Blake, J.R. and Otto, S.R. (2004). Bubble interactions near a free surface. *Engineering Analysis with Boundary Elements*, **28**(4), pp. 295–313.

Pirlo, R.K. Wu, P., Liu, J. and Ringeisen, B. (2012). PLGA/hydrogel biopapers as a stackable substrate for printing HUVEC networks via BioLP™. *Biotechnology and Bioengineering*, **109**(1), pp. 262–273.

Raof, N.A. Schiele, N.R., Xie, Y., Chrisey, D.B. and Corr D.T. (2011). The maintenance of pluripotency following laser direct-write of mouse embryonic stem cells. *Biomaterials*, **32**(7), pp. 1802–1808.

Ringeisen, B.R Kim, H., Barron, J.A., Krizman, D.B., Chrisey, D.B., Jackman, S., Auyeung, R.Y. and Spargo ,B.J. (2004). Laser printing of pluripotent embryonal carcinoma cells. *Tissue Engineering*, **10**(3–4), pp. 483–491.

Robinson, P.B., Blake, J.R., Kodama, T., Shima, A. and Tomita, Y. (2001). Interaction of cavitation bubbles with a free surface. *Journal of Applied Physics*, **89**(12), pp. 8225–8237.

Saunders, R.E., Gough, J.E. and Derby, B. (2008). Delivery of human fibroblast cells by piezoelectric drop-on-demand inkjet printing. *Biomaterials*, **29**(2), pp. 193–203.

Schiele, N.R. Corr, DT, Huang Y, Raof NA, Xie Y, Chrisey DB. (2010). Laser-based direct-write techniques for cell printing. *Biofabrication*, **2**(3), p. 032001.

Schiele, N.R. Koppes, R.A., Corr, D.T., Ellison, K.S., Thompson, D.M., Ligon, L.A., Lippert, T.K.M. and Chrisey, D.B. (2009). Laser direct writing of combinatorial libraries of idealized cellular constructs: biomedical applications. *Applied Surface Science*, **255**(10), pp. 5444–5447.

Serra, P. Duocastella, M., Fernandez-Pradas, J.M. and Morenza, J.L. (2009). Liquids microprinting through laser-induced forward transfer. *Applied Surface Science*, **255**.

Voldman, J. (2006). Engineered systems for the physical manipulation of single cells. *Current Opinion in Biotechnology*, **17**(5), pp. 021012–021022.

Wang, W. Huang, Y., Grujicic, M. and Chrisey, D.B. (2008). Study of impact-induced mechanical effects in cell direct writing using smooth particle hydrodynamic method. *Journal of Manufacturing Science and Engineering*, **130**(2), pp. 021012-10.

Wu, P.K. and Ringeisen, B.R. (2010). Development of human umbilical vein endothelial cell (HUVEC) and human umbilical vein smooth muscle cell (HUVSMC) branch/stem structures on hydrogel layers via biological laser printing (BioLP). *Biofabrication*, **2**(1), p. 014111.

Xiu-Mei, L. Jie, H., Jian, L. and Xiao-Wu, N. (2008). Growth and collapse of laser-induced bubbles in glycerol–water mixtures. *Chinese Physics B*, **17**(7), pp. 2574–2579.

Young, D. Auyeung, R.C.Y., Piqué, A., Chrisey, D.B. and Dlott, D.D. (2002). Plume and jetting regimes in a laser based forward transfer process as observed by time-resolved optical microscopy. *Applied Surface Science*, **197–198**, pp. 181–187.

6
Scaffolding hydrogels for rapid prototyping based tissue engineering

R. A. SHIRWAIKER, M. F. PURSER and R. A. WYSK,
North Carolina State University, USA

DOI: 10.1533/9780857097217.176

Abstract: Scaffolding hydrogels provides the ability to pattern cell suspensions directly within 3D configurations of hydrated polymer networks to mimic the physical and biological characteristics of natural extracellular matrices. This chapter first reviews the developments, key characteristics, and applications of some commonly used and emerging hydrogel biomaterials. It then discusses the compatibility between hydrogels and rapid prototyping (RP) scaffolding processes, and highlights some of the recent emerging trends in scaffolding biomaterials research.

Key words: hydrogels, biopaper, scaffolds, organ printing.

6.1 Introduction

The interdisciplinary field of tissue engineering focuses on the design and fabrication of functional biological substitutes for damaged human tissue or organs using the synergistic application of principles and methods from engineering and life sciences. The underlying premise of tissue engineering is that new tissues can be grown *in vitro* by regenerating the appropriate types of cells utilizing scaffold templates to guide tissue development (Langer and Vacanti, 1993). All traditional tissue engineering technologies have three fundamental components: cells, biomaterials and scaffolds, and bioreactor-based processes (Fig. 6.1).

Scaffolds refer to the broad set of 3D biomaterial configurations that serve as an alternative extracellular matrix (ECM) for the structure, support, and collocation of biological materials including cells, nutrients, and bioactive molecules. In addition to providing the physical architecture for 3D growth, they also present stimuli to the cells, allowing them to respond as if they were in their natural *in vivo* tissue environment (Maher *et al.*, 2009). Materials used for tissue engineering scaffolds need to meet some specific requirements, including biocompatibility, biodegradability, mechanical properties, and manufacturability in order to fulfill the goals of tissue

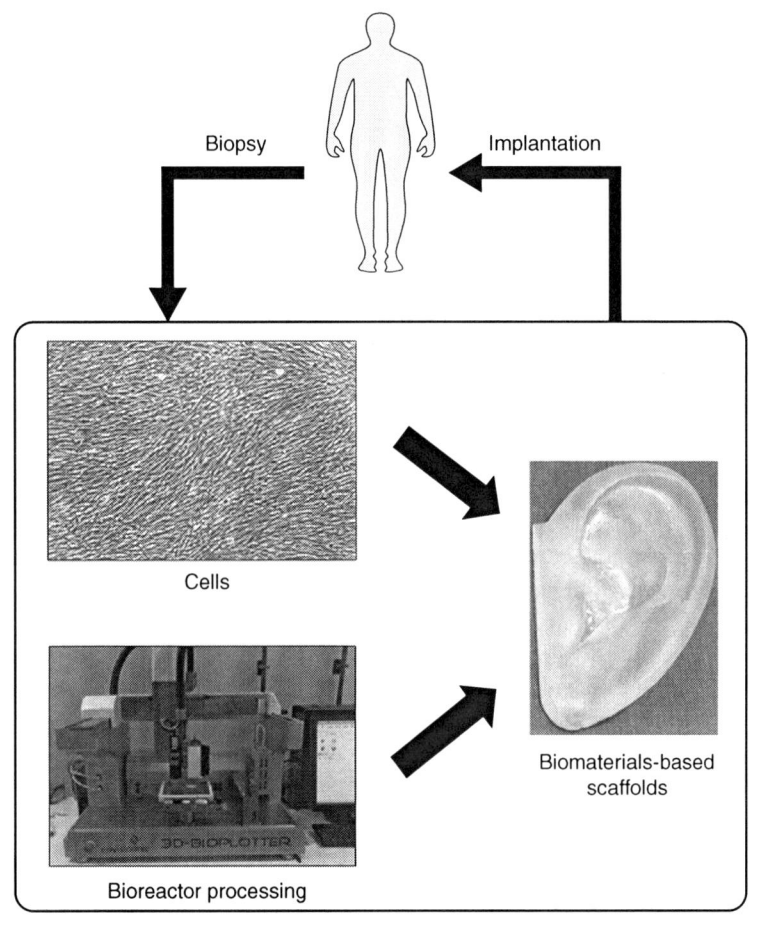

6.1 Tissue engineering of cells noting the major components of the process (cells, biomaterials and scaffolds, and bioreactor-based processes); a template for tissue formation by allowing cells to adhere, grow and proliferate.

regeneration. In addition, scaffolding processes should enable the fabrication of these materials into 3D forms that satisfy the scaffold design requirements for the specific tissue under consideration.

Recent advances in tissue engineering strategies have resulted in the emergence of the concept of direct tissue/organ bioprinting, a principal biomedical application of rapid prototyping (RP) technology. It provides a paradigm shift from the traditional scaffold-based approach – seeding cells and nutrients into a preformed 3D polymer matrix – by enabling the simultaneous deposition of cells and biocompatible scaffolding gels (known as hydrogels) with the principles of self-assembly (Mironov *et al.*, 2006). The former approach does not allow for well-defined cell distribution in that it does not facilitate the placement of cells at specific locations within the solid synthetic scaffolds (Nakamura et al., 2010). The direct bioprinting approach in which an autologous or autogenic cell suspension 'bioink' is printed onto the stimuli-sensitive 'biopaper' – a group of specially designed, processible and biomimetic tissue fusion-permissive hydrogels – allows these drawbacks to be overcome. These hydrogels not only allow for easy processing with various RP techniques, but they can also sustain cell encapsulation by providing a highly hydrated microenvironment that is amenable to nutrient diffusion and present biochemical, cellular, and physical stimuli to direct cellular processes (Fedorovich et al., 2007). With the evolution of the knowledge and understanding of the characteristics and behavior of natural ECMs, contemporary research efforts have been focusing on the development and applications of 3D hydrogels (materials, structure, and RP processes) to mimic various functional ECMs.

This chapter focuses on the characteristics and applications of hydrogels as biopaper for tissue engineering scaffolds. The remainder of the chapter is organized as follows: Section 6.2 provides a general overview of traditional scaffolding biomaterials used in tissue engineering and the need for/evolution of hydrogels. Sections 6.3 and 6.4 provide a review of some of the most commonly used hydrogel materials in tissue engineering and their applications in scaffold-based RP, respectively. The chapter concludes in Section 6.5 with a brief discussion about the future prospects in scaffolding biomaterials.

6.2 Biomaterials in tissue engineering

Biomaterials play a central role in tissue engineering by providing the physical support and structure interface necessary for the interactions with biological systems. Some of the important aspects of biomaterials in scaffold-based tissue engineering are discussed in this section.

6.2.1 Background of scaffolding materials

Scaffolds provide the support for cell growth and define the ultimate shape of the cultured functional tissues. Scaffold materials play an important role in tissue regeneration by addressing critical physical (e.g., mechanics, degradation, gel formation), mass transport (e.g., diffusion), and biological (e.g., cell adhesion and signaling) characteristics that are often unique to each application. Inorganic materials – metals such as titanium, stainless steel and their alloys, and ceramics such as hydroxyapatite – have traditionally been used for hard tissue applications such as orthopedic implants, while natural and synthetic polymers such as polycaprolactone (PCL) have been

Scaffolding hydrogels for rapid prototyping 179

primarily used for soft tissue scaffolds. These scaffolds can be either fully dense or trabecular, and degradable or non-degradable, depending on the intended use (Ramakrishna *et al.*, 2001). Figure 6.2 shows examples of two different types of tissue engineering scaffolds – Ti6Al4V tibial wedge and silicone auricular implant – fabricated by electron beam melting and 3D-bioplotting RP processes, respectively.

As such, for virtually all tissue types, a number of requirements must be maintained when defining a tissue engineering process and the related materials used in the scaffold (O'Brien, 2011). Some of these important characteristics are summarized below:

- *Biocompatibility*: Key in any tissue engineering is that all materials must be biocompatible. That is, all scaffold materials must serve as an adherent surface for cells where the cells can congregate/migrate onto the scaffold surface for 'normal function' and eventually begin to proliferate to become normal tissue. Once implanted in a body, the engineered tissue structure must interact with the total system, produce minimal immune response, avert severe inflammatory response, minimize rejection by the body, and maximize healing.

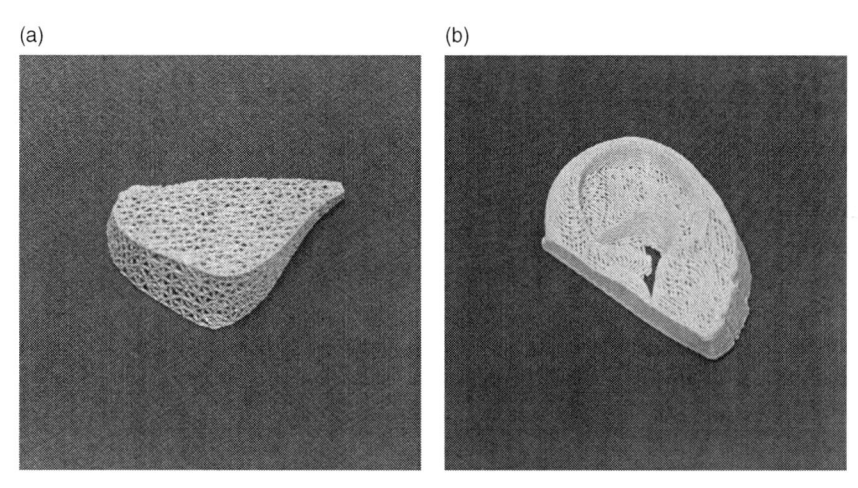

6.2 Examples of tissue engineering scaffolds fabricated by RP methods at North Carolina State University. (a) Metal (Ti6Al4V) tibial wedge designed by Materialise Inc. using the Mimics Innovation Suite and fabricated by Electron Beam Melting process. (*Source*: Courtesy: Materialise Inc., Belgium, and Rapid Prototyping Laboratory – Dr Ola Harrysson and Mr Tim Horn.) (b) Polymer (silicone) auricular implant fabricated by 3D Bioplotting process. (*Source*: Courtesy: Biomedical Manufacturing Laboratory – Dr Rohan A. Shirwaiker and Dr Molly Purser.)

- *Biodegradable/bioabsorable*: The vision of tissue engineering is to create functional tissue so that the body performs as it originally did. Given this goal, any synthetic materials introduced into the system must be purged from the system so that they will not impede the performance of the system in any way. For this to happen, all synthetic materials must be biodegradable in that they break down chemically into components that can be absorbed into the system. This will allow the cells to produce their own matrix structure and maintain their shape and function without the synthetic support. The by-product(s) from the chemical degradation must be non-toxic and able to exit the body in some natural manner without interfering with other organs or body functions.
- *Mechanical properties*: The mechanical component of tissue engineering is as important as any other aspect of the process. Scaffolds need to have the requisite mechanical properties and must have both micro and bulk geometries similar to what is needed for each application. For example, in the case of traditional metal scaffolds – titanium-based implants – both the bone and the implant scaffold must function collectively as the original bone did, and both must be able to tolerate the required stresses and strains associated with the physical processes of mammals. Hence, it must have immediate strength and the capability to integrate rapidly with the bone. For biodegradable scaffolds, the scaffold must degrade at the same rate as the growth of the regenerated tissue in order to transition the mechanical properties.
- *Scaffold architecture and geometry*: The architecture used in scaffolds is critical to the performance of any tissue engineered product. Generally, tissue engineered products are grouped into 2D and 3D architectures. For creating a product like skin, a 2D structure is adequate to form the entire dermis. The same is true for the manufacturing of a bladder, where bladder tissue is first produced as a 2D structure and then sewn together as a 3D bladder vessel. For other types of tissue, like bone and most organs, the entire 3D geometry is important. Generally, scaffolds should have the requisite porosity to ensure cellular penetration and adequate diffusion of nutrients to cells within the structure. A porous interconnected structure is typically required to allow diffusion of waste products from the scaffold and the products of the scaffold degradation (in the case of biodegradable scaffolds) to exit without interference with other organs and surrounding tissues.
- *Manufacturing technology*: For any tissue engineered product to be commercially viable, it must be produced efficiently and economically. This implies that the scaffold component of any tissue engineered product should be quickly produced from inexpensive materials. Today, RP technologies are providing the ability to consistently and economically fabricate viable scaffolds.

Natural polymers were the first biodegradable biomaterials to be used clinically (Nair and Laurencin, 2007). As Dhandayuthapani *et al.* (2011) have noted, natural materials, owing to their bioactive properties, have better interactions with the cells, allowing them to enhance cell performance in biological systems. These polymers can be classified as proteins (e.g., silk, collagen, gelatin, fibrinogen), polysaccharides (e.g., cellulose, amylose, dextran, glycosaminoglycans (GAGs)), or polynucleotides (e.g., DNA, RNA) (Yannas, 2004). Synthetic polymers represent the largest group of biodegradable polymers today, since they generally exhibit predictable and reproducible physicochemical and mechanical properties necessary to facilitate restoration of structure and function of damaged or diseased tissues (Gunatillake *et al.*, 2006). Polylactic acid (PLA), polyglycolic acid (PGA), and their copolymers are among the most popular synthetic polymers in tissue engineering (Ma, 2004). Polyhydroxyalkanoates (PHAs) represent another class of synthetic biopolymers that are being increasingly considered for applications in tissue engineering (Chen and Wang, 2002; Dhandayuthapani *et al.*, 2011).

6.2.2 Evolution towards hydrogel-based scaffolds

With the advent of direct tissue/organ bioprinting, processible and biomimetic hydrogels have become central to tissue engineering materials. Hydrogels are a class of highly hydrated polymer materials (natural and synthetic, water content $\geq 30\%$ by weight) consisting of hydrophilic homopolymer, copolymer, or macromer chains (Drury and Mooney, 2003; Park and Lakes, 1992; Slaughter *et al.*, 2009). Their inherent structural and compositional framework is similar to naturally occurring ECM, and they can often be processed under relatively mild conditions or formed *in situ*. In fact, hydrogels lend themselves well to RP methods because they can transform into a crosslinked 3D network in response to physical stimuli (e.g., change in pH, temperature, stresses) or chemical stimuli (e.g., enzymatic reaction, crosslinking agent, photopolymerization) (Patel and Mequanint, 2011). Scaffolding hydrogels have potential advantages of biocompatibility, cell-controlled degradability, and intrinsic cellular interaction (Jhon and Andrade, 1973). They allow for the uniform distribution and encapsulation of cells, growth factors, and other bioactive compounds while allowing rapid diffusion of hydrophilic nutrients and metabolites of incorporated cells.

In common with other tissue engineering materials, hydrogels used for direct bioprinting are also expected to possess all the desirable scaffolding characteristics described in Section 6.2.1. These properties, including the mechanical performance and degradation behavior, are directly related to the mechanisms and dynamics of gel formation, which in turn are controlled by the intrinsic properties of the main chain polymer and the crosslinking mechanisms (i.e., amount, type, and size of crosslinking molecules) as well as the environmental conditions. The crosslinking characteristics also govern the practicality of RP fabrication, incorporation of viable cells and bioactive molecules within the hydrogel, and the mass transport properties throughout the scaffold, all of which are extremely critical for cell viability and nutrition in direct bioprinting. As Imani *et al.* (2011) note, several studies have demonstrated the inability of single polymers alone to satisfy different demands in tissue engineering in terms of both characteristics and functionality. Hence, different polymers are often mixed to create more suitable hydrogel blends that may demonstrate synergistic properties for practical scaffolding applications (Liu *et al.*, 2005).

Biocompatible hydrogel blends are currently used in cartilage wound healing, bone regeneration, wound dress, and as carriers for drug delivery (Peppas and Khare, 1993). Hydrogels with growth factors can act directly to support the development and differentiation of cells in the newly formed tissues (Tabata, 2003). In part due to their high water content, hydrogels are often favorable for promoting cell migration, angiogenesis, and rapid nutrient diffusion (Bryant and Anseth, 2001). The hydrogel scaffolds have received intensive study for their use in the engineering of replacement connective tissues, primarily due to their biochemical similarity with the highly hydrated GAG components of connective tissues. Examples of hydrogel-forming polymers of natural origin are collagen (Wallace and Rosenblatt, 2003), fibrin (Eyrich *et al.*, 2007), alginate (Kong *et al.*, 2003), hyaluronic acid (HA) (Solchaga *et al.*, 2002), gelatin (Kim *et al.*, 2004), and chitosan (Suh and Matthew, 2002). Examples of synthetic polymers include poly(propylene fumarate) (PPF)-derived copolymers (Behravesh and Milos, 2003), poly(ethylene glycol) (PEG)-derivatives, and poly(vinyl alcohol) (PVA) (Bryant *et al.*, 2004).

Some commonly used hydrogel-forming biomaterials are discussed in the following section.

6.3 Review of commonly used hydrogel-forming scaffolding biomaterials

Hydrogels used for scaffolds can be categorized as either naturally derived or synthetic materials. Many naturally derived hydrogels are popular due to their innate biocompatibility, while synthetic gels are known for their ability to be more uniform, tunable, and translatable to large-scale production. The more widely used hydrogels for RP scaffolds include the naturally derived hydrogels collagen, fibrin, alginate, and HA, in addition to the synthetic gels Pluronic® F-127, poly(ethylene)-based hydrogels, fumaric acid-based hydrogels, and PVA.

6.3.1 Collagen

Collagen is the most widely used natural material for tissue engineered scaffolds. It is the most abundant fibrous protein in mammalian tissue ECM and can account for as much as 25% of the total protein content of a mammal (Drury and Mooney, 2003). Although there are several forms of collagen, in its basic form it consists of a self-aggregating triple helix structure – polypeptide chains held together by both hydrogen and covalent bonds. The collagen strands are then capable of self-aggregating to form stable fibers (Lee *et al.*, 2001). Collagen (in fiber form) possesses good tensile strength, is easy to process, dissolves readily in acidic aqueous solutions, and can form a gel at physiological temperatures.

Collagen is popular in tissue engineering due to its inherent ability to improve cell adhesion, proliferation, and function (Ulery, *et al.*, 2011). It is naturally degraded in the body by enzymatic digestion, specifically by collagenase and serine proteases, and naturally promotes cell adhesion. Since collagen is a natural polymer, there can be a high variability in the chemical and mechanical properties depending on the source. The main sources of natural collagen are animal and cadaver and, often, the processing can induce an immunologic response *in vivo* (Ulery, *et al.*, 2011). As with most hydrogels, the mechanical strength is low, and can often be enhanced by various crosslinking methods: chemical (e.g., glutaraldehyde, formaldehyde) or physical (e.g., ultraviolet (UV) irradiation, heating) (Jin and Dijkstra, 2010). Collagen is often also combined with other natural polymers such as alginate (Sang *et al.*, 2010), HA (Park *et al.*, 2002), and chitosan (Lin *et al.*, 2009), as well as synthetic polymers such as PCL (Xu *et al.*, 2013) and poly(lactic-co-glycolic acid) (PLGA) (Liu *et al.*, 2010) to improve the mechanical strength.

Collagen scaffolds have found applications in adipose tissue, bladder, blood vessel, bone cartilage, cornea, heart, intervertebral disk, ligament, liver, muscle, nerve, nucleus pulposus, pancreas, and skin (Fedorovich *et al.*, 2007).

6.3.2 Fibrin

Fibrin is a biopolymer of the monomer fibrinogen and a critical physiological component of blood that is responsible for hemostasis (Eyrich *et al.*, 2007). Fibrinogen comprises two sets of three polypeptide chains – Aα, Bβ, and γ chains – that are held together by 29 sulfide bonds (Ahmed *et al.*, 2008). Fibrinogen and thrombin are combined to form a fibrin hydrogel. The fibrin monomer self-associates to form insoluble fibrin. The blood coagulation factor XIIIa covalently crosslinks with the γ chains in the fibrin polymer to produce a fibrin network that is stable and resists protease degradation (Ahmed et al., 2008).

Advantages of fibrin-based scaffolds include a high cell-seeding density, uniform cell distribution, and inherent cell adhesion capabilities (Christman et al., 2004). In addition, fibrin is readily available: it can be produced from the blood of the patient eliminating a foreign body response, or purified fibrinogen and thrombin can be commercially purchased. However, fibrin, like most hydrogels, has a low mechanical stiffness, high shrinkage and rapid degradation. The loss of shape and volume of fibrin hydrogels can be overcome by incorporation of fixing agents such as poly-L-lysine. Mechanical properties can be enhanced by incorporation with other scaffolding biopolymer such as polyurethane, PCL, β-tricalciumphosphate (β-TCP), β-TCP/PCL and PEG (Ahmed et al., 2008). The rapid degradation rate can be moderated by optimizing factors such as pH, calcium ions, and fibrinogen concentrations, using a lower cell density, or adding protease inhibitors. Biologically active peptides, such as fibronectin, vitronectin, laminin, and collagen can also be incorporated into fibrin.

Fibrin-based hydrogels have been used in inkjet printing as well as syringe deposition methods due to their ability to easily and quickly crosslink with thrombin at room temperature. Rapid prototyped 3D scaffolds of fibrin have found applications in adipose, bone, cardiac, cartilage, liver, ocular, skin, tendon, and ligament tissues (Ahmed et al., 2008).

6.3.3 Alginate

Alginate is a hydrophilic linear polysaccharide copolymer of D-mannuronic acid (M) and L-guluronic acid (G) derived from brown seaweed and bacteria (Johnson et al., 1997). The arrangement of the M and G monomers within the polysaccharide (MMMM, GGGG, or GMGM) depends on the source from which it is isolated and affects the characteristics of the material (Smidsrød and Skjakbraek, 1990). Alginate forms a gel under mild conditions, has a low toxicity, and can easily be modified with peptides to improve cell adhesion (Chung and Burdick, 2008). In general, alginate-based gels are formed by the interaction of divalent cations (e.g., Ca^{++}, Ba^{++}) with the G monomers to form ionic bridges between different polymer chains. The M/G ratio and the concentration of the divalent cations within the gel have a significant bearing on its crosslinking density, mechanical properties and pore sizes; the higher the M/G ratio, the smaller the average pore size is (Drury and Mooney, 2003; Smidsrød and Skjakbraek, 1990).

While alginate-based hydrogels are known to be biocompatible and non-immunogenic, they possess two critical disadvantages from the scaffold tissue engineering perspective – low mechanical strength and slow degradation rate (Chung and Burdick, 2008). Alginates cannot be enzymatically broken down and have poorly regulated degradation. Gamma irradiation and partial oxidation with sodium periodate can be used to control and promote the degradation. In addition, cell adhesion to alginates is weak, unless the material is modified with adhesion promoting molecules (e.g., peptides). Interactions between these molecules and the cells also improve the strength of alginate-based hydrogels (Drury et al., 2005).

Alginate hydrogels have been used in cartilage tissue scaffolds when combined with other materials such as covalently crosslinking with PEG (Drury and Mooney, 2003). They have also been used as a bioink for inkjet biofabrication (Arai et al., 2011).

6.3.4 Hyaluronic acid

HA (also known as hyaluronan) is a high molecular weight, linear polysaccharide consisting of repeating disaccharide units of (1–3) and (1–4)-linked b-d-glucuronic acid and N-acetyl-b-d-glucosamine. It is a simple form of GAG, one of the main components of the ECM found in cartilage, skin, and the vitreous humor (Drury and Mooney, 2003; Slaughter et al., 2009). Hydrogels of HA can be formed by covalent crosslinking with hydrazide derivatives (Prestwich et al., 1998), by esterification (Mensitieri et al., 1996), or by annealing (Fujiwara et al., 2000). HA can also be combined with collagen and alginate to form composite hydrogels.

HA hydrogels are important in RP-based scaffold applications owing to their ability to be deposited with incorporated cells and crosslink *in situ*. They are naturally degraded by hyaluronidase or by reactive oxygen intermediates, allowing cells to regulate the clearance of the material in a controlled manner. While HA is highly viscous and can be difficult to handle, acid or base treatments can be used to produce a lower molecular weight gel with a specific compressive modulus, degradation time, and swelling ratio (Jin and Dijkstra, 2010). HA has been shown to affect cell proliferation, differentiation, inflammation, and wound repair (Chen and Abatangelo, 1999). Cells incorporated within HA-hydrogels have demonstrated their ability to retain their phenotype and secrete ECM *in vivo*.

HA-based products are commercially available, such as Healon® (Abbott Laboratories, Inc.) which is a FDA-approved product used for eye surgery and corneal transplant. Hyalograft® C (Anika Therapeutics, Inc.) is a HA-based mesh scaffold seeded with autologous chondrocytes that is currently in FDA clinical trials for articular cartilage repair (http://www.anikatherapeutics.com). Hyaluronan combined with PVA and calcium chloride ($CaCl_2$) has been used as a biopaper with alginate as the bioink (Arai et al., 2011). Blends of HA with other natural gel-forming materials such as gelatin, alginate, and collagen have also been utilized for tissue engineering applications.

6.3.5 Pluronic® F-127

Pluronics® or poloxamers are tri-block copolymers of poly(ethylene oxide)-poly(propylene oxide)-poly(ethylene oxide) (PEO-PPO-PEO) (Lippens *et al.*, 2013). This group of synthetic polymers is thermoreversible in aqueous solutions. The sol-gel transition is governed by the composition, molecular weight, and concentration of each constituent block polymer. The hydrophilic ethylene oxide and the hydrophobic propylene oxide give Pluronics an amphiphilic structure – meaning it has a polar, water-soluble group attached to a nonpolar water-insoluble hydrocarbon chain. Amphiphilic block copolymer molecules self-assemble into micelles (a packed chain of molecules) in aqueous solution. Micelle formation is temperature dependent and affects the degradation properties of the biomaterial: below a certain characteristic temperature known as the critical micelle temperature, both the ethylene and propylene oxide blocks are hydrated and the PPO block becomes soluble.

Pluronic F-127, also known as Poloxamer 407, is often used in tissue engineering because of the commercial availability of a consistent product that will undergo a sol-gel transition near physiological temperature and pH (Klouda and Mikos, 2008; Lippens *et al.*, 2013). A disadvantage of Pluronic is its fast degradation rate *in vivo*. To overcome this problem, Pluronic is frequently crosslinked with another α-hydroxy or amino acid in order to alter the chemical structure of its depsipeptide unit. In a recent study, crosslinking with alanine amino acid resulted in a tailored degradation rate that could be achieved using UV irradiation, without any harmful side-effects on the sol-gel transition or the critical micelle temperature (Lippens *et al.*, 2013).

In terms of applications, while Pluronics are known to inhibit surface–tissue adhesion for many cell types (Klouda and Mikos, 2008), they have been successfully used for scaffolding applications that involve hematopoietic stem cells (Higuchi *et al.*, 2006) and lung tissue (Cortiella *et al.*, 2006).

6.3.6 Poly(ethylene)-based hydrogels

Poly(ethylene)-based hydrogels are the most widely used synthetic hydrogels for tissue engineering applications – with several FDA-approved medical applications – primarily due to their versatile chemistry that allows for tailored properties (Drury and Mooney, 2003). They are prepared by the polymerization of ethylene oxide by condensation, and classified into PEG and poly(ethylene oxide) (PEO) based on their molecular weights.

Materials with higher molecular weight (usually above 10 kDa) are classified as PEOs, while the ones with lower molecular weight are PEG (Fedorovich et al., 2007; Slaughter et al., 2009). These materials are non-ionic, hydrophilic, biocompatible, and are limited in their natural abilities of protein binding and cell adhesion.

PEG is often used in tissue engineering applications to encapsulate cells into hybrid polymer/gel scaffolds. PEG repels non-specific protein adsorption and cell adhesion and can protect encapsulated cells from undesirable interactions with the scaffold polymers (e.g., inflammatory cell adhesion, capsule formation) (Lin et al., 2009). However, this also limits the absorption of necessary bioactive molecules such as ECM proteins that are needed for cell survival (Lin et al., 2009). To overcome this limitation, PEG chains are modified with bioactive peptides and growth factors such as the adhesion peptide arginine-glycine-aspartic acid (RGD), pancreatic β-cell, and human mesenchymal stem cells (MSCs). PEG possesses high permeability (up to 95% water in its total mass) that facilitates nutrient waste exchange necessary for cell survival. However, the high permeability can also allow for small cytotoxic molecules to pass through, which is detrimental to cells (Lin et al., 2009).

By customizing PEG, the hydrogel can be tailored to enhance cell survival, proliferation, secretory properties, and even differentiation of cells (Slaughter et al., 2009). The crosslinking of PEG hydrogels can also be tailored to the application. Modification of each end of the polymer with either acrylates or methacrylates allows for crosslinking via photopolymerization when mixed with the appropriated photoinitiator (Drury and Mooney, 2003). Crosslinking can be done with either UV or natural light. Thermoreversible hydrogels can be formed from block copolymers with PLA and PGA (Drury and Mooney, 2003; Jin and Dijkstra, 2010). PLA and PGA are aliphatic, biocompatible polymers that are hydrophobic and not soluble in water. Water soluble PEG-PLA, PEG-PGA or PEG-PLGA can be synthesized through ring-opening polymerization (Jin and Dijkstra, 2010). Changing the ratio of the polymers and the PEG length controls the transition temperature (Fedorovich et al., 2007). Degradation rates of these hydrogels can be tailored by combining with matrix metalloproteinases that form cleavable linkages in order to facilitate the cell-responsive degradation as well as releasing growth factors beneficial for tissue regeneration (Lin et al., 2009).

6.3.7 Fumaric acid-based hydrogels

Hydrogels based on fumaric acid are another popular group of synthetic biopolymers that include poly(propylene fumarate) (PPF), poly(propylene fumarate-co-ethylene glycol) (PPF-co-EG) and oligo(poly(ethylene glycol fumarate) (OPF). These materials are based on fumaric acid, which is non-toxic and naturally formed in the Krebs cycle (Jin and Dijkstra, 2010).

PPF is an unsaturated linear polyester which is water-insoluble. It is generally crosslinked with N-vinyl pyrrolidone (NVP) using benzoyl peroxide (BP) as an initiator, or by covalent bonding through addition polymerization, resulting in water soluble compounds that can be processed through the kidneys (Temenoff et al., 2007). PPF-co-EG is a block copolymer of PPF and PEG. The additions of PEG chains make the block copolymer water soluble (Temenoff et al., 2007). By varying the amounts of PPF and PEG, mechanical properties, degradation rate, and cell adhesion of the hydrogel can be altered to suit the application (Ketlow and Mikos, 2007). Similarly, the bioactive properties can also be modified; for example, the addition of RGD resulted in enhanced cell adhesion, migration, and the ability to differentiate into mineralized tissue of MSCs *in vitro* (Behravesh and Mikos, 2003). From an RP direct bioprinting perspective, Fisher et al. (2004) have developed a method to produce a thermoreversible P(PF-co-EG) that is liquid at temperatures below 25°C and a gel above 35°C allowing for the addition of cells at room temperature and the formation of a hydrogel at physiological temperatures.

OPF comprising alternating PEG and single fumarate moieties has also been explored for tissue engineering applications. OPF, like P(PF-co-PEG), is water-soluble and possesses both greater hydrophilicity and swelling ratio (Temenoff et al., 2007). Tailoring of tensile mechanical properties, swelling characteristics, and cell attachment properties can be achieved by synthesizing with PEG of different molecular weights (Kretlow and Mikos, 2007). Park *et al.* (2005) have modified OPF to deliver both cells and growth factors, and demonstrated how hydrogels incorporating growth factors produced a greater amount of GAG than hydrogels without growth factors.

6.3.8 Poly(vinyl alcohol)

PVA is a synthetic polymer that is derived from poly(vinyl acetate) through partial or full hydrolysis to eliminate the acetate groups (Baker et al., 2012; Slaughter et al., 2009). It is biocompatible, water soluble, chemical resistant, and non-adhesive to cells and proteins. Biological molecules can be attached to enhance cell adhesion and migration through the available pendant alcohol groups (Slaughter et al., 2009). The amount of hydroxylation during synthesis determines its physical characteristics as well as the chemical and mechanical properties. For example, greater degrees of hydroxylation and polymerization will lower the solubility of PVA in water (Baker et al., 2012).

To form a hydrogel, PVA can be crosslinked either physically, by repeated freeze–thaw cycles, or chemically, with aldehydes. However, neither of these methods allows for the physiological environment necessary for cell survival (Fedorovich *et al.*, 2007). PVA grafted with photosensitive groups can be photopolymerized in a biologically compatible environment using a non-toxic photoinitiator and short exposure to UV light (Schmedlen *et al.*, 2002). With this method, cells can be seeded within the hydrogel for direct bioprinting. The flexibility, elasticity, and diffusion properties of PVA-based hydrogels can be modified by controlling the degree of crosslinking (Baker, 2012; Schmedlen *et al.*, 2002). This is important because the elastic property of PVA can be used to transmit mechanical stimuli to seeded cells to enhance matrix synthesis (Slaughter *et al.*, 2009).

PVA is often used for articular cartilage tissue engineering because it is stronger than most synthetic gels, has a low coefficient of friction and has similar tensile and compressive strength of human cartilage (Schmedlen *et al.*, 2002).

The basic structural formulae of the hydrogel-forming synthetic polymers described above are shown in Fig. 6.3. The important hydrogel materials and their crosslinking and RP processing-based characteristics are highlighted in Table 6.1. In this table, nozzle-based systems imply systems

6.3 Basic structural formulae of some commonly used synthetic hydrogel-forming polymers used in RP-based tissue engineering: (a) Pluronic F-127, (b) PEG, (c) PEO, (d) P(PF-co-EG), (e) OPF, (f) PVA.

Table 6.1 Crosslinking and RP processing-based characteristics of hydrogel-forming polymers commonly used in tissue engineering

Hydrogel	Crosslinking mechanisms				RP Processes			References
	Thermo	Photo	Chemical	Enzymatic	Nozzle based	Laser based	Inkjet based	
Collagen	x				x		x	Smith et al., 2004; Boland et al., 2006
Fibrin				x	x		x	Schantz et al., 2005; Xu et al., 2006
Alginate			x	x	x		x	Sakai and Kawakami, 2007; Fedorovich et al., 2008; Nishiyama et al., 2009
HA		x	x	x	x	x		Segura et al., 2005; Hutmacher et al., 2004; Zhang et al., 2007
Pluronic® F-127	x	x			x		x	Lee et al., 2004; Smith et al., 2004; Biase et al., 2011
Poly(ethylene)-based	x	x	x		x	x	x	Fedorovich et al., 2007; Tsang et al., 2007; Maher et al., 2009; Sanjana and Fuller, 2004
Fumaric acid-based	x	x	x			x		Temenoff et al., 2007; Lee et al., 2007
PVA		x	x			x		Fedorovich et al., 2007; Schemedlen et al., 2002; Chua et al., 2004

involving extrusion of hydrogels based on pressure-actuated, piezoelectric, solenoid-actuated, and volume-actuated designs (e.g., fused deposition modeling (FDM) and 3D-bioplotting). Laser based systems are the ones that use light or radiation energy to crosslink liquid hydrogels, requiring the use of photocrosslinkable polymers (e.g., stereolithography, solid ground curing system). Inkjet printing systems refer to the ones that eject the liquid polymer as micro-droplets under pressure. The crosslinking in these systems occurs when the liquid polymer dries or cures on the biopaper after deposition (Billet et al., 2012).

6.4 Applications of scaffolding hydrogels

Rapid prototyping of hydrogel scaffolds is challenging because of the comparatively low mechanical strength and stability of the gels. The trend in research is in methods of crosslinking and using hybrid scaffolds that combine several different polymers to achieve the desired degradation, cell behavior, and mechanical properties while maintaining cell viability and functionality. Some of the applications of hydrogels in RP-based tissue engineering are presented in this section.

Tsang et al. (2007) used PEG to fabricate a multilayer hydrogel construct to support hepatic tissue using a photopatterning technique. The PEG hydrogel was modified to include the RGD peptide sequence to increase cell adhesion. The PEG-based polymer solution was created by combining the PEG monoacrylate with poly(ethylene glycol) diacrylate (PEGDA) to obtain the desired concentration and molecular weight. The photoinitiator 2,2-dimethoxy-2-phenyl-acetophenone was added, enabling crosslinking when exposed to 365 nm UV light for 90–110 s through a glass cover slip. Hepatocytes were then isolated from adult Lewis rats and co-cultured with fibroblast in Dulbecco's modified Eagles Medium (DMEM) with high glucose and supplemented with 5% bovine calf serum and penicillin and streptomycin. For photopatterning, the cells were suspended in the polymer solution at a concentration of $5-15 \times 10^6$ cells/mL. A construct, consisting of three layers of the polymer/cell hydrogel, was placed in a perfusion bioreactor to allow for nutrient flow through the scaffold. Results of imaging after 3 days in the bioreactor showed viable hepatocytes throughout all layers of the patterned hydrogel scaffold. Culture medium collected over 12 days showed higher levels of liver specific markers than un-patterned solid hydrogel constructs. Photopolymerizable PEGDA with incorporated RGD enabled the proliferation and ECM production of hepatocytes. This study demonstrates the feasibility of three-dimensional hepatic tissue fabricated by photopatterning, perhaps leading to a tissue engineered implantable liver (Tsang et al., 2007)

A study by Cohen et al. (2006) used a similar linear actuated syringe free form fabrication system. The difference was that the alginate crosslinking could be done prior to deposition. A low-viscosity, high G-content alginate was used to encapsulate chondrocyte cells harvested from the femeropatellar groove of 1- to 3-day old calves. The cell/alginate mixture was mixed in a 2:1 ratio with calcium sulfate ($CaSO_4$) to initiate the crosslinking of the alginate. The optimal printing time was approximately 15 min after the crosslinker was added to alginate to allow for sufficient crosslinking density. Cell viability was 94% after printing, and cell distribution was uniform throughout the scaffold. GAG and hydroxyproline content increased over an 18 week incubation period of the scaffold: after 18 weeks, the scaffold still held the same overall shape. The elastic modulus increased with time after fabrication, and was reported to be higher than in an alginate scaffold prepared by photopolymerization. Several shapes, including a crescent, a disk with two types of batches of alginate each stained a different color, and a model of an ovine meniscus from a computed tomography (CT) scan were printed to estimate the error in the printing process. The ability of alginate to be crosslinked when in contact with non-cytotoxic calcium ions at room temperature is an advantage in this fabrication process and does not decrease the cell viability.

A limitation of hydrogel scaffolds is their poor mechanical properties which inhibit use in load-bearing applications such as bone or cartilage. Xu et al. (2013) have recently fabricated scaffolds composed of both synthetic polymers with high stiffness and hydrogels to provide a favorable environment for cells using a hybrid electrospinning/inkjet printing method. In this method, flexible mats of electrospun synthetic polymers were alternately layered with inkjet printed hydrogel with suspended chondrocyte cells. The material for the electrospinning was PCL and Pluronic F-127 dissolved in acetone. Pluronic F-127 was used to reduce the viscosity of the PCL as well as increase the hydrophilicity of the PCL. Chondrocytes were isolated from rabbit ear cartilage, and cultured in media in an incubator until the sixth passage. Cells were then resuspended in a mixture of fibrinogen, collagen, and phosphate buffered saline (PBS) at a concentration of $3-4 \times 10^6$ cells/mL. The layered scaffold was fabricated by first electrospinning the PCL/F-127 polymer directly into a Petri dish containing PBS. Since PCL/F-127 will rapidly absorb water from the fibrinogen/collagen, which will desiccate the cells, the electrospinning was done wet. The cell/collagen/fibrinogen was subsequently inkjet printed onto the mesh PCL/F-127 mat. In order to crosslink the fibrinogen, a layer of thrombin was printed on top of the cell/collagen/fibrinogen layer and allowed to react for 15 min. This three-layer process was repeated until five layers were completed. The outer two layers of the PCL/F-127 were 300 μm thick while the inner layer was only 150 μm thick. This allowed for the outer layers to provide mechanical stability to the scaffold, while the thinner inner layer allowed for cell interaction between layers. As a comparison group, the alginate was used as a replacement for the fibrinogen/collagen. The alginate was cross-linked with $CaCl_2$ and then immediately inkjet printed before the hydrogel was completely gelled. Mechanical testing showed that the scaffold with the electrospun PCL, collagen/fibrin, and thrombin had a higher Young's modulus and ultimate yield stress than the PCL/F-127-only scaffold and the alginate scaffold. Seeded scaffolds were incubated and cultured in DMEM with 10% fetal bovine serum (FBS) for 4 weeks. Cell viability after 1 week *in vitro* culture was 82%. *In vivo* scaffold explanted after 8 weeks showed dense and well-organized collagen formation with type II collagen throughout, GAGs and rounded chondrocytes were visible and appeared typical of elastic cartilage. The results of this work suggest that a hybrid scaffold containing both a mechanically stable region and a hydrogel region to support the cells is feasible. The use of an electrospun flexible mesh provides support for the 3D organization for cell–cell signaling and could also provide a moldable scaffold to conform to a defect.

Shim *et al.* (2012) also used a hybrid scaffold design with PCL for mechanical properties and an alginate hydrogel to provide an environment for the cells; however, the scaffold was fabricated by a multi-head bioprinter. The system consisted of two heated heads for deposition of the polymers and for unheated heads for the hydrogels. The molten polymer was extruded using pneumatic pressures of up to 400 kPa, while the hydrogels were extruded using a plunger system driven by a stepper motor. PCL with a molecular weight of 70–90 kDa was extruded in a grid pattern. Two cell types, chondrocytes and osteoblasts, were used in order to demonstrate the feasibility of scaffolds with multiple cell types. Cells were suspended in sodium alginate and DMEM. The cell/alginate hydrogel was dispensed in every other pore produced by the PCL grid. The empty pores provided areas for oxygen and nutrient transport. A 20-layer scaffold with osteoblasts in the lower region and chondrocytes in the upper region was produced (Fig. 6.4).

After incubation of the cell seeded scaffold for 7 days, the viability of chondrocytes was 94% and the osteoblasts was 96%. However, the chondrocytes did not proliferate as well as the osteoblasts. It was suggested that the alginate might not promote chondrocyte adhesion, proliferation, and differentiation and suggested that bioactive proteins added to the hydrogel may improve chondrocyte proliferation. This process combines both a mechanically stable structure with various cell types suspended in a hydrogel and demonstrates that a complex tissue structure could be fabricated on one machine in a relatively short time period with a patient-specific geometry.

In contrast to constructing mechanically stable scaffolds, researchers at Wake Forest Institute for Regenerative Medicine have developed a novel skin repair printer that would allow for *in situ* wound repair to treat burn

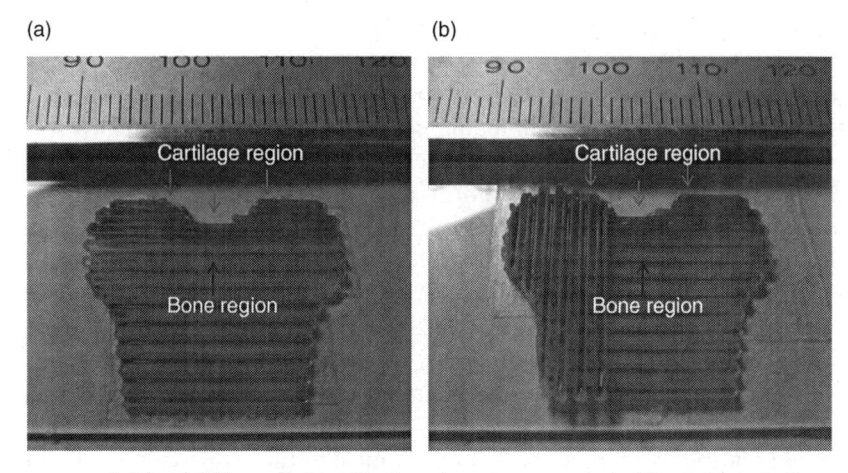

6.4 (a,b) PCL scaffold with chondrocytes suspended in alginate in the upper region, and osteoblasts in alginate in the lower region printed in select pores between the PCL strands (Shim *et al.*, 2012).

victims (Albanna *et al.*, 2012). Fibroblasts and keratinocytes were isolated from porcine skin and suspended in fibrinogen and collagen. The fibroblast/fibrinogen/collagen hydrogel is first printed directly onto the wound. This is followed by thrombin to crosslink the fibrinogen into a hydrogel. Next the keratinocytes/fibrinogen/collagen hydrogel are printed. The animals that received the treatment showed an approximate 90% reduction in wound size, 80% re-epithelialization and 40% contracture of original wound size. The cell encapsulated hydrogel in this case could be directly applied to the wound. The fibrinogen and thrombin remained viscous enough to be directly printed, and crosslinking occurred after the printing process to create the gel.

These are just a few of the examples of RP-specific applications of hydrogels in tissue engineering. Several other RP- and non-RP-specific examples are available and have been reviewed in the literature (Drury and Mooney, 2003; Fedorovich *et al.*, 2007).

6.5 Conclusion

Scaffolds and their materials have transformed significantly in the short, two-decade period since they arrived on the research scene. They have evolved to provide collocation of cells and supporting biologics, along with sufficient strength for surgical procedures and structure for post-surgery ambulation.

Recently, hydrogels have received attention because of their innate structural and compositional similarities to the ECM and their extensive framework for cellular proliferation and survival. They are also attractive because their degradation properties can be tailored to mimic those of natural tissues. However, one of the main limitations of hydrogels when it comes to rapid prototyping of 3D architectures is their lack of mechanical stability. This limits their applications, especially in bone and cartilage tissue engineering. A recent trend in research is the concept of hybrid scaffolds that synergistically utilize a biodegradable synthetic polymer framework for mechanical stability along with hydrogels to provide cells with a microenvironment physiologically similar to natural ECM. For soft tissues, various combinations of polymers and crosslinkers are being investigated in order to fabricate such scaffolds with sufficient stability while also promoting cell proliferation. Manufacturing technology has also evolved to a point where synthetic and biological materials can be interspersed into adjacent layers in the scaffold building process. The development of new hydrogel materials and hybrid strategies such as these is expected to be an active topic of exploration. Alongside this, investigations into the ability of decellularized components to provide the basic scaffolding structure for simple tissue geometries, like bladders, as well as for complex geometries, like vascular networks, will continue to be a critical biomaterial focus in tissue engineering research.

Eventually, these advances in scaffolding materials combined with the evolution of efficient, repeatable, reliable, and economical RP processes will together lead to the rapid progress toward engineering complex organs. In the near future, these hybrid processing approaches – merging of sophisticated biological protocols with state-of-the-art engineering practices – will begin to produce more complicated organs than we are capable of producing today. It is our vision that within a generation, we will be able to produce, or at least repair, organs that do not function as required.

6.6   References

Ahmed, T., Dare, E. and Hincke, M. (2008) 'Fibrin: A versatile scaffold for tissue engineering applications', *Tissue Eng Pt B*, **14**(2), 199–215. doi:10.1089/ten.teb.2007.0435.

Albanna, M., Murphy, S., Zhao, W., El-Amin, I., Tan, J., Dice, D., Kang, H., Jackson, J., Atala, A. and Yoo, J. (2012) 'In situ bioprinting of skin for reconstruction', *J Am Coll Surgeons*, **215**(3), S87. doi:10.1016/j.jamcollsurg.2012.06.235.

Arai, K., Iwanaga, S., Toda, H., Genci, C., Nishiyama, Y. and Nakamura, N. (2011) 'Three-dimensional inkjet biofabrication based on designed images', *Biofabrication*, **3**(3). doi:10.1088/1758-5082/3/3/034113.

Baker, M., Walsh, S., Schwartz, Z. and Boyan, B. (2012) 'A review of polyvinyl alcohol and its uses in cartilage and orthopedic applications', *J Biomed Mater Res Part B*, **100B**(5), 451–1457. doi:10.1002/jbm.b.32694.

Behravesh, E. and Mikos, A. (2003) 'Three-dimensional culture of differentiating marrow stromal osteoblasts in biomimetic poly(propylene fumarate-co-ethylene glycol)-based macroporous hydrogels', *J Biomed Mater Res A*, **66A**(3), 698–706. doi:10.1002/jbm.a.10003.

Biase, M., Saunder, R., Tirelli, N. and Derby, B. (2011) 'Inkjet printing and cell seeding thermoreversible photocurable gel structures', *Soft Matter*, **7**, 2639–2646. doi:10.1039/C0SM00996B.

Billet, T., Vandenhaute, M., Schelfhout, J., Van Vlierberghe, S. and Dubruel, P. (2012) 'A review of trends and limitations in hydrogel-rapid prototyping for tissue engineering', *Biomaterials*, **33**(26), 6020–6041. doi.org.prox.lib.ncsu.edu/10.1016/j.biomaterials.2012.04.050.

Boland, T., Xu, T., Damon, B. and Cui, X. (2006) 'Application of inkjet printing to tissue engineering', *Biotechnol J*, **1**(9), 910–917. doi:10.1002/biot.200600081.

Bryant, S. and Anseth, K. (2001) 'The effects of scaffold thickness on tissue engineered cartilage in photocrosslinked poly(ethylene oxide) hydrogels', *Biomaterials*, **22**(6), 619–626. doi.org/10.1016/S0142-9612(00)00225-8.

Bryant, S., Davis-Arehart, K., Luo, N., Shoemaker, R., Arthur, J. and Anseth, K. (2004) 'Synthesis and characterization of photopolymerized multifunctional hydrogels: water-soluble poly(vinyl alcohol) and chondroitin sulfate macromers for chondrocyte encapsulation', *Macromolecules*, **37**(18), 6726–6733. doi:10.1021/ma0499324.

Chen, L. and Wang, M. (2002) 'Production and evaluation of biodegradable composites based on PHB-PHV copolymer', *Biomaterials*, **23**(13) 2631–2639.

Chen, W. and Abatangelo, G. (1999) 'Functions of hyaluronan in wound repair', *Wound Rep Reg*, **7**(2), 79–89. doi:10.1046/j.1524-475X.1999.00079.x.

Christman, K., Fok, H., Sievers, R., Fang, Q. and Lee, R. (2004) 'Fibrin glue alone and skeletal myoblasts in a fibrin scaffold preserve cardiac function after myocardial infarction', *Tissue Eng*, **10**(3–4), 403–409. doi:10.1089/107632704323061762.

Chua, C., Leong, K., Tan, K., Wiria, F. and Cheah, C. (2004) 'Development of tissue scaffolds using selective laser sintering of polyvinyl alcohol/hydroxyapatite biocomposite for craniofacial and joint defects', *J Mater Sci-Mater M*, **15**, 1113–1121.

Chung. C. and Burdick, J. (2008) 'Engineering cartilage tissue', *Adv Drug Delivery Rev*, **60**(2), 243–262. doi.org.prox.lib.ncsu.edu/10.1016/j.addr.2007.08.027.

Cohen, D., Malone, E., Lipson, H. and Bonassar, L. (2006) 'Direct freeform fabrication of seeded hydrogels in arbitrary geometries', *Tissue Eng*, **12**(5), 1325–1335. doi:10.1089/ten.2006.12.1325.

Cortiella, J., Nichols, J., Kojima, K., Bonassar, L., Dargon, P., Roy, A., Vacant, M., Niles, J. and Vacanti, C. (2006) 'Tissue-engineered lung: an in vivo and in vitro comparison of polyglycolic acid and Pluronic F-127/somatic lung progenitor cell constructs to support tissue growth', *Tissue Eng*, **12**(5), 1213–1225. doi:10.1089/ten.2006.12.1213.

Dhandayuthapani, B., Yoshida, Y., Maekawa, T. and Kumar S. (2011) 'Polymeric scaffolds in tissue engineering applications: a review', *Int J Polym Sci*, **2011**, 1–11. doi:10.1155/2011/290602.

Drury, J. and Mooney, D. (2003) 'Hydrogels for tissue engineering: scaffold design variables and applications', *Biomaterials*, **24**(24), 4337–4351. doi.org/10.1016/S0142-9612(03)00340-5.

Drury, J., Boontheeku, T. and Mooney, D. (2005) 'Cellular cross-linking of peptide modified hydrogels', *J Biomech Eng Trans*, **127**(2), 220–228. doi:10.1115/1.1865194.

Eyrich, D., Brandl, F., Appel, B., Wiese, H., Maier, G., Wenzel, M., Staudenmaier, R., Goepferich, A. and Blunk, T. (2007) 'Long-term stable fibrin gels for cartilage engineering', *Biomaterials*, **28**(1), 55–65. doi.org/10.1016/j.biomaterials.2006.08.027.

Fedorovich, N., Alblas, J., de Wijn, J., Hennink, W., Verbout, A. and Dhert, W. (2007) 'Hydrogels as extracellular matrices for skeletal tissue engineering: state-of-the-art and novel application in organ printing', *Tissue Eng*, **13** (8), 1905–1925. doi:10.1089/ten.2006.0175.

Fedorovich, N., Dewijn, J., Verbout, A., Alblas, J. and Dhert, W. (2008) 'Three-dimensional fiber deposition of cell-laden, viable, patterned constructs for bone tissue printing', *Tissue Eng Pt A*, **14**(1), 127–133. doi:10.1089/ten.a.2007.0158.

Fisher, J., Jo, S., Mikos, A. and Reddi, A. (2004) 'Thermoreversible hydrogel scaffolds for articular cartilage engineering', *J Biomed Mater Res A*, **71A**(2), 268–274. doi:10.1002/jbm.a.30148.

Fujiwara J., Takahashi, M., Hatakeyama, T. and Hatakeyama, H. (2009) 'Gelation of hyaluronic acid through annealing', *Polym Int*, **49**(12), 1604–1608. doi:10.1002/1097-0126(200012)49:12<1604::AID-PI558>3.0.CO;2-4.

Gunatillake, P., Mayadunne, R. and Adhikari, R. (2006) 'Recent developments in biodegradable synthetic polymers', in El-Gewely, M., *Biotechnology Annual Review*, Elsevier, Amsterdam, 301–347.

Higuchi, A., Aoki, N., Yamamoto, T., Gomei, Y., Egashira, S., Matsuoka, Y., Miyazaki, T., Fukushima, H., Lee, Y., Jyujyoji, S. and Natori, S. (2006) 'Bio-inert surface of Pluronic-immobilized flask for preservation of haematopoietic stem cells', *Biomacromolecules*, **7**(4), 1083–1089. doi:10.1021/bm050868n.

Hutmacher, D., Sittinger, M. and Risbud, M. (2004) 'Scaffold-based tissue engineering: rational for computer-aided design and solid free-form fabrication system', *Trends Biotechnol*, **22**(7), 354–362. doi.org/10.1016/j.tibtech.2004.05.005.

Imani, R., Emami, S., Sharifi, A., Moshtaq, P., Baheiraei, N. and Fakhrzadeh, H. (2011) 'Evaluation of novel 'biopaper' for cell and organ printing application: an in vitro study', *J Diabetes Metab Disorders*, **10**, 1–13.

Jhon, M. and Andrade, J. (1983) 'Water and hydrogels', *J Biomed Mater Res*, **7**(6), 509–522. doi:10.1002/jbm.820070604.

Jin, R. and Dijkstra, P. (2010) 'Hydrogels for tissue engineering applications' in R. Ottenbrite, Park, K., Okano, T. and Peppas, N., *Biomedical Applications of Hydrogels Handbook*, Springer Science, New York.

Johnson F., Craig, D. and Mercer, A. (1997) 'Characterization of the block structure and molecular weight of sodium alginates', *J Pharm Pharmacol*, **49**(7), 639–643. doi:10.1111/j.2042–7158.1997.tb06085.x.

Ketlow, J. and Mikos, A. (2007) 'From material to tissue: biomaterial development, scaffold fabrication, and tissue engineering', *AIChE J*, **54**(12), 3048–3067. doi:10.1002/aic.11610.

Kim, U., Park, J., Li, C., Jin, H., Valluzzi, R. and Kaplan, D. (2004) 'Structure and properties of silk hydrogels', *Biomacromolecules*, **5**(3), 786–792. doi:10.1021/bm0345460.

Klouda, L. and Mikos, A. (2008) 'Thermoresponsive hydrogels in biomedical applications – a review', *Eur J Pharm Biopharm*, **68**(1), 34–45. doi.org.prox.lib.ncsu.edu/10.1016/j.ejpb.2007.02.025.

Kong, H., Smith, M. and Mooney, D. (2003) 'Designing alginate hydrogels to maintain viability of immobilized cells', *Biomaterials*, **24**(22), 4023–4029. doi.org/10.1016/S0142-9612(03)00295-3.

Langer, R. and Vacanti, J. (1993) 'Tissue engineering', *Science*, **260**(5110), 920–926.

Lee, C., Singla, A. and Lee, Y. (2001) 'Biomedical applications of collagen', *Int J Pharm*, **221**(1–2), 1–22. doi.org/10.1016/S0378-5173(01)00691-3.

Lee, J., Yoon, J., Lee, D. and Park, T. (2004) 'Photo-crosslinkable, thermo-sensitive and biodegradable Pluronic hydrogels for sustained release of protein', *J Biomater Sci Polymer Edn*, **15**(12), 1571–1583. doi:10.1163/1568562042459751.

Lee, J., Lan, P., Kim, B., Lim, G. and Cho, D. (2007) '3D scaffold fabrication with PPF/DEF using micro-stereolithography', *Microelectron Eng*, **84**(5–8), 1702–1705. doi.org/10.1016/j.mee.2007.01.267.

Lin, Y., Tan, F., Marra, K., Jan, S. and Liu, D. (2009) 'Synthesis and characterization of collagen/hyaluronan/chitosan composite sponges for potential biomedical applications', *Acta Biomater*, **5**(7), 2591–2600. doi:10.1016/j.actbio.2009.03.038.

Lippens, E., Swennen, I., Gironès, J., Declercq, H., Vertenten, G., Vlaminck, L., Gasthuys, F., Schacht, E. and Cornelissen, R. (2013) 'Cell survival and proliferation after encapsulation in a chemically modified Pluronic® F127 hydrogel', *J Biomater Appl*, **27**(7), 828–839, doi:10.1177/0885328211427774.

Liu J., Lin, S. and Li, L. (2005) 'Release of theophylline from polymer blend hydrogels', *Int J Pharm*, **298**(1), 117–125. doi.org/10.1016/j.ijpharm.2005.04.006.

Liu, S., Kau, Y., Chou, C., Chen J., Wu, R. and Yeh, W. (2010) 'Electrospun PLGA/collagen nanofibrous membrane as early-stage wound dressing', *J Membrane Sci*, **355**(1–2), 53–59. doi.org.prox.lib.ncsu.edu/10.1016/j.memsci.2010.03.012.

Ma, P. (2004) 'Scaffolds for tissue fabrication', *Mater Today*, **7**(5) 30–40. doi.org/10.1016/S1369-7021(04)00233-0.

Maher, P., Keatch, R., Donnelly, K., Mackay, R. and Paxton, J. (2009) 'Construction of 3D biological matrices using rapid prototyping technology', *Rapid Prototyping J*, **15**(3), 204–210. doi.org/10.1108/13552540910960307.

Mensitieri, M., Ambrosio, L., Nicolais, L., Bellini, D. and O' Regan, M. (1996) 'Viscoelastic properties modulation of a novel auto-cross-linked hyaluronic acid polymer', *J Mater Sci Mater Med*, **7**, 695–698.

Mironov, V., Reis, N. and Derby B. (2006) 'Review: bioprinting: a beginning', *Tissue Eng*, **12**(4), 631–634. doi:10.1089/ten.2006.12.631.

Nair, L. and Laurencin, C. (2007) 'Biodegradable polymers as biomaterials', *Prog Polym Sci*, **32**(8–9), 762–798. doi.org/10.1016/j.progpolymsci.2007.05.017.

Nakamura, M., Iwanaga, S., Henmi, C., Arai, K. and Nishiyama, Y. (2010) 'Biomatrices and biomaterials for future developments of bioprinting and biofabrication', *Biofabrication*, **2**(1), 035001 (6 pp), doi:10.1088/1758-5082/2/1/014110.

Nishiyama, Y., Nakamura, M., Henmi, C., Yamaguchi, K., Mochizuki, S., Nakaqawa, H. and Takiura, K. (2009) 'Development of a three-dimensional bioprinter: construction of cell supporting structures using hydrogel and state-of-the-art inkjet technology', *J Biomech Eng*, **131**(3), 035001 (6 pp) 035001-1-035001-6. doi.org/10.1115/1.3002759.

O'Brien, F. (2011) 'Biomaterials & scaffolds for tissue engineering', *Mater Today*, **14**(3), 88–95. doi.org/10.1016/S1369-7021(11)70058-X.

Park, J. and Lakes R. (1992) *Biomaterials: An Introduction*, 2nd edn. Plenum Press, New York.

Park, S., Park, J., Kim, H., Song, M. and Suh, H. (2002) 'Characterization of porous collagen/hyaluronic acid scaffold modified by 1-ethyl-3-(3-dimethylaminopropyl) carbodiimde crosslinking', *Biomaterials*, **23**(4), 1205–1212. doi.org.prox.lib.ncsu.edu/10.1016/S0142-9612(01)00235-6.

Park, H., Temenoff, J., Holland, T., Tabata, Y. and Mikos, A. (2005) 'Delivery of TGF-beta1 and chondrocytes via injectable biodegradable hydrogels for cartilage tissue engineering applications', *Biomaterials*, **26**(34), 7095–7103. doi.org.prox.lib.ncsu.edu/10.1016/j.biomaterials.2005.05.083.

Patel, A. and Mequanint K. (2011) 'Hydrogel biomaterials', in Fazel-Rezai, R. *Biomedical Engineering – Frontiers and Challenges*, ISBN: 978-953-307-309-5, Available from: http://www.intechopen.com/books/biomedical-engineering-frontiers-and-challenges/hydrogel-biomaterials.

Peppas, N. and Khare, A. (1993) 'Preparation, structure and diffusional behavior of hydrogels in controlled release', *Adv Drug Deliver Rev*, **11**(1–2), 1–35. doi.org/10.1016/0169-409X(93)90025-Y.

Prestwich, G., Marecak, D., Marecek, J., Vercruysse, K. and Ziebell, M. (1998) 'Controlled chemical modification of hyaluronic acid: synthesis, applications, and biodegradation of hydrazide derivatives', *J Control Release*, **53**(1–3), 92–103. doi.org.prox.lib.ncsu.edu/10.1016/S0168-3659(97)00242-3.

Ramakrishna, S., Mayer, J., Wintermantel, E. and Leong, E. (2001) 'Biomedical applications of polymer-composite materials: a review', *Compos Sci Technol*, **61**(9), 1189–1224. doi.org/10.1016/S0266-3538(00)00241-4.

Sakai, S. and Kawakami, K. (2007) 'Synthesis and characterization of both ionically and enzymatically cross-linkable alginate', *Acta Biomater*, **3**(4), 495–501. doi.org/10.1016/j.actbio.2006.12.002.

Sang, L., Wang, X., Chen, Z., Lu, J., Gu, Z. and Li, X. (2010) 'Assembly of collagen fibrillar networks in the presence of alginate', *Carbohyd Polym*, **82**(40), 1264–1270. doi.org.prox.lib.ncsu.edu/10.1016/j.carbpol.2010.07.005.

Sanjana, N. and Fuller, S. (2004) 'A fast flexible ink-jet printing method for patterning dissociated neurons in culture', *J Neuro Sci Meth*, **136**(2), 151–163. doi.org/10.1016/j.jneumeth.2004.01.011.

Schantz, J., Brandwood, A., Hutmacher, D., Khor, H. and Bittner, K. (2005) 'Osteogenic differentiation of mesenchymal progenitor cells in computer designed fibrin-polymer-ceramic scaffolds manufactured by fused deposition modeling', *J Mater Sci-Mater M*, **16**, 807–819.

Schmedlen, R., Masters, K. and West, J. (2002) 'Photocrosslinkable polyvinyl alcohol hydrogels that can be modified with cell adhesion peptides for use in tissue engineering', *Biomaterials*, **23**(22), 4325–4332. doi.org.prox.lib.ncsu.edu/10.1016/S0142-9612(02)00177-1.

Segura, T., Anderson, B., Chung, P., Webber, R., Shull, K. and Shea, L. (2005) 'Crosslinked hyaluronic acid hydrogels: a strategy to functionalize and pattern', *Biomaterials*, **26**(4), 359–371. doi.org/10.1016/j.biomaterials.2004.02.067.

Shim, J., Lee, J., Kim, J. and Cho, D. (2012) 'Bioprinting of a mechanically enhanced three-dimensional duel cell-laden construct for osteochondral tissue engineering using a multi-head tissue/organ building system', *J Micromech Microeng*, **22**(8), 1–11. doi:10.1088/0960-1317/22/8/085014.

Slaughter, B., Khurshid, S., Fisher, O., Khademhosseine, A. and Peppas, N. (2009) 'Hydrogels in regenerative medicine', *Adv Mater*, **21**(32–33), 3307–3329. doi:10.1002/adma.200802106.

Smidsrød, O. and Skjakbraek, G. (1990) 'Alginate as immobilization matrix for cells', *Trends Biotechnol*, **8**, 71–78. doi.org.prox.lib.ncsu.edu/10.1016/0167-7799(90)90139-O.

Smith, C., Stone, A., Parkhill, R., Stewart, R., Simpkins, M., Kachurin, A., Warren, W. and William, S. (2004) 'Three-dimensional bioassembly tool for generating viable tissue-engineered constructs', *Tissue Eng*, **10**(9–10), 1566–1576. doi:10.1089/ten.2004.10.1566.

Solchaga, L., Gao, J., Dennis, J., Awadallah, A., Lundberg, M., Caplan, A. and Goldberg, M. (2002) 'Treatment of osteochondral defects with autologous bone marrow in a hyaluronan-based delivery vehicle', *Tissue Eng*, **8**(2), 333–347. doi:10.1089/107632702753725085.

Suh, J. and Matthew, H. (2000) 'Application of chitosan-based polysaccharide biomaterials in cartilage tissue engineering: a review', *Biomaterials*, **21**(24), 2589–2598. doi.org/10.1016/S0142-9612(00)00126-5.

Tabata, Y. (2003) 'Tissue regeneration based on growth factor release', *Tissue Eng*, **9**(4), 5–15. doi:10.1089/10763270360696941.

Temenoff, J., Kasper, F. and Mikos, A. (2007) 'Fumarate-based macromers as scaffolds for tissue engineering applications', in Ashammakhi, N., Reis, R. and Chiellini, E. *Topics in Tissue Engineering*, Vol. 3.

Tsang, V., Chen, A., Cho, L., Jadin, K., Sah, R., DeLong, S., West, J. and Bhatia, S. (2007) 'Fabrication of 3D hepatic tissues by additive photopatterning of cellular hydrogels', *FASEB J*, **21**(3), 790–801. doi:10.1096/fj.06-7117com.

Ulery, B., Nair, L. and Laurencin, C. (2011) 'Biomedical applications of biodegradable polymers', *J Polm Sci*, **49**(12), 832–864. doi: 10.1002/polb.22259.

Wallace, D and Rosenblatt, J. (2003) 'Collagen gel systems for sustained delivery and tissue engineering', *Adv Drug Deliver Rev*, **55**(12), 1631–1649. doi.org/10.1016/j.addr.2003.08.004.

Xu, T., Gregory, C., Molnar, P., Cui, X., Jalota, S., Bhaduri, S. and Boland, T. (2006) 'Viability and electrophysiology of neural cell structures generated by the inkjet printing method', *Biomaterials*, **27**, 3580–3588. doi:10.1016/j.biomaterials.2006.01.048.

Xu, T., Binder, K., Albanna, M., Dice, D., Zhao, W., Yoo, J. and Atala, A. (2013) 'Hybrid printing of mechanically and biologically improved constructs for cartilage tissue engineering applications', *Biofabrication*, **5,** 015001 (10 pp), doi:10.1088/1758-5082/5/1/015001.

Yannas, V. (2004) 'Classes of materials used in medicine: natural materials', in Ratner, B., Hoffman, A., Schoen, F. and Lemons, J., *Biomaterials Science – An Introduction to Materials in Medicine*, Elsevier Academic Press, San Diego, Calif, USA, 127–136.

Zhang, T., Yan, Y., Wang, X., Xiong, Z., Lin, F., Wu, R. and Zhang, R. (2007) 'Three-dimensional gelatin and gelatin/hyaluronan hydrogel structures for traumatic brain injury', *J Bioact Compat Pol*, **22**(1), 19–28. doi:10.1177/0883911506074025.

7
Bioprinting for constructing microvascular systems for organs

T. XU, J. I. RODRIGUEZ-DEVORA,
D. REYNA-SORIANO and B. MOHAMMOD,
University of Texas at El Paso, USA, L. ZHU and K. WANG,
Sun Yat-sen University, China and Y. YUAN, Medprin
Regenerative Technologies, China

DOI: 10.1533/9780857097217.201

Abstract: Regenerative medicine/tissue engineering (TE) has emerged as a potential solution to overcome the current shortage of organs/tissue for transplantation. However, generating microvascularized tissue still challenges TE for the creation of suitable organs. In addressing this, numerous approaches over the last decade culminated in one main strategy emerging: bioprinting. The combination of medical imaging, bio-computer aided design (bio-CAD) systems, and modern fabrication approaches promises to overcome the current limitations in constructing organs with microvascular systems. Specifically, appropriate bioprinting strategies have been proposed to fabricate biomimetic organs/tissue microvasculature. Bioreactor systems have been incorporated into bioprinting to enhance maturation of the bioprinted organ/tissue. Despite important advances in microvascular bioprinting, further work is required to enhance the interface of medical imaging and bio-CAD, optimize bioprinting systems, and accelerate the maturation process of bioprinted implants.

Key words: bioprinting, microvasculature, biofabrication, tissue engineering, regenerative medicine.

7.1 Introduction

Human biological systems depend highly on arterial networks and functionalized vasculature composed of endothelial cells, connective tissues, and epithelial cells, without which the transportation of nutrients, oxygen, and waste would not be possible. To date, vascular grafts for diseased large vessels (millimeter range) are well established; however, for microvasculature (small vessels in the micrometer range) investigations remain in the early stages and clinical translation is not foreseen based on the current knowledge and information available. Tissue engineers Langer and Vacanti[1] pioneered solutions to this bioengineering challenge by opening the doors for other scientists to pursue this field. Kaplan and his group[2] recently published a review identifying the main strategies to follow, classified as: material functionalization, scaffold design, microfabrication, bioreactor development, endothelial cell seeding, modular assembly, and *in vivo* testing systems. Their strategies included theoretical models and measurement methods which advanced microvascularization towards a quantitative assessment of the diffusion and consumption of biological substances.

Tissue engineering is currently limited by its ability to adequately vascularize tissue *in vitro* or *in vivo*. Lack of nutrient perfusion and mass transport restricts construct development to less than clinically relevant dimensions and also limits *in vivo* integration. The development of tissue-engineered organs has a strong dependency on host vasculature for oxygen, nutrients, and waste removal.[3,4] Although some avascular tissues, such as heart valve leaflets, can receive nutrients by diffusion, the majority of tissues and organs depend on a complex microvascular system. Without an intrinsic capillary network, the maximum viable thickness of engineered tissues is approximately 150–200 µm due to oxygen diffusion limitations;[5,6] this highlights the need for incorporating microvascularization in engineered tissue constructs prior to implantation. Microvascularization *in vitro* can maintain cell viability during tissue growth, induce structural organization, and promote angiogenesis after implantation. While many strategies have been investigated to resolve this challenge, microvasculature printing emerges as one possible solution to building precise human microvasculature with the help of suitable cell- and material-additive bioinks. To pursue successful microvasculature bioprinting, three sequential steps need to be considered: pre-printing preparation (biomimetic modeling), printing (fabrication strategy), and post-printing (implantation and maturation).

7.2 Biomimetic model for microvasculature printing

The creation of a biomimetic model is required as the first step towards microvasculature printing. On the basis of anatomical and physiological principles of natural blood vessels, biomimetic models of simple and complex vascular networks for engineered tissue can be developed. Many researchers have proposed angiogenesis as a means of establishing a network of blood vessels within a tissue-engineered organ.[7,8] For a solid organ such as the liver, even if an angiogenesis network could be established in a tissue *in vitro*, by the time the network inosculates (uniting blood vessels, nerve fibers, or ducts) with the host vascular system, the majority of the parenchymal cells of the liver would be impaired or become necrotic. Based on the principle of minimum work required to achieve flow through a network, in the early stages of biomedical

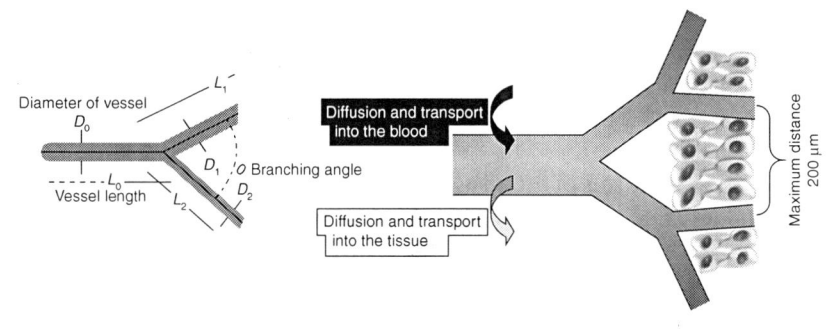

7.1 Geometry of vascular network.[7,8]

engineering (as shown in Fig. 7.1), Murray derived a relationship between parent and daughter diameters of branching networks in the body that has been shown to be valid for blood vessels as well as airway structures.[9,10] Vascular networks may be optimally designed with diameters that conform to this relationship so that shear stress can be maintained within a narrow range.

In general, the sequence of biomimetic modeling is structured in the following main stages: (i) development of a computer modeling representation of a three dimensional (3D) organ (bio-blueprint model), which provides detailed informative data of a specific organ of interest; and (ii) identification of the biological and mechanical arrangements of the organ of interest to be printed.

7.3 The bio-blueprint for microvasculature printing

The bio-blueprint depicts the graphic features of an organ structure along with the required biological data for organ anatomy, tissue heterogeneity, and vascular networking. It can be used to introduce and facilitate organ manufacturing design. To acquire critical geometry and cellular structure of living organs, bio-blueprint is associated with cross sections of each layer of tissue, providing detailed informative data on individual cells. The bio-blueprint serves the following basic functions:

- describes anatomy, geometry, and internal architecture of an organ of interest, including the tissue heterogeneity, the individual tissue geometry, and the boundary distinction within the organ;
- defines a vascular network and 3D topology of an organ of interest; and
- provides a necessary database for organ/tissue geometry, heterogeneity, and the associated vascular network that can be used for tool path generation of 3D cell and organ printing.

204 Rapid prototyping of biomaterials

The bio-blueprint is generated using medical imaging data in order to replicate organ/tissue anatomy, including detailed internal and external morphology, geometry, vascularization, and tissue identification. Bio-blueprint is a software representation containing bio-information, physical and material information, and anatomic and geometric information of the printed tissue or organ of interest. Figure 7.2 shows a blueprint of the heart, depicting only five cross-sections and side-views; however, the real life bio-blueprint consists of a huge number of cross-sections (equal to the number of cell layers). The blueprint will use color to specify the type of cells in each layer to print using the bioprinting method.

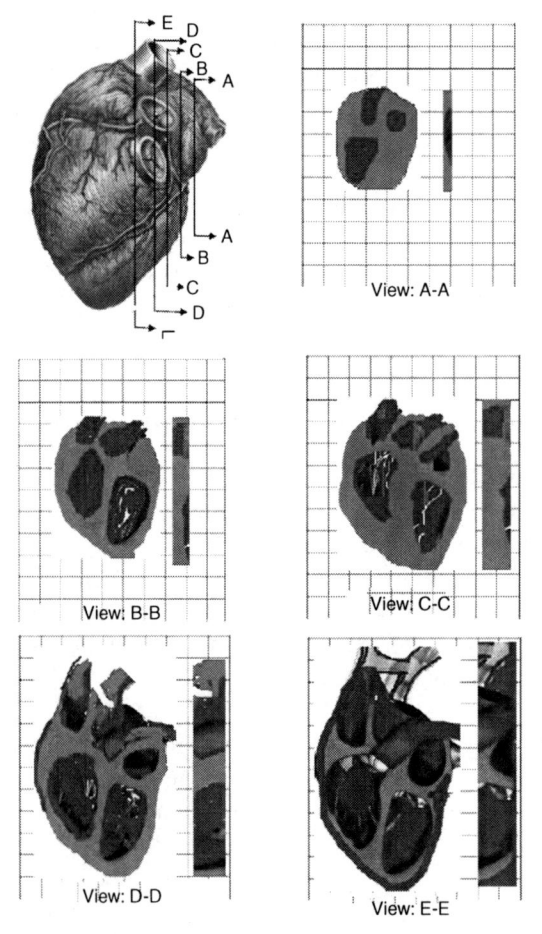

7.2 Blueprint of heart depicting the cross-section and corresponding side views.

7.3.1 Data acquisition

Recent advances in imaging technologies and reverse engineering techniques have helped extensively in biomedical engineering applications ranging from clinical medicine through customized medical implant design to tissue engineering. The primary imaging modalities capable of producing 3D views, such as computed tomography (CT), magnetic resonance imaging (MRI), optical microscopy, and single photon emission computed tomography (SPECT) are used for anatomic data acquisition to construct a bio-computer aided design (bio-CAD) model. The bio-CAD model provides the biological data needed for organ anatomy, tissue heterogeneity, and vascular networking. It can be used to introduce and facilitate the design or manufacturing intent. In general, an image-based bio-CAD modeling process involves the following four major steps: (i) non-invasive image acquisition, (ii) imaging process and 3D reconstruction to form voxel-based volumetric image representation, (iii) construction of a CAD-based model, and (iv) viewing operations and display.

Computed tomography (CT) and micro-CT

CT provides nondestructive 3D visualization and characterization, creating images that map the variation of X-ray attenuation within objects, which relates closely to density. X-Ray CT measures the spatially varying X-ray attenuation[11] to show internal structures. CT or micro-CT scans require exposure of a sample to small quantities of ionizing radiation, the absorption of which is detected and imaged. This results in a series of 2D images displaying a density map of the sample. Stacking these images creates a 3D representation of the scanned area. In particular, the latest development of micro-CT technology has been successfully used to quantify the microstructure–function relationship of tissues and the designed tissue structures, including the characterization of the microarchitecture of tissue scaffolds to help with the design and fabrication of tailored tissue microstructures, and determine tissue morphologies and internal physical activities.

Magnetic resonance imaging (MRI)

MRI provides images for soft tissues as well as for hard tissues and, as such, is vastly superior in differentiating soft tissue types and recognizing border regions of tissues of similar density. MRI shows excellent contrast between varieties of soft tissues; however, the variety of surfaces presents a challenge to 3D surface construction and to the techniques required for selective extraction and display.

Optical microscopy

Optical microscopy has limited application in 3D bio-tissue modeling due to the intensive data manipulation required. For example, to examine a sample with high resolution using optical microscopy it must be physically sectioned onto very thin slides. The division into these slides is labor intensive and the resulting images of the target organ would be a series of thousands of 2D images that must be digitally stacked into 3D columns. Due to excessive computation and a memory-intensive process it is a significant challenge to train computers to identify individual cells by their visual characteristics, even with the aid of complex staining. However, to date, differentiating tissue down to the level of the individual cell may still be only possible using optical microscopy.

Single photon emission computed tomography (SPECT)

SPECT measures the emission of gamma rays. The sources of these rays are radioisotopes distributed within the body. SPECT can show the presence of blood in structures with a much lower dose than that required by CT.

7.3.2 3D microvasculature system

A feature-primitive reconstruction method for vascular networks is used to generate a 3D biological microvascular system for organ growth. In this primitive feature modeling approach, the characteristic parameters of the basic vascular primitives (e.g., conjugate or segment) are determined from patient-specific CT/MRI images, and further use of Boolean operation algebra forms a high-level vessel assembly. The vascular feature primary are represented as surfaces which are mathematically described using special polynomial functions, such as non-uniform rational B-splines (NURBS); the parameters in the NURBS equations can be determined by measuring the spatial positions of the vascular CT/MRI images at different projections.

7.3.3 Microvasculature modeling

In using the bio-blueprint approach for organs, the model must be capable of producing 3D views of anatomy, differentiating heterogeneous tissue types, displaying the vascular structure, and generating computational tissue models for other downstream uses, such as analysis and simulation. Modeling simulation is used to manipulate the images generated and create a mathematical or CAD-based medical model.[12] There are three types of functionalities which have to be integrated into the modeling simulation algorithm: (i) the accurate representation of biological structures by the computer model; (ii) the possibility of simulating all surgical actions; and (iii) the capability of obtaining certain parametric information such as tissue volume and anatomic distances.

The bio-blueprint model is usually constructed through either segmentation or volumetric representation. 2D segmentation (i.e., slices) obtained from image data undergoes independent processing, including inner and outer contours of the living tissue, using a conjugate gradient (CG) algorithm.[13] The contours are stacked in three dimensions to create a solid model, usually through skinning operations. 3D segments of the image data set are able to identify within the voxels binding the organ and extract a 'tiled surface' (a discrete representation made of connected polygons) from them. The microvascular modeling should result in efficient and realistic estimation of tissue behaviors and interactive forces.

Modern CAD systems use boundary representation, in which a solid object is defined by its surfaces. These surfaces are mathematically described using NURBS[14] functions. NURBS facilitate operations such as intersection and closure of the boundary surface to construct the computer model using fewer numbers of digitized points, which would significantly decrease the size of the files. In the last few years some commercial programs, for example, SurgiCAD by Intergraph ISS, USA; Med-Link, by Dynamic Computer Resources, USA; and Mimics and MedCAD, by Materialise, Belgium, are being used to convert image data to models. However, none of these programs has been widely applied in biomechanical engineering due to intricately specific and expensive clinical applications, and the inability to generate a sufficiently sophisticated model.

7.3.4 Mechanical and biological arrangements

Design of 3D tissue scaffolds for microvasculature printing should consider the complex hierarchy and structural heterogeneity of the host tissue and/or scaffold environment. Other than important factors of porosity, pore size, interconnectivity, and transport of nutrients that would enable the ingrowth of new cells and cell-tissue formation, the designed scaffolds should also be able to have mechanical properties compatible with the host environment as well as the required mechanical strength after implantation.

The 3D tissue scaffolds must have certain characteristics of their own in order to function as a true tissue/organ which satisfies the biological, mechanical, and geometrical constraints. Such characteristics include: (i) biological requirements – the scaffold must facilitate cell attachment and distribution, growth of regenerative tissue, and facilitate the transport of nutrients and signals. This requirement can be achieved by controlling the porosity of the structure, by providing appropriate interconnectivity inside the structure, and by selecting appropriate biocompatible materials; (ii) mechanical requirement – the scaffold must provide structural support at the replacement site while the tissue regenerates to occupy the space defined by the scaffold structure.

7.4 Microvasculature printing strategies

A number of methods of biomimetic microvascular fabrication have been developed that are capable of emulating their hierarchical and branching channel designs, especially in terms of 3D form. These techniques include direct-ink writing,[15] soft lithography,[14] direct laser ablation,[16] stereolithography,[17] two-photon polymerization,[18] jet-based printing,[19] and electrostatic discharge.[20] In each case there are two critical points to producing a viable construct: (i) the ability to measure and model the oxygen requirement throughout the engineered tissue and (ii) the point of functional anastomosis to connect the graft-originated prevascular structure with host blood vessels.

7.4.1 Generating 2D microvascularized tissue constructs

To generate microvascularized tissue constructs, 2D vascularized patterns are fabricated on various biomaterials (silicon, glass, or polymer) normally starting from a single channel that branches out multiple times into thinner and thinner channels. In the early days of microvascular research, Borenstein and Vacanti adopted photolithography techniques to generate 2D vascular and capillary networks on silicon and Pyrex to culture endothelial cells and hepatocytes with desired proliferative and functional capabilities.[15] Later, soft lithographic techniques were applied to mold polydimethylsiloxane (PDMS) onto silicon wafer with the intention of bifurcating enclosed vascular networks of capillary diameters.[21] Even though endothelial cells successfully seeded, PDMS is not biodegradable and has limited biocompatibility; it is therefore unsuitable as a biomaterial for implantation. To address this limitation, King et al. produced highly branched microfluidic scaffolds from biodegradable poly(L-lactic-co-glycolic acid) (PLGA)[22] on PDMS molds. Although this quick, versatile approach is inexpensive, the potential limitations of a PLGA vascular network are brittleness and mechanical weakness with cytotoxic by-products, as well as inflammatory immune responses which are generated during implantation.[23] To overcome these limitations, Wang et al. easily synthesized a novel biocompatible, tough, and biodegradable elastomer known as poly(glycerol sebacate) (PGS).[24] Moreover, Fidkowski et al. applied the novel PGS to fabricate capillary networks.[25] As shown in Fig. 7.3, PGS has been imprinted with a capillary network pattern using microfabricated silicon wafers produced by standard

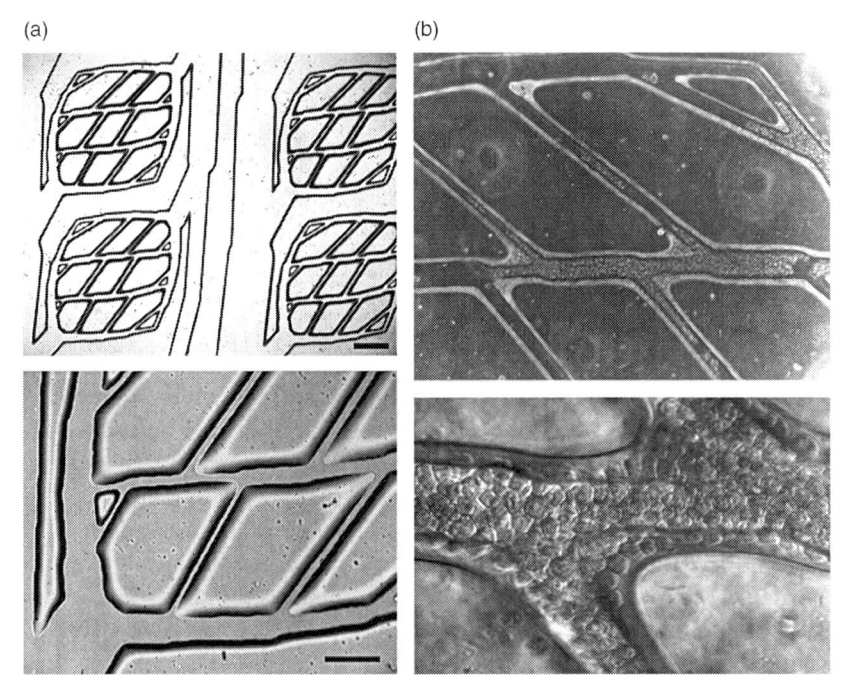

7.3 (a) Photomicrographs of patterned PGS capillary network 500 μm (top), 200 μm (bottom). (b) Photomicrographs of endothelialized capillary network, ×10 (top) ×40 (bottom).[26]

lithography as a mold. Two PGS layers were attached to yield flow rates of approximately 150 and 1000 cm/s within the largest and smallest channels, respectively. Human umbilical vein endothelial cells (HUVECs) cells were seeded to PGS network devices without the use of adhesion proteins and maintained proliferation and functionality. Although these capillaries can be incorporated into functional devices, this method is limited to generating only monolayer capillary networks. In addition to PLGA, PGS, and polycaprolactone (PCL),[26,27] silk fibroin protein (from the Bombyx mori silkworm)[28] has also been used for vascularized microfluidic scaffolds.

7.4.2 Generating 3D vascularized tissue constructs

Fabricating 3D vascularized microfluidic scaffolds is of vital importance for the success of tissue engineering applications and for better physiological replication than 2D constructs; however, construction in 3D remains a challenge. So far, the most commonly used approach is to stack and assemble 2D vascularized films into large 3D devices suitable for transplantation. In the early stages of microvascular fabrications, some scientists developed 3D scaffold stacking or bonding together with 2D vascularized polymer films using PLGA,[22] PGS,[27] and PDMS;[25] however, building 3D microvascular networks by stacking 2D layers is a cumbersome process requiring multiple fabrication and masking steps that are difficult to scale up. To overcome this challenge 3D microvascular network systems were introduced through direct printing processes. Therriault et al. demonstrated the fabrication of 3D microvascular networks through direct-write assembly of a paraffin-based organic ink[30] via robotic deposition. In addition, Lim et al. demonstrated a faster and more flexible alternative method to fabricate multiple-level microfluidic channels using maskless laser direct micromachining.[31] A multi-width and multi-depth microchannel was fabricated to generate biomimetic vasculatures whose channel diameters changed according to Murray's law across multiple branching generations, mimicking physiological flow patterns with minimum flow resistances and more gradual changes in the flow velocities across different generations of branching compared to channels of uniform depth.

In addition to polymers, hydrogels have recently become more widely used as 3D scaffolds due to their superior properties (biodegradability, biocompatibility, and resemblance to the natural extra cellualr matrix (ECM)). Commonly used hydrogels include natural hydrogels (i.e., collagen, hyaluronic acid (HA), and alginate), synthetic hydrogels (i.e., polyethylene glycol-diacrylate (PEGDA), and polyvinyl alcohol (PVA)),[16] and hybrid natural-synthetic composites.[32] Collagen, in particular, has shown promise with platelet-derived growth factor (PDGF) and has been molded into scaffolds that enhance capillary formation in a dermal wound healing model.[33] In a recent study, collagen–HA 3D scaffolds were produced by mixing varying concentrations of HA, an inhibitor of angiogenesis, into collagen gels with the aid of micropatterning techniques.[34] By implanting endothelial cell spheroids into these gels, microchanneling techniques can be used for enhancing migration, and offer dominant tools to organize microvascularization for tissue engineering.

Fibrin produced from the blood of a patient and used as an autologous scaffold for tissue engineering can potentially serve as a biopaper scaffolding material.[32] Human capillaries and the entire inner lining of a cardiovascular system are composed of endothelial cells. Cui and Boland[35] printed human microvascular endothelial cells (HMVEC) on fibrin polymerized from fibrinogen and thrombin solutions as bioink for microvasculature construction. An inkjet printer was used to seed HMVECs together with fibrin to form the microvasculature (fibrin channel). As shown in Fig. 7.4, the printed cells were aligned and proliferated to form a confluent lining along with the fibrin scaffold to show the functionality of the printed human microvasculature. The fabricated microvasculature also showed increased integrity after being cultured for 3 weeks.

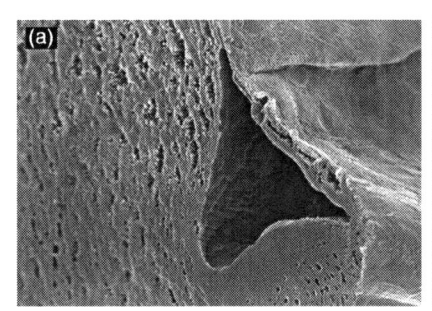

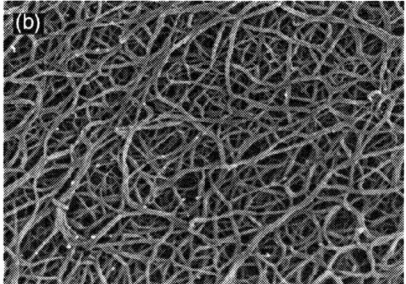

7.4 SEM images of printed fibrin fiber.[29] (a) Channel structure at the cross-section. (b) Nano-sized fibers on the printed fibrin scaffold surface.

Stroock et al. introduced a functional 3D microfluidic structure fabricated from calcium alginate hydrogel[36] by arraying the channels in appropriate dimensions. These microfluidic channels enabled an efficient exchange of solutes within the interior of the hydrogel scaffolds and created a quantitative control of the soluble environment experienced by the cells in their 3D environment. Alginate scaffolds loaded with vascular endothelial growth factor (VEGF) have also been shown to promote microvascularization.[37]

Ling et al. built highly porous mammalian cell-laden microfluidic channels from agarose hydrogels by directly encapsulating cells within the microfluidic channels.[38] Using standard soft lithographic techniques, molten agarose was molded against an SU-8 patterned silicon wafer to build microfluidic channels of different dimensions. Cells embedded within the microfluidic molds are well distributed and media was pumped through the channels allowing for the exchange of nutrients and waste products. While most cells were found to be viable upon initial device fabrication, only those cells near the microfluidic channels remained viable after 3 days, demonstrating the importance of a perfused network of microchannels for delivering nutrients and oxygen to maintain cell viability in large hydrogels. Skardal et al. also developed 3D microvasculature grafts using agarose.[39] Two four-armed polyethylene glycol (PEG) derivatives with different PEG chain lengths, TetraPEG8 and TetraPEG13, were synthesized to form gel. After gel formation, human hepatoma cells (HepG2 C3A) and human intestinal epithelial cells (Int 407) were encapsulated with gel and printed layer-by-layer on agarose with the aid of the Fab@Home printing system. These microvascular structures maintained viability in culture for up to 4 weeks which offered hope that functional blood vessel structures and branched vascular networks may soon be accessible using a bioprinting strategy.

Future 3D scaffold designs may be improved by using computer-aided tissue engineering to model, design, and fabricate scaffolds with proper microvascular channel elements of different geometry.[26] Computer-generated engineering scaffolds can match native features of vascularization with accurate positioning and morphology.[26] Overall, these scaffolding techniques offer fine control over microvascularization potentials and a multitude of options in scaffold design and engineering to create functional tissue outcomes. One of the mechanisms to be tested is omnidirectional printing. As shown in Fig. 7.5, omnidirectional printing involves 3D microvascular branched networks which are fabricated using a three-axis robotic deposition stage (ABL9000, Aerotech® Inc.) to print a 3D pattern into a physical gel reservoir. The 3D patterns are designed using commercially available CAD software (AutoCAD 2010, Autodesk) and translated into G-code instructions for the deposition stage using a custom Visual Basic program.[40]

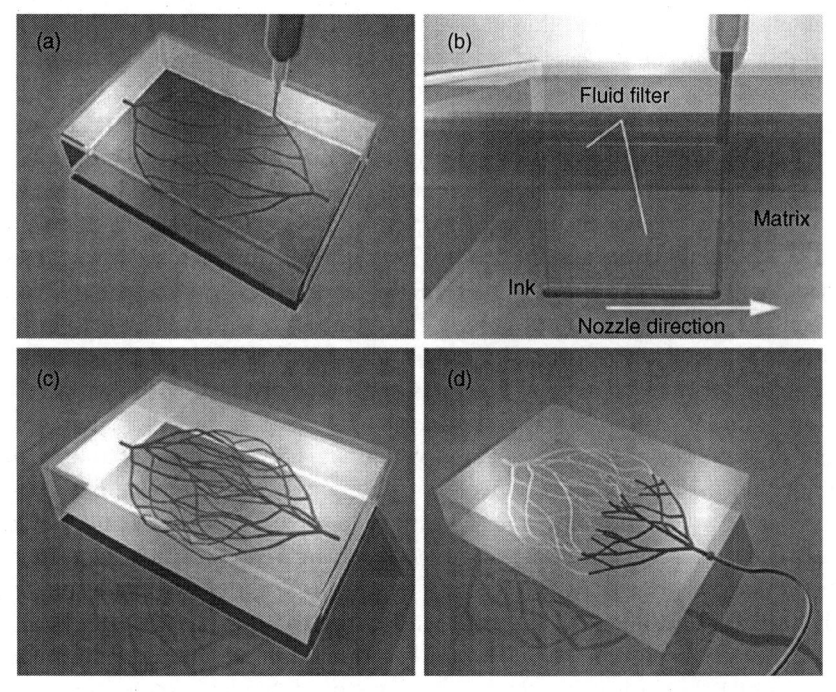

7.5 Schematics of omnidirectional printing of 3D microvascular networks within a hydrogel reservoir. (a) Deposition of a fugitive ink into a physical gel reservoir allows hierarchical, branching networks to be patterned. (b) Voids induced by nozzle translation are filled with liquid that migrates from the fluid capping layer. (c) Subsequent photopolymerization of the reservoir yields a chemically cross-linked, hydrogel matrix. (d) The ink is liquified and removed under a modest vacuum to expose the microvascular channels.[40]

7.4.3 Cell sources for engineered vasculature

Appropriate cell sources are also a key factor in the fabrication of microvasculture. One technique involves isolating and growing endothelial cells on polymer scaffolds *in vitro*, followed by *in vivo* implantation. This method has been tested in an ovine model, where expanded pulmonary arterial cells were grown on polyglactin/PLGA tubular scaffolds for 1 week before transplantation into pulmonary arteries of lamb.[39] Over a period of 24 weeks, the vascular grafts showed growth and development of the endothelial lining and the production of ECM components, such as collagen and elastic fibers. Tubular structures were also prepared without the use of any exogenous material.[22] Another method of artificial vessel production is to culture vascular cells on polymer scaffolds and to subject the scaffold to a pulsating flow.[29] This biomimetic system involves seeding bovine aortic smooth muscle cells into hollow tubular polyglycolic acid (PGA) scaffolds, followed by injection of bovine aortic endothelial cells into the lumen. Compared with native arteries, the engineered arteries demonstrate similarities in wall thickness and collagen content after 8 weeks of culture in a bioreactor. Endothelial progenitor cells can also function as a cell source for engineering blood vessels. These cells normally circulate in the blood, migrating to sites of trauma or ischemia.[41] When endothelial progenitor cells from the peripheral blood of sheep were grown on decellularized porcine iliac vessels, the grafts exhibited endothelial cell morphology, contractile activity, and patency for 150 days.[15] These studies show that endothelial cells grown on matrices have both the structural and functional capabilities of blood vessels *in vitro*.[42]

Hibino *et al.*[21] showed that bone marrow cells can be used directly in vascular grafts without culturing. Bone marrow cells were also used to engineer vascular patches on decellularized tissue matrices.[22] In addition, Poh *et al.*[23] demonstrated the use of human vascular cells isolated from elderly patients to engineer blood vessels. To overcome the limited replicative capacity of these adult somatic cells, telomerase was overexpressed in the cells and enabled the successful culture of autologous blood vessels. It is also important to study the behavior of embryonic stem cells (ESC)-derived endothelial cells in engineered vessels, as human derived embryonic stem cells (hESCs) could provide an unlimited source of proliferating 'young' human endothelial cells. Endothelial cells have the ability to organize into vascular networks *in vitro*. Isolated hESC-derived endothelial cells can form tube-like structures *in vitro* and can form blood-carrying microvessels *in vivo* (Fig. 7.6a and 7.6b).[12] Seeding hESCs on polymeric scaffolds shows the added potential of these cells to produce vasculogenesis. During differentiation of the cells into epithelial or mesenchymal cells, and formation of specific tissue structures (affected by the addition of specific growth factors), some of the cells differentiate into endothelial cells, which organize into vessel-

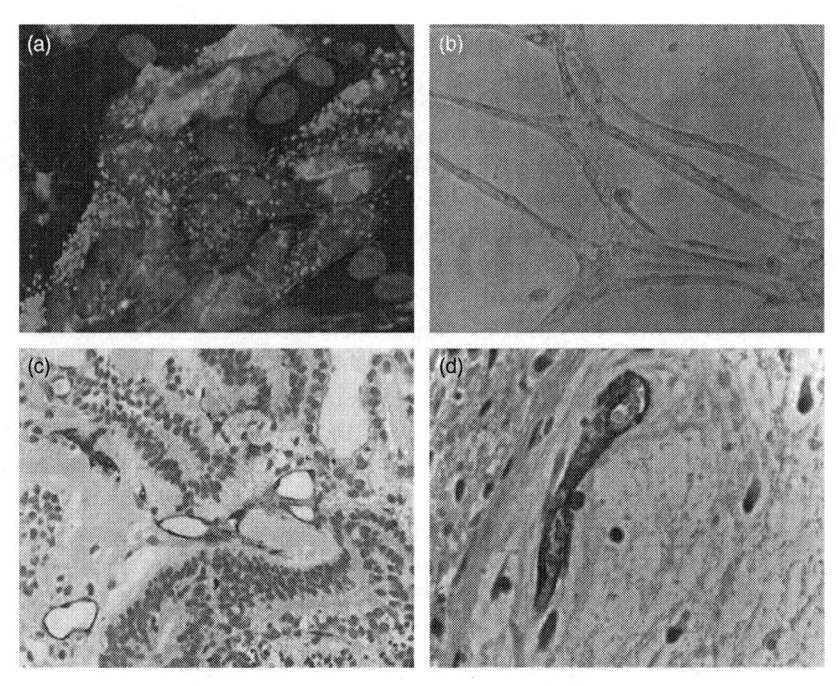

7.6 The vasculogenic potential of hESC-derived endothelial cells. (a) Isolated CD31+ cells grown in culture and immunofluorescently stained using anti-CD31 antibodies and anti-von Willebrand factor antibodies to detect these endothelial markers and DAPI for nuclear staining. (b) CD31+ cells forming tube-like structures on Matrigel™. (c) The vascularization of hESC-derived three-dimensional tissue constructs. Differentiating hESC cells grown on biodegradable polymer scaffolds form endothelial vascular networks within the engineered tissue-like structures. (d) Implantation of hESC-derived engineered tissue constructs. Human endothelial cells (from the implanted tissue) form into blood-carrying microvessels in mice (as shown by staining endothelial cells using anti-human specific CD31 antibody). (*Source*: Reproduced from Reference 42.)

like structures throughout the tissue (Fig. 7.6c). This observation indicates that the three-dimensional culture of the ESCs can promote the formation of massive three-dimensional vascular networks that interact closely with the surrounding tissue.[25,43] Upon implantation into immunodeficient mice (to prevent rejection of the human cells implanted), the donor endothelial cells within the implants appeared to form vessel structures and to anastomose with the host vasculature (Fig. 7.6d).[26] Gerecht-nir *et al.*[30] later showed that hESCs cultured on alginate scaffolds form well-vascularized embryoid bodies (EBs). In mouse ESC studies, Flk-1+-derived cells contribute to adult neovascularization when injected subcutaneously into tumor-bearing mice.[31] In addition, endothelial cells selected from growing ESCs (using the puromycin resistance gene under the control of a vascular endothelial promoter tie-1) incorporated into the neovasculture of mice tumors.[16,42]

7.5 Microvasculature post-printing stage

7.5.1 Maturation

Bioprinting is not sufficient for the creation of functional vasculatures and tissue constructs which are immediately suitable for implantation. It takes time for bioprinted tissue to mature into functional tissue constructs and post-processing is probably the most crucial step in organ printing technology. Effective post-processing or accelerated tissue maturation will require the development of new types of bioreactors and more efficient accelerated tissue maturation technologies, as well as methods of non-invasive and non-destructive biomonitoring. One of the latest is the conceptual design of novel irrigation dripping, tripled perfusion bioreactors with temporally removable porous mini-tubes suitable for bioprinting.[44] The design is a triple perfusion bioreactor because it has three circuits: one for maintaining a wet environment around the bioprinted construct, the second for media perfusion through an intra-organ branched vascular tree, and the third and most essential circuit, for temporal perfusion. The last type of perfusion is undertaken using extremely strong, thin, porous, non-biodegradable, removable mini-tubes that serve as temporal supports and artificial microchannels. The main goal of the proposed dripping irrigation circuit system is to maintain the irrigation until the intra-organ branched vascular system has matured sufficiently for the initiation of intravascular perfusion. Moreover, this temporal perfusion system can be used for the delivery of cells, soluble extracellular matrix molecules, and maturogens. The rational design behind such a bioreactor, especially the level of porosity and distance between mini-tubes, must be based on systematic mathematical modeling and computer simulation of interstitial flow. The identification of appropriate materials and coatings of these mini-tubes, and the optimal way to retrieve the inert mini-tubes without severe tissue injury, are other important engineering challenges.

7.5.2 Implantation

Long-term survival and function of engineered tissue substitutes crucially depend on the rapid establishment of an adequate blood supply post-implantation.[45] After implantation, the host body undergoes fibroblast migration, neovascularization, and, finally, vessel maturation. A successful microvascular system depends on the graft material and how it is arranged. The material would have to be non-thrombogenic, must have similar compliance to native vessels to avoid intimal hyperplasia[46] at its arterial end, and must be sufficiently porous to allow nutrient exchange at the capillary level. Endothelialization of vascular prostheses also depends on the biomaterial used and the pore morphology.

To establish connections between prevascularized tissues and host vasculature, geometry connections, flow rates, and ease of integration are critical to the ultimate viability of the implanted tissue. If the printed organ is implanted with no connection to the host blood vessels, oxygen and nutrient supplies become a diffusion-limiting process, resulting in the loss of cell viability at the core of the engineered tissue. By designing a prevascularized tissue with connections to host vasculature, the cells within the construct are immediately supplied with the required nutrients from the moment of implantation. Most of the strategies aim to vascularize tissue constructs based on the stimulation of angiogenesis and blood vessel ingrowth into the implants;[47–50] however, because physiological growth of blood vessels is not normally faster than 5 mm/h, these strategies cannot prevent cell death within the center of large 3D tissue constructs.

Inosculation within the blood vessels of the host microvasculature after implantation is a promising alternative that may significantly accelerate and improve microvascularization of tissue constructs.[51] Matthias *et al.* found

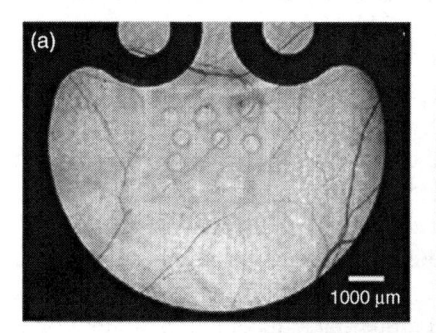

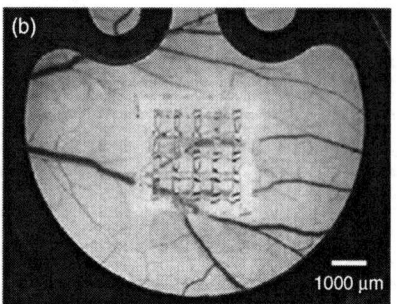

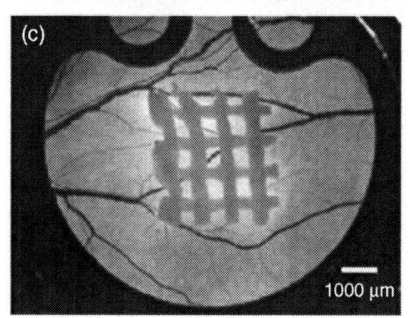

7.7 Three-dimensional scaffolds (3 mm × 3 mm × 1 mm) directly after implantation onto the striated muscle tissue in the dorsal skinfold chamber of balb/c mice: (a) prepared from an isogeneic calvarial bone block; or plotted from (b) PLGA and (c) hydrogel.

that the preformed blood externally inosculating with the host microvasculature can contribute to sprouting angiogenesis in the center of the scaffolds, further increasing the microvessel density of the implants.[52] In the beginning, implantation inosculation is dependent on the formation of capillary sprouts in the prevascularized implants and the host tissue that grow toward each other to form the vascular interconnections. On the basis of this hypothesis, scaffolds were implanted into the flank of donor mice to allow the ingrowth of a granulation tissue with many new blood vessels. Experiments show that noncultivated, prevascularized tissue constructs were successfully transferred directly from the flank of donor mice into the dorsal skinfold chambers of recipient animals.[53] Granulation tissue from the surrounding host tissue had grown into the scaffold pores and contained completely new microvascular networks (Fig. 7.7). Following the implantation of the tissue constructs into dorsal skinfold chambers of recipient mice, fluorescence microscopy showed the inosculation of the implant microvascular networks with blood vessels of the host microvasculature *in vivo*.

7.6 Future trends

Future directions of this research should focus on the creation of new strategies to bioprint complex microvascular networks within 3D tissue constructs *in vitro* before implantation. With rapid development of tissue engineering and rapid prototyping technologies, tremendous opportunities have been created for building microvascularized tissue engineering constructs using bioprinting systems. Microvasculature will continue to be one of the most important and challenging directions in the tissue engineering field; however, some obstacles need to be overcome in order to build microvasculature systems for clinical applications in the future. Although microvasculature improves cell viability and generates capillary-like structures within a tissue construct, there is still a fundamental disconnect between defining the metabolic needs of a tissue through quantitative measurements of oxygen and nutrient diffusion and the potential ease of integration into host vasculature for subsequent *in vivo* implantation. Full and proper endothelialization of the vasculature systems with specific geometries and materials needs to be accomplished for successful results. For microvasculature structures to mimic the capillaries in a scalable manner, precise fabrication of the bio-blueprint model is a must. Another obstacle which needs to be overcome is how to enrich the complexity of the biomimetic model by involving engineering of cellular matrix from multiple cell types and controlling cell–matrix and cell–cell interactions. As part of these challenges, the designing of new suitable bioinks to increase the variety of cells and biomaterials is required in order to enhance the interaction between the bio-blueprint model and the organ/microvasculature bioprinting process.

7.7 Acknowledgements

The authors would like to thank NIH-National Heart, Lung, and Blood Institute (NHLBI) (grant #1SC2HL107235–01), National Science Foundation (grant #CBET0936238), and Department of Education (grant #P116V090013) for funding this project. Dr Lei Zhu and Dr Kun Wang want to thank the Science and Technology Program of Guangdong Province of China (grant #2009B06070045) for financial support.

7.8 References

1. Langer, R. and J.P. Vacanti (1993), Tissue engineering. *Science*, **260**(5110): pp. 920–926.
2. Lovett, M., K. Lee, A. Edwards and D.L. Kaplan (2009), Vascularization strategies for tissue engineering. *Tissue Engineering Part B Reviews*, **15**(3): pp. 353–370.
3. Langer, R.S. and J.P. Vacanti (1999), Tissue engineering: the challenges ahead. *Scientific American*, **280**(4): pp. 86–89.
4. Vacanti, J.P. and R. Langer (1999), Tissue engineering: the design and fabrication of living replacement devices for surgical reconstruction and transplantation. *Lancet*, **354**: pp. Si32–Si34.
5. Colton, C.K. (1995), Implantable biohybrid artificial organs. *Cell Transplantation*, **4**(4): pp. 415–436.
6. Folkman, J. and M. Hochberg (1973), Self-regulation of growth in 3 dimensions. *Journal of Experimental Medicine*, **138**(4): pp. 745–753.
7. Mohebi, M.M. and J.R. Evans (2002), A drop-on-demand ink-jet printer for combinatorial libraries and functionally graded ceramics. *Journal of Recombinant Chemistry*, **4**(4): pp. 267–274.
8. Koike, N., D. Fukumura, O. Gralla, P. Au, J.S. Schechner and R.K. Jain (2004), Tissue engineering: creation of long-lasting blood vessels. *Nature*, **428**(6979): pp. 138–139.
9. Murray, D. (1926), The physiological principle of minimum work: I. The vascular system and the cost of blood volume. *Proceedings of National Academy of Sciences U S A*, **A**(12): p. 207.
10. Emerson, D.R., K. Cieslicki, X. Gu and R.W. Barber (2006), Biomimetic design of microfluidic manifolds based on a generalised Murray's law. *Lab Chip*, **6**(3): pp. 447–454.
11. Bates, R.H.T., K.L. Garden and T.M. Peters (1983), Overview of computerized-tomography with emphasis on future-developments. *Proceedings of the IEEE*, **71**(3): pp. 356–372.
12. Foroutan, M., B. Fallahi, S. Mottavalli and M. Dujovny (1998), Stereolithography: application to neurosurgery. *Critical Reviews in Neurosurgery*, **8**(4): pp. 203–208.
13. Viceconti, M., M. Casali, B. Massari, L. Cristofolini, S. Bassini and A. Toni (1996), The 'standardized femur program' proposal for a reference geometry to be used for the creation of finite element models of femur. *Journal of Biomechanics*, **29**(9): p. 1241.
14. Kaihara, S., J. Borenstein, R. Koka, S. Lalan, E.R. Ochoa, M. Ravens, H. Pien, B. Cunningham and J.P. Vacanti (2000), Silicon micromachining to tissue engineer branched vascular channels for liver fabrication. *Tissue Engineering*, **6**(2): pp. 105–117.

15. Chrisey, D.B. (2000), Materials processing – The power of direct writing. *Science*, 289(5481): pp. 879–881. Kaihara, S., *et al.*, Silicon micromachining to tissue engineer branched vascular channels for liver fabrication. *Tissue Engineering*, 2000. 6(2): pp. 105–117.
16. Lim, D., Y. Kamotani, B. Cho, J. Mazumder and S. Takayama (2003), Fabrication of microfluidic mixers and artificial vasculatures using a high-brightness diode-pumped Nd:YAG laser direct write method. *Lab on a Chip*, 3(4): pp. 318–323.
17. Bertsch, A., S. Heimgartner, P. Cousseau and P. Renaud (2001), Static micromixers based on large-scale industrial mixer geometry. *Lab on a Chip*, 1(1): pp. 56–60.
18. Kawata, S., H.B. Sun, T. Tanaka and K. Takada (2001), Finer features for functional microdevices – Micromachines can be created with higher resolution using two-photon absorption. *Nature*, 412(6848): pp. 697–698.
19. Jakab, K., C. Norotte, F. Marga, K. Murphy, G. Vunjak-Novakovic and G. Forgacs (2010), Tissue engineering by self-assembly and bio-printing of living cells. *Biofabrication*, 2(2).
20. G. Jans, J.V. Sloten, R. Gobin, G. Van der Perre, R. Van Audekercke and M. Mommaeits (1999), Computer aided craniofacial surgical planning implemented in CAD software. *Computer Aided Surgery*, 4: pp. 117–128.
21. Richardson, T.P., M.C. Peters, E.B. Ennett and D.J. Mooney (2001), Polymeric system for dual growth factor delivery. *Nature Biotechnology*, 19(11): pp. 2001–1034.
22. King, K.R., C.C.J. Wang, M.R. Kaazempur-Mofrad, J.P. Vacanti and J.T. Borenstein (2004), Biodegradable microfluidics. *Advanced Materials*, 16(22): p. 2007–2012.
23. Yang, Y. and A.J. El Haj (2006), Biodegradable scaffolds – delivery systems for cell therapies. *Expert Opinion on Biological Therapy*, 6(5): pp. 485–498.
24. Wang, Y., A.G., Sheppard, B.J. and Langer, R. (2002), A tough biodegradable elastomer. *US National Library of Medicine*, 20: pp. 602–606.
25. Fidkowski, C.M.R. Kaazempur-Mofrad, J. Borenstein, J.P. Vacanti, R. Langer and Y.D. Wang (2005), Endothelialized microvasculature based on a biodegradable elastomer. *Tissue Engineering*, 11(1–2): pp. 302–309.
26. Zhang, G. and L.J. Suggs (2007), Matrices and scaffolds for drug delivery in vascular tissue engineering. *Advanced Drug Delivery*, 61(14): p. 1386.
27. Tabata, Y. (2003), Tissue regeneration based on growth factor release. *Tissue Engineering*, 9: pp. S5–S15.
28. Shea, L.D., E. Smiley, J. Bonadioand D.J. Mooney (1999), DNA delivery from polymer matrices for tissue engineering. *Nature Biotechnology*, 17: pp.551–554.
29. Borenstein, J.T. *et al.* (2002), Microfabrication technology for vascularized tissue engineering. *Biomedical Microdevices*, 4(3): pp. 167–175.
30. Leclerc, E., Y. Sakai and T. Fujii, Cell culture in 3-dimensional microfluidic structure of PDMS (polydimethylsiloxane). *Biomedical Microdevices*, 5(2): pp. 109–114.
31. Therriault, D., S.R. White and J.A. Lewis (2003), Chaotic mixing in three-dimensional microvascular networks fabricated by direct-write assembly. *Nature Materials*, 2(4): pp. 265–271.
32. Bettinger, C.J. K.M. Cyr, A. Matsumoto, R. Langer, J.T. Borenstein and D.L. Kaplan (2007), Silk fibroin microfluidic devices. *Advanced Materials*, 19: pp. 2847–2850.
33. Jain, R.K. *et al.* (2005), Engineering vascularized tissue. *Nature Biotechnology*, 23(7): pp. 821–823.
34. Elcin, Y.M., V. Dixit, and T. Gitnick, Extensive in vivo angiogenesis following controlled release of human vascular endothelial cell growth factor:

implications for tissue engineering and wound healing. *Artificial Organs*, **25**(7): pp. 558–565.
35. Cui, X.F. and T. Boland (2009), Human microvasculature fabrication using thermal inkjet printing technology. *Biomaterials*, **30**(31): pp. 6221–6227.
36. Flores-Ramirez, N. *et al.* (2005), Characterization and degradation of functionalized chitosan with glycidyl methacrylate. *Journal of Biomaterials Science-Polymer Edition*, **16**(4): pp. 473–488.
37. Friess, W. (1998), Collagen – biomaterial for drug delivery. *European Journal of Pharmaceutics and Biopharmaceutics*, **45**(2): pp. 113–136.
38. Cabodi, M. N.W. Choi, J.P. Gleghorn, C.S.D. Lee, L.J. Bonassar and A.D. Stroock (2005), A microfluidic biomaterial. *Journal of the American Chemical Society*, **127**(40): pp. 13788–13789.
39. Ling, Y. *et al.* (2007), A cell-laden microfluidic hydrogel. *Lab on a Chip*, **7**(6): pp. 756–762.
40. Wu, W., A. DeConinck and J.A. Lewis (2011), Omnidirectional printing of 3D microvascular networks. *Advanced Materials*, **23**(24): pp. H178–H183.
41. Shea, L.D. E. Smiley, J. Bonadio and D.J. Mooney (1999), DNA delivery from polymer matrices for tissue engineering. *Nature Biotechnology*, **17**(6): pp. 551–554.
42. Levenberg, S. (2005), Engineering blood vessels from stem cells: recent advances and applications. *Current Opinion in Biotechnology*, **16**(5): pp. 516–523.
43. Wei, Y. and R.F. Sandenbergh (2007), Effects of grinding environment on the flotation of Rosh Pinah complex Pb/Zn ore. *Minerals Engineering*, **20**(3): pp. 264–272.
44. Mironov, V., V. Kasyanov and R.R. Markwald (2011), Organ printing: from bioprinter to organ biofabrication line. *Current Opinion in Biotechnology*, **22**(5): pp. 667–673.
45. Mironov, V., V. Kasyanov and R.R. Markwald (2008), Nanotechnology in vascular tissue engineering: from nanoscaffolding towards rapid vessel biofabrication. *Trends in Biotechnology*, **26**(6): pp. 338–344.
46. Peirce, S.M. and T.C. Skalak (2003), Microvascular remodeling: a complex continuum spanning angiogenesis to arteriogenesis. *Micro Circulation*, **10**(1): pp. 99–111.
47. Laschke, M.W. *et al.* (2006), Angiogenesis in tissue engineering: breathing life into constructed tissue substitutes. *Tissue Engineering*, **12**(8): pp. 2093–2104.
48. Richardson, T.P., M.C. Peters, A.B. Ennett and D.J. Mooney (2001), Polymeric system for dual growth factor delivery. *Nature Biotechnology*, **19**(11): pp. 1029–1034.
49. Nillesen, S.T., P.J. Geutjes, R. Wismans, J. Schalkwijk, W.F. Daamen and T.H. van Kuppevelt (2007), Increased angiogenesis and blood vessel maturation in acellular collagen-heparin scaffolds containing both FGF2 and VEGF. *Biomaterials*, **28**(6): pp. 1123–1131.
50. Laschke, M.W. *et al.* (2008), Incorporation of growth factor containing Matrigel promotes vascularization of porous PLGA scaffolds. *Journal of Biomedical Material Research A*, **85**(2): pp. 397–407.
51. Schumann, P. *et al.* (2009), Consequences of seeded cell type on vascularization of tissue engineering constructs in vivo. *Microvascular Research*, **78**(2): pp. 180–190.
52. Laschke, M.W., B. Vollmar and M.D. Menger (2009), Inosculation: connecting the life-sustaining pipelines. *Tissue Engineering Part B Reviews*, **15**(4): pp. 455–465.
54. Rücker, M., B. Vollmar and M.D. Menger (2006), Angiogenic and inflammatory response to biodegradable scaffolds in dorsal skinfold chambers of mice. *Biomaterials*, **27**(29): 5027–5038.

8
Feasibility of 3D scaffolds for organs

T. BURG and K. BURG, Clemson University, USA

DOI: 10.1533/9780857097217.221

Abstract: The feasibility of using rapid-prototyping techniques to produce 3D tissue is directly tied to the feasibility of creating a 3D support structure that promotes the life-cycle of the cellular component. The material used to fabricate this scaffold, the manner in which the material is assembled, the evolution of the material over time, the biocompatibility or cellular affinity of the material, and the mechanical properties of the material are all degrees-of-freedom that the organ designer must use. Rapid-prototyping techniques provide additional design options to exploit biomaterials to create unique scaffold configurations and co-fabricate cellular and acellular components. The promise of 3D organs is thus linked to managing the material design and assembly process.

Key words: engineered tissue voxel, scaffold feasibility, tissue fabrication error.

8.1 Introduction

The central task of creating three-dimensional scaffolds to support organ fabrication can be reduced to defining the *right process to place the right biomaterials with the right characteristics at the right place at the right time*. In this chapter, the role of rapid fabrication techniques in addressing each part of this complex objective is discussed relative to the feasibility of assembling the biomaterials and biological components to produce a three-dimensional (3D) organ. The term biofabrication will be used generically to refer to the full process of assembling 'biomaterials', cellular materials, and biochemical agents to produce an organ; such a definition will encompass both independent fabrication of the scaffold and later addition of the cellular component, as well as co-assembly of the scaffold and cellular components. The scope of biofabricated organs discussed will span from tissue test systems, functional *in vitro* units for treatment discovery and biological mechanism study, to *in vivo* replacement organs, such as hearts or kidneys. The successful organ scaffold must support and promote the appropriate microenvironment, such as cell-to-cell signaling and cell adhesion opportunities, and the appropriate macro-behavior, such as integration with native vascular networks. Further, the scaffold must evolve with the organ and provide different functions as the construct matures toward the ultimate goal of a completely cellularized organ. Though daunting, there has been consistent movement toward overcoming these challenges; review of the current work in two-dimensional (2D) and 3D tissue fabrication shows encouraging progress and supports the assumption that three-dimensional organ fabrication is an imminent reality.

8.2 Overview of organ fabrication

The most general perspective of organ fabrication must start with a definition of an organ as a constellation of living cells that perform a function. It is apparent that native organs, such as heart or lungs, meet this definition and that a replica of such organs *via* biofabrication techniques would also meet this definition. However, to expand the discussion, and include the most likely immediate potential of rapid prototyping techniques, the definition of an organ will be indicated to specifically include tissue test systems; that is, *in vitro* constellations of living cells that can reproduce or mimic a biological function. Using this definition of an organ, the opportunities for organ fabrication products are shown in Fig. 8.1. The most obvious, and most likely longest-horizon, product is an implantable replacement organ. In the replacement scenario, cells would be extracted from the donor, expanded, integrated into a support matrix (i.e., scaffold), cultured toward functional maturity, and the resulting construct would be implanted in the patient. A potential enhancement on this replacement paradigm is to fabricate the replacement organ *in situ*; that is, assemble the necessary scaffolding and biological material at the repair site.

The tissue test system concept relies on the same core technologies as the organ replacement scenario; however, the scale and criteria for success are different. A tissue test system is a heterotypic, engineered, tissue culture system that has almost immediate utility to guide therapy discovery and biology exploration and, in the longer term, to serve as a tool to personalize medical treatments. The short-term vision for a tissue test system is

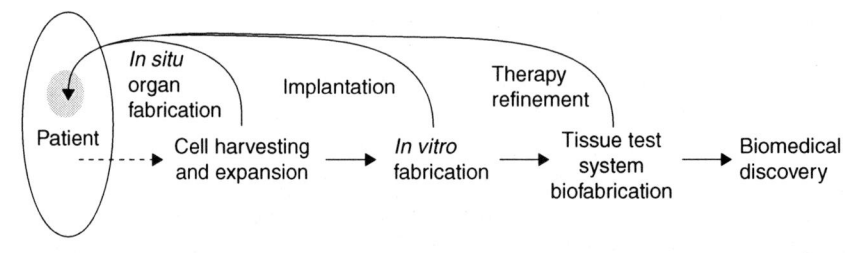

8.1 Potential of biofabrication to produce organ products.

a 3D tissue that is small enough to survive in a bioreactor with diffusion-limited exchange of wastes, nutrients, and intercellular signals (i.e., without vascularization). A significant potential advantage lies in the ability to more readily observe a tissue test system as compared to an *in vivo* system: it is much easier to isolate observations in a controlled laboratory environment. In addition, if the tissue test system can be faithfully reproduced, then multiple questions can be asked and answered in parallel. Ultimately, the processing methods and instrumentation could be used to build other complementary organ modules that could be interconnected to model more complex phenomena. The approach can, for example, be used to produce a modular, functional, validated, 3D microphysiological organ system for drug efficacy and toxicity testing that will reduce the need for animal models. Challenging questions that must be asked are: 'How can one verify that the heterotypic, engineered, tissue culture systems are similar to *in vivo* human tissues structurally and functionally?', 'What are the accepted benchmarking methods?', 'What are the "functional" parameters that will be used to make that determination, in order to validate the system and its potential clinical relevance?', and 'How closely and in how many aspects must a test system approximate native tissue in order to provide tangible value?' Obviously, such 3D systems must be selected with careful respect to a driving clinical or scientific question; that is, there will be no 'one size fits all' 3D system appropriate for all applications.

As an example of the potential utility for organ test systems, consider that breast cancer may take many years to develop (5–30 years) and any epidemiological studies conducted to study this disease process will take even longer (Kim *et al.*, 2004). Each year, approximately 191 000 women in the United States are diagnosed with breast cancer (DHHS, 2010). The majority of breast tumors are detected early and treated with combinations of radiation, chemotherapy, and surgical resection. When chemotherapy is administered prior to surgical resection, the pathology report post-surgery suggests whether or not the chemotherapeutic regimen was effective and if a specific adjuvant chemotherapy would benefit the patient following surgery. Currently, the clinical assessment of cancer progression is mostly based on the prior outcome of patients with similar breast cancer presentation (Pantel and Woelfle, 2004). Tools to select the most effective chemotherapeutic option for a specific patient are still lacking. Indeed, although recently developed molecular analyses add pertinent information to the American Joint Committee on Cancer Staging by incorporating additional information about the receptor status and genetic parameters of the tumors, these analyses are only informative for a subset of patients (Oakman *et al.*, 2010). Overall the choice of a chemotherapy protocol is made on the basis of general data rather than on the specific tumor characteristics of the patient (Shoemaker, 2006). Although this approach has been beneficial to many patients, it falls short of the goal of an individualized diagnosis and treatment plan, largely because of the heterogeneity of the primary breast tumor. Therefore, there is an urgent need for alternate methods that integrate patient-specific information into a semi-automated process or system to suggest the most suitable chemotherapy for that patient. *In vitro* test systems, essentially cultures of the tumor cells of a particular patient, have shown promise in matching the numerous and diverse breast tumors with the optimal post-surgery chemotherapy (Shrivastav *et al.*, 1980). Thus an *in vitro* organ that reproduces the salient features of the breast tissue environment relevant to tumor progression to allow study of the basic mechanisms in a compressed time frame has enormous medical potential.

In order to systematically approach this complex fabrication problem, the fabricated organ concept may be distilled into basic elements. The basic concept of a fabricated tissue voxel as the building block for complex organs or tissue test system is depicted in Fig. 8.2. The tissue voxel is shown to illustrate the idea of tissue constructed from repeated blocks of biomaterials, biological elements, and biochemical agents. The voxel represents achievement of the end goal of producing a functional arrangement, or a constellation, of cells – the voxel is shown as a minimum divisible unit of the whole that contains the necessary simulated biology from which the function of the whole is derived. Different types of voxels will exist within any tissue type. The cell and biomaterial interactions within the voxel and the cell and biomaterial connections between voxels define the function of the organ as a whole. The ability to create an organ will depend on how well the fabrication technique reproduces the microenvironment, including biomechanical forces, cellular signaling, nutrient supply, and waste removal, and the macro-environment, including biomechanical forces and vascular and nervous system connections. These complex and conflicting criteria

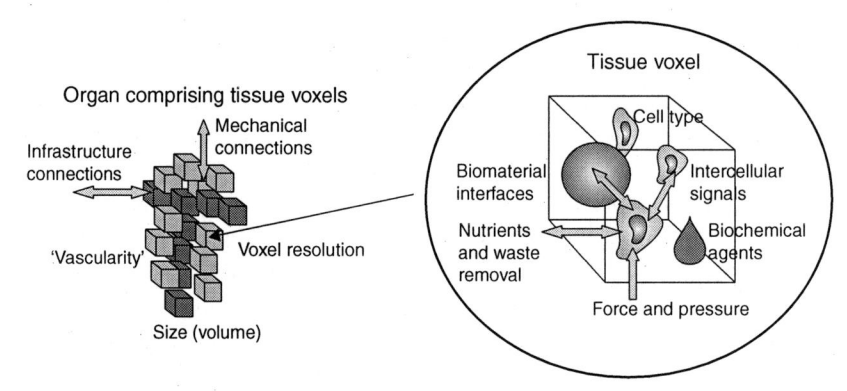

8.2 Tissue can be decomposed as voxels to provide a starting point for defining requirements of biofabrication tools.

defining tissue structure and function will dictate the requirements of the biomaterial scaffold.

The voxel description of a tissue does not dictate a specific scaffold and cellular assembly technique but rather indicates the end product of the fabrication process; for example, a woven matrix with cells seeded *via* perfusion may be used to create voxels. However, it is clear that the fabrication techniques for the tissue scaffold must be harmonized with the techniques to place cellular and biochemical agents. That is, co-design of the biomaterial scaffold and the cellular component is a requirement (without requiring co-fabrication). An interesting variable is the initial starting condition from which the organ will develop into its final form. At one extreme is the hypothesis that if the cells are close together then they will self-assemble into the desired tissue structure (Jakab *et al.*, 2010), while the alternative hypothesis is that the cells and biomaterials must start in an exact spatial configuration in order for the tissue to develop in the appropriate ordered arrangement. The former hypothesis has spawned tissue construction using tissue spheroids while the latter has spanned numerous rapid-prototyping approaches using cellular and acellular components. In reality, exact placement and replication is not realistic, nor is self-assembly in the absence of control inputs. Rather, the convergence of the two ideas, where cells and biomaterials are placed 'close enough' in controlled (if not exact) relative locations, will lead to the next generation of 3D tissue structures; the evolution of the field will eventually allow definition of 'close enough'.

The concept of developing a heterogeneous injectable implant, that is, a cellular composite material, was introduced to the tissue engineering community to restore tissue to a defect site. The original composite design comprised cells seeded on degradable beads in a degradable gel (Burg *et al.*, 2000b). The advantage to the composite aspect was that, unlike classic homogenous injectables (e.g., gels), a composite material accommodates anchorage-dependent cells and allows one to change the building blocks to tailor a system to a particular clinical or patient need. Subsequently, the potential of the composite approach was envisioned and translated to building heterogeneous benchtop tissues that require different 'zones' or microenvironments (Burg and Boland, 2003). Again, the classic 3D approach, for example in cancer benchtop systems, was to suspend cells in a homogeneous environment. It was postulated that tissues are complex structures and therefore a heterogeneous 3D environment is warranted. Hence in this composite scenario, a voxel may include a bead and anchorage-dependent cells or may include cells and a gel.

8.3 The right place: physical properties of the scaffold

The primary function of the organ scaffold is to spatially arrange the cellular material, that is, the cellular constellation. Figure 8.2 suggests two scales of spatial scaffold requirements: bulk properties and voxel-level properties. The most obvious bulk property is specific size, shape, and volume of the organ – in a replacement organ these features will be specific to a patient. This seemingly innocuous requirement may already limit the type of scaffold material and fabrication technique. For example, the liver, the second largest organ in the body, has an average width of 22 cm, a vertical height of 16 cm, a front-to-back thickness of 11 cm and an average weight of 1.5 kg (Wolf, 1990). A biofabricated liver implant must connect to the existing infrastructure of the replacement site. For example, the entire liver structure is permeated by a network of blood capillaries, bile capillaries, and lymph capillaries that must make exterior connections.

Finally, there are mechanical connections to support and hold an organ; for example, the falciform ligament attaches the liver to the anterior body wall and must be incorporated into a replacement organ. Considering only these mechanical constraints, a large group of biomaterials such as gels are unlikely candidate materials for liver biofabrication. Embedded in the bulk property consideration is the need to tailor the organ, and hence the scaffold dimensions, to match the differing requirements of each specific site. Considering this further, it is expected that the scaffold will support the microenvironment at each voxel in a predictable manner and hence the size and shape of the scaffold should not produce unknown compression and change the effective modulus within the organ – the anchorage and support of the organ in the body, for example, the liver support ligament, or even in a bioreactor may make this seemingly simple constraint a challenging problem.

The need to distribute nutrients and oxygen and collect waste throughout a structure is a major challenge in scaling from 2D cell cultures, where all cells are by default within diffusion lengths of the medium, to 3D organ cultures, where the access distance to the medium must be created for all locations within the organ. The simplest approach would be to create a network of interconnected cells, for example, a homogenous, open-cell foam; however, to build an organ the size of a liver will require a purposeful network of channels to support organ development.

The voxel diagram suggests that the mechanical quality of the organ scaffold is important (the placement and self-assembly of the cellular component is discussed later). That is, in order for each voxel, and hence the whole tissue, to behave in a predictable manner the voxels must be similar. If a voxel pattern is defined with millimeter resolution, such as from a scan of a histology slide, the scaffold fabrication technique must have sub-millimeter resolution in order to produce an approximate copy of the desired pattern. The approximation of continuous biology with discrete fabrication components will always suffer from quantization error. Two other types of scaffold fabrication error stem from limitations in the accuracy of placing biomaterials at a desired location and limitations in the precision (repeatability) of placing these materials. The accuracy will affect how well the scaffold meets an overall design objective and the precision and repeatability of the fabrication instrumentation will affect the uniformity and hence cellular-level behavior of specific voxel types. Thus to create the tissue voxels the scaffold fabrication process must be accurate and repeatable to sub-voxel resolution. Note that there is much debate about how much quantization error and fabrication error can be tolerated in building a useful tissue structure.

8.4 The right time: temporal expectations on the scaffold

A successful organ scaffold will dynamically change to meet the needs of the cellular components and the organ as a whole. In most cases, the scaffold is a temporary structure that will facilitate the initial arrangement of cells and then help orchestrate the development of the organ; as such, the temporal properties of the scaffold should be considered a tool that can transform its material and physical properties to promote and direct the organ development. For example, the scaffold strength and mass loss can be tailored to match the needs and capability of the evolving tissue. Several considerations are illustrated in the simple representation of this process in Fig. 8.3. The actual modulus of the scaffold may be prescribed so that the scaffold does not shield the developing organ from mechanical stresses that might actually be important for organ development – the graph is drawn to suggest

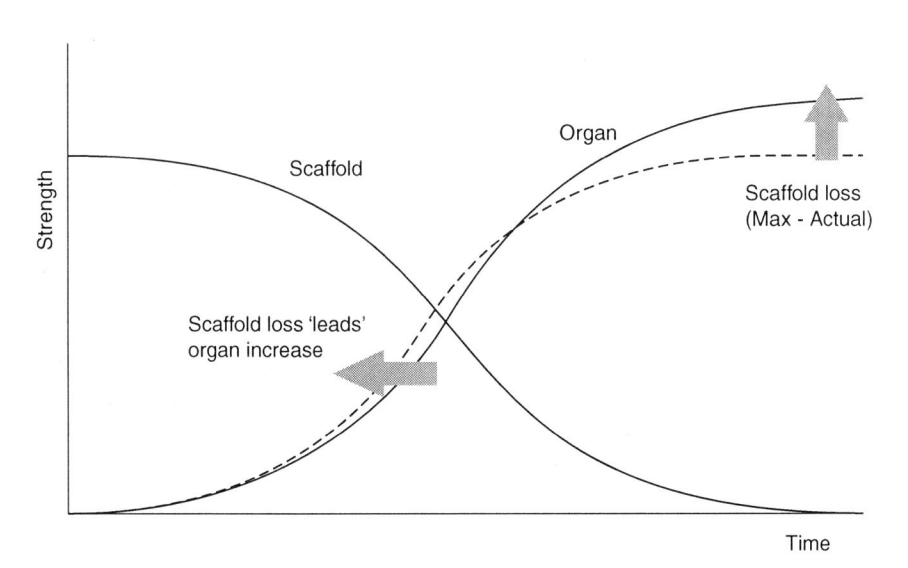

8.3 Scaffold strength can be designed to decrease over time as the organ develops and begins to develop its own support structure.

that it may be advantageous for the scaffold loss to lead the development of organ strength. A number of other material properties may be manipulated to evolve over time.

The scaffold may supply varying biochemical factors or drugs as the organ tissue evolves. A drug-eluting biomaterial may help exert biochemical influences to modify the behavior of the cellular component. As an example, it may be beneficial to inhibit some of the cells so that they remain dormant until other parts of the tissue have grown to a certain capacity. Waiting for vasculature to develop into the inner core of the structure requires the cells at the core of the organ to maintain a quiescent state until the support infrastructure with nutrient delivery and waste removal is ready. At the early stages the scaffold may need to support cell attachment in order to set or define the initial cellular constellation 'map'; however, as the organ grows, the scaffold could promote the migration and interconnection of cells as the cellular component forms the organ tissue. An evolving scaffold may be designed to provide a directed pathway for the migration of cells by continuously revealing surface modifications that lure the cells towards target locations. As a second example, the surface texture or modulus may need to evolve in order to promote stem cell differentiation (Stevens and George, 2005; Chaubey *et al.*, 2008). These examples suggest that the biomaterial should evolve in surface texture, surface chemistry, etc., as a function of time in order to fully exploit the scaffold capabilities. The temporal behavior of the organ scaffold can provide an additional degree of freedom with which to tailor organ development. Thus, smart temporal design can help the scaffold builder address additional needs of the developing organ tissue.

The evolution of a scaffold with time may also have negative consequences. Sensitivity of the cellular components to degradation by-products requires that the structure is absorbed or dissociated, in a biocompatible manner, as the tissue volume expands. The degradation profile of the scaffold (if the biomaterial is degradable) must match the capability of the cellular component to accommodate environmental changes from this breakdown. For example, a polylactide material may have a bolus release of acid that could be detrimental to the surrounding cells; depending on the timing of this release relative to the maturity of the surrounding tissue, mature tissue may be more or less robust to environmental fluctuations.

8.5 The right biomaterials: scaffold fabrication effects on non-scaffold components

Conceptually the biofabrication process can be divided into two broad categories seen in Fig. 8.4; either the scaffold is fabricated independently before the cellular component is added or the scaffold is fabricated concurrently with the cellular component. In some instances, the scaffold may

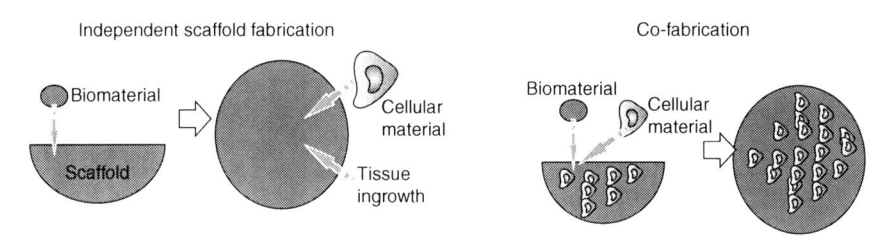

8.4 Tissue scaffolds may be fabricated before any cellular component is added or fabricated concurrently with the deposition of the cellular component.

comprise multiple layers of prefabricated forms concurrently placed with cells (e.g., a 3D cellular fibrous form, or a form comprising multiple cellular sheets); hence these instances involve both pre-fabrication and fabrication concurrent with seeding. In the first category, the structure is fabricated as a standalone biomaterial structure and then the cellular components are added or the scaffold is implanted with the expectation of tissue ingrowth. In the second, the biomaterials and cellular components are assembled at the same time. In either case, a rapid prototyping approach to scaffold fabrication is possible; however, both approaches create challenges to ensuring that (i) the cellular component can be delivered to the desired location in the scaffold and (ii) the scaffold fabrication process, or residue from the process, does not negatively affect the cellular component.

Regardless of the biofabrication approach, the effect of the scaffold fabrication on the cellular component must be considered. In short, the environment of the cell must not be changed beyond healthy bounds; that is, it must not be toxic. The fabrication process may result in changes in cell behavior (e.g., differentiation of cells, expression of particular genes); these changes may not be detrimental, but the fabrication process must be characterized such that the changes are predictable. Some of the major potential effects of the fabrication process on the cellular component are shown in Fig. 8.5. The most serious challenge to an independently fabricated scaffold may be the accessibility of the interior of the scaffold to the cells; for example, it would be impossible to directly seed or place cells in the interior of a closed-cell foam. Even creating the desired voxels throughout an open cell matrix with a post-fabrication seeding technique such as perfusion in a bioreactor is unrealistic; that is, it would be difficult to evenly distribute cells throughout the scaffold. Additionally, the remnants of the fabrication process, especially solvents or unpolymerized agents, may affect the cellular component. Thus, a pre-fabricated organ scaffold, built independent of cells by rapid prototyping techniques, has the potential advantages of having tailored geometry and material properties but may sacrifice the possibility of precise initial cell deposition throughout the tissue structure.

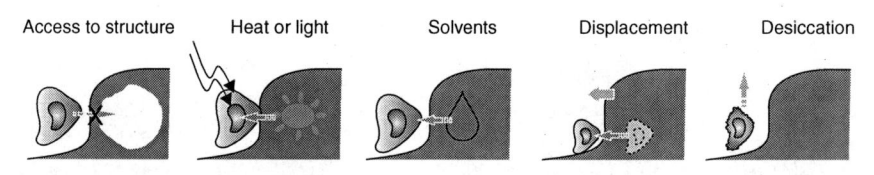

8.5 Sources of interaction during fabrication that have potential detrimental effects on cellular behaviors.

Perhaps more challenging is the concept of co-fabrication of the scaffold and cellular component. In most cases the processes that make a biomaterial 'flowable' for integration into a scaffold, such as heat, solvents, or pressure, are incompatible with the cellular component. For example, the deposition of a polymer which is heated to its melting point for deposition may cause excessive thermal stress on the cellular components; thus, a biomaterial with a low enough melting point or a process to protect the cells from the heat is needed. Similarly, any solvents used to dissolve the scaffold material may also damage the cellular component. It is worth a reminder that the search for a compatible biomaterial deposition process must not sacrifice the structural goals for the scaffold; thus there is a limited range of materials and processes that can be used in this manner. Other approaches to forming the scaffold may introduce additional considerations; for example, the intensity and wavelength of light used to photocure a polymer may damage the cellular component.

The seemingly more mundane requirements that the cellular component can be placed as desired and then stay in that position actually present significant challenges to any rapid biofabrication system. If the polymer is deposited as a (biofabrication compatible) flowable liquid, then it is likely that the spreading of the polymer will cause movement of the cells. Finally, the major consideration in a co-fabrication system is the amount of time for which the organ must remain in the fabrication process. The amount of time that the cells must spend in the construct, not supported by oxygenation or transport processes, is critical. A material fabrication process adequate for independent fabrication of the scaffold may be too slow for co-fabrication with the cellular component. Special misting or other systems may be needed to maintain the cellular component during manufacture. Additionally, the cells may take a certain amount of time to attach and thus limit the speed at which a construct can be assembled.

8.6 The right characteristics: material types

Implantable materials must be biocompatible; the ideal would be a material that serves as a scaffold in the short term and disappears with time, degrading as incorporated cells and invading cells form tissue. Naturally-derived materials are inherently variable, as their characteristics vary with changes in source and processing. Batch to batch variability can be enormous. Naturally-derived materials are more likely to elicit an inflammatory response than synthetic materials, which can be designed in a readily controlled, repeatable manner. Synthetic materials tend not to have the same high level of good or bad cellular affinity; however, they can be specifically designed to meet defined structural integrity or biocompatibility specifications. Naturally-derived materials are generally remodeled at a faster rate than synthetic materials; many of the remodeling processes of naturally-derived materials are enzymatically driven and are therefore quite variable depending on the local levels of relevant enzymes. Synthetic materials are degraded or absorbed largely by hydrolysis and therefore may be tuned to meet specific degradation needs.

Additionally, the cellular materials would ideally be built in place in a minimally invasive manner such as with laparoscopic means. The goal of a minimally invasive mode of delivery limits the materials selection to those with low viscosity or flow. That is, the material selection is limited to those materials that can be readily extruded or ejected from a small diameter bore size. If the cellular material is to have a designated form, as opposed to simply supplying bulking value, it must have a means of setting or maintaining the form. Thus materials that gel upon exposure to certain conditions (e.g., light, pH, temperature, etc.) would be of greatest interest. The local microenvironment makes this selection tricky as a wound site may have very different properties and conditions than the same site under healthy conditions. A material may have to incorporate very different features for the short and the long term. Also, the mode of gelation may require delivery of the gelling agent, and so could potentially mean a second laparoscopic tool (light source, heat probe, for example).

Materials for test systems can take on relatively any form or shape and are limited by the means of fabrication. The range of materials in question encompasses distribution of size, viscosity, form, charge, and chemistry, which are all factors influencing mechanisms of biofabrication. While biocompatibility is not relevant to an *in vitro* system, cellular affinity and toxicity are important considerations. It may not be necessary for a benchtop material to degrade, particularly if the interest is in long-term retention of the tissue system. It is interesting to note that naturally-derived materials that degrade in the body *via* enzymatic means may have a significantly lower degradation rate or may not degrade in a benchtop system, where the enzymes are present in low quantities or are even absent. Clearly, materials used in tissue test systems encompass a wider pool than those designed for implantation.

A practical consideration in the biofabrication of an organ is the means of sterilizing the material. Traditional sterilization processes may change the biomaterial or leave residual chemicals that affect the cellular component.

For example, autoclave may affect the scaffold material with temperature and exposure to water, a dry heat sterilization may eliminate the exposure to water at the expense of prolonged high temperatures, gas sterilization such as ethylene oxide may leave residuals harmful to the cellular component, and gamma radiation may affect the chemical structure of a polymer. Consideration should also be given to regulations and standards which will significantly limit the use of some material types.

8.7 The right process: biofabrication

Many approaches have been suggested to build cellular systems. The conventional, sequential approach, where the material is built first and then the cells are added, results in inconsistent incorporation of the cellular components. Loading of a volume of cells onto or into a biomaterial scaffold by cell seeding methods is either static, such as in a well-plate, or dynamic, for example, in a stir flask, or in a bioreactor (Burg et al., 2000a; Burg et al. 2002). These loading approaches results in inconsistent incorporation of the cellular components. For example, static, dynamic, and bioreactor seeding of identical scaffolds will result in different cell distributions throughout the scaffold volume (Burg et al., 2002). Static seeding most often results in a distribution of cells around a scaffold rather than throughout a scaffold, while bioreactor seeding most often results in the other extreme. Generally, the goal is to achieve uniform cell deposition on a surface or within an open-cell volume (Xu and Burg, 2007), specifically within a homogenous volume. This approach does not allow spatial control of the seeded cells and results in random placement of cells on the construct (Burg et al., 2000a). Additionally, it does not allow building of heterogeneous structures. Ideally, one would want to control placement of each cell in the appropriate material microenvironment. We generally term this concept micro-biofabrication, where the intent is to build cells into materials.

A wide range of culture systems, from 2D to 3D to animal models, have been investigated; animal models provide a native, living system for cell growth, but they are expensive and complex, and are often ill-suited for elucidating answers to very specific scientific questions (Yamada and Cukierman, 2007). Additionally, extrapolation of results derived in an animal model to a human is generally not realistic. In many instances, the biochemical and biomechanical environment in an animal is vastly different from that in a human. For example, unlike murine mammary tissue in which epithelial structures are surrounded by fat, human parenchymal components are in direct contact with collagen-based stromal tissue (Ronnov-Jessen et al., 1996). Thus, information gleaned from a murine system regarding breast cancer may be very difficult to extrapolate or even be irrelevant to a human system.

The scaffold provides the major environmental factor that contributes to the maintenance of cell function integrity *in vitro*. Traditional cell culture methods involve the seeding, growth, differentiation, and long-term culture of primary and immortalized cells on rigid plastic substrates. Though a multitude of cell types can actually remain viable on these artificial surfaces through many successive passages, their ultimate behavior in monolayer culture may not successfully recapitulate *in vivo* activity. For example, in 1977, Emerman and colleagues demonstrated that primary murine mammary epithelial cells cultured on rigid plastic surfaces always failed to secrete casein (a milk protein), even when the cells were supplemented with lactogenic hormones. Many examples have been given since that initial finding, all pointing to the need for a 3D environment. More recently, it has been shown that even the transition to a 3D environment is not simple – the choice of 3D environment strongly affects the behavior of cells and therefore the suitability of a system to answer a particular scientific question (Booth *et al.*, 2013). For example, it was determined when testing specific breast cancer cells that collagen/agarose systems were ill-suited to questions regarding migration and metastasis, but were amenable to questions regarding primary tumor growth. Matrigel™, on the other hand, was ill-suited to questions regarding primary tumor growth, but well-suited to questions regarding migration and metastasis. While this may seem obvious, particularly to the biomaterials research community – that is, biomaterial specifications and characteristics influence cellular behavior – this point has not been reflected in the research of the biology-focused scientific community, where scientists continue to rely on one 'go-to' material to answer all questions. Hence, there is a need to provide biofabrication systems that will accommodate a range of materials in order to answer a range of scientific questions.

8.8 Conclusion

The promise of 3D organ systems is remarkable; however, the reality is that building even the simplest of tissues presents a series of complex problems. Fortunately, the scaffold features including material type, fabrication methods, *in situ* evolution, and integration with the biofabrication process, provide a great number of degrees-of-freedom for the organ designer. The downside to such freedom and flexibility is the complexity of setting and optimizing all of these parameters; scaffold designers are just beginning to exert sufficient control over the design process to create 3D tissue structures. However, critics question the validity of 3D systems, seemingly unimpressed by anything short of cloning. In response, we suggest that the feasibility of creating fully functioning 3D organs should be extrapolated from the current successes in designing organ test systems. As researchers seek to answer specific scientific questions by developing and demonstrating simplified tissue test systems, each with its own set of bounds and assumptions, the goal of whole tissue creation comes closer. As with any

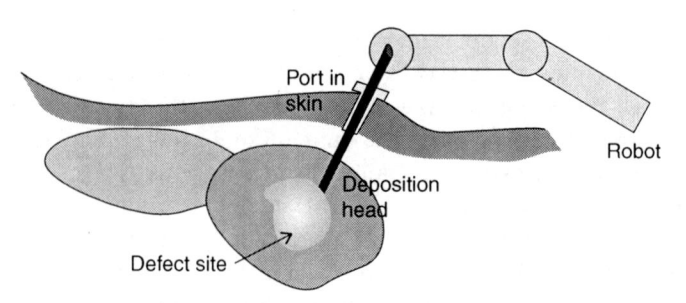

8.6 The ultimate medical objective would be to repair any defect with minimal impact on the patient.

new, complex idea, the process towards realizing it in its entirety is slow, iterative, and often incremental.

We propose the organ repair/replacement paradigm in Fig. 8.6 as the ultimate challenge to organ designers; specifically, producing a 3D organ or organ repair *in situ* using minimally invasive surgical techniques. Reaching this milestone will surely signal that the next generation of medical treatment has arrived.

8.9 Sources of further information and advice

Relevant professional organizations, reflecting biomaterials design, cellular engineering, and biofabrication, respectively, include the Society For Biomaterials (www.biomaterials.org), the Tissue Engineering and Regenerative Medicine International Society (www.termis.org), and the IEEE Engineering in Medicine & Biology Society (www.embs.org). Additionally, the newly published *Principles of Tissue Engineering* book overviews many methods, including the development of breast cancer test systems.

8.10 References

Department of Health and Human Services (DHHS), Atlanta (GA), Centers for Disease Control and Prevention and National Cancer Institute (2010). United States Cancer Statistics: 1999–2006 Incidence and Mortality Web-based Report.

Booth, B., Park, J. and Burg K.J.L. (2013) 'Evaluation of normal and metastatic mammary cells grown in different biomaterial matrices: establishing potential tissue test systems', *J Biomater Sci: Polym Ed*, **24**, 6, 758–768.

Burg, K. J. L. and Boland, T. (2003) 'Bioengineered devices: Minimally invasive tissue engineering composites and cell printing', *IEEE Eng Med Biol*, **22**(5): 84–91.

Burg, K.J.L., Holder Jr, W.D., Culberson, C.R., Beiler, R.J., Greene, K.G., Loebsack, A.B., Roland, W.D., Eiselt, P., Mooney, D.J. and Halberstadt, C.R. (2000a) 'Comparative study of seeding methods for three-dimensional polymeric scaffolds', *J Biomed Mater Res*, **51**, 4, 642–649.

Burg, K.J.L., Austin, C.E., Culberson, C.R., Greene, K.G., Halberstadt, C.R., Holder, Jr, W.D., Loebsack, A.B. and Roland, W. D. (2000b) 'A novel approach to tissue engineering: Injectable composites', *Transactions of the 2000 World Biomaterials Congress*, Kamuela, HI, 5/2000.

Burg, K.J.L., Delnomdedieu, M., Beiler, R.J., Culberson, C.R., Greene, K.G., Halberstadt, C.R., Holder, Jr, W.D., Loebsack, A.B., Roland, W.D. and Johnson, G.A. (2002) 'Application of magnetic resonance microscopy to tissue engineering: a polylactide model', *J Biomed Mater Res*, **61**, 3, 380–390.

Burg, K.J.L., Inskeep, B. and Burg, T.C. (2013) 'Breast tissue engineering: Reconstruction implants and three-dimensional tissue test systems', in Lanza, R., Langer R. and Vacanti, J. (Eds) *Principles of Tissue Engineering*, 4th Edition. Elsevier: Amsterdam, The Netherlands.

Chaubey, A., Ross, K.J., Leadbetter, M.R. and Burg, K.J.L. (2008), 'Surface patterning: Tool to modulate stem cell differentiation in an adipose system', *J Biomed Mater Res*, **84B**, 70–78.

Emerman, J.T., Enami, J., Pitelka, D.R. and Nandi, S. (1977) 'Hormonal effects on intracellular and secreted casein in cultures of mouse mammary epithelial cells on floating collagen membranes', *Proc Natl Acad Sci USA* **74**, 10, 4466–4470.

Jakab, K., Norotte, C., Marga, F., Murphy, K., Vunjak-Novakovic, G. and Forgacs, G. (2010) 'Tissue engineering by self-assembly and bio-printing of living cells', *Biofabrication* **2**, 2.

Kim, J.B., Stein, R. and O'Hare, M.J. (2004) 'Three-dimensional *in vitro* tissue culture models of breast cancer – a review', *Breast Cancer Res Tr*, **85**, 281–291.

Lanza, R., Langer R. and Vacanti, J. (Eds) (2013) *Principles of Tissue Engineering*, 4th Edition. Elsevier/Academic Press : Amsterdam, The Netherlands.

Oakman, C., Santarpia L. and Di Leo, A. (2010) 'Breast cancer assessment tools and optimizing adjuvant therapy', *Nat Rev Clin Oncol*, **7**, 725–732.

Pantel, K. and Woelfle, U. (2004) 'Micrometastasis in breast cancer and other solid tumors', *J Biol Regul Homeost Agents*, **18**, 2, 120–125.

Ronnov-Jessen, L., Petersen, O.W. and Bissell, M.J. (1996) 'Cellular changes involved in conversion of normal to malignant breast: importance of the stromal reaction', *Physiol Rev* **76**, 1, 69–125.

Shoemaker, R.H. (2006) 'The NCI60 human tumour cell line anticancer drug screen', *Nat Rev Cancer*, **6**, 813–823.

Shrivastav, S., Bonar, R.A., Stone, K.R. and Paulson, D.F. (1980) 'An in vitro assay procedure to test chemotherapeutic drugs on cells from human solid tumors', *Cancer Res*, **40**, 4438–4442.

Stevens, M. and George, J. (2005) 'Exploring and engineering the cell surface interface', *Science*, **310**, 5751, 1135–1138.

Wolf, D.C. (1990) 'Evaluation of the size, shape, and consistency of the liver', in Walker, H.K., Hall, W.D., and Hurst, J.W. (Eds) *Clinical Methods: The History, Physical, and Laboratory Examinations*. 3rd edition, Boston, Butterworth, 478–481.

Xu, F. and Burg, K.J.L. (2007) 'Three-dimensional polymeric systems for cancer cell studies', *Cytotechnology*, **54**, 3, 135–43.

Yamada, K.M. and Cukierman, E. (2007) 'Modeling tissue morphogenesis and cancer in 3D', *Cell* **130**, 4, 601–610.

9
3-D organ printing technologies for tissue engineering applications

H.-W. KANG, C. KENGLA, S. J. LEE, J. J. YOO and
A. ATALA, Wake Forest School of Medicine, USA

DOI: 10.1533/9780857097217.236

Abstract: Organ printing technology was developed to allow construction of biological substitutes mimicking structures and functions of native tissues or organs. This technology enables precise placement of various cell types and biomaterials in a single three-dimensional (3-D) architecture. Among technologies developed for tissue engineering and regenerative medicine, organ printing technology is one of the most attractive and powerful for use in constructing a structure that mimics a real 3-D tissue or organ. The hypothesis driving organ printing development is that by precisely placing cells in relation to each other, an environment that encourages physiologically relevant cues can be created, resulting in a tissue construct with functionality. This chapter discusses organ printing technologies and applications in tissue engineering and regenerative medicine.

Key words: organ printing, inkjet printing, extrusion-based printing, tissue engineering, regenerative medicine.

9.1 Introduction

The aim of tissue engineering and regenerative medicine is to meet the demand for replacing tissues/organs with those bioengineered from the cells of an individual patient (Nerem, 2006). With the advances in modern medicine come increases in lifespan and age-related illness. We live in an era of increasing demand for organ transplantation, but the supply of donor organs has barely risen. Thus, there is a significant shortage of tissues and organs for transplant (Atala, 2009). There are many advanced tissue engineering technologies which are producing remarkable success stories; however, the realization of the tissue engineering dream remains unfulfilled. Typical approaches utilizing cells and/or biomaterials have significant limitations in spatial organization, and are thus limited to primarily hollow, two-dimensional (2-D), or avascular tissue and organ structures.

Three-dimensional printing technology for building a tissue or organ construct is a recent advanced tissue engineering approach that has the potential to overcome the limitations of other approaches. An area of active research, 3-D printing technology is demonstrating the feasibility of producing complex tissues and organs at sizes which are anatomically and clinically applicable. Through spatial combinations of cells and biomaterials in complex, meaningful, 3-D arrangements, we can better harness the regenerative capacity innate to cells and thereby generate needed tissues or organs. To achieve this goal, various types of 3-D printing methods exist and several have been utilized for the purpose of organ printing. Organ printing technology can be supported by three broad methods based on working principles: inkjet-, extrusion-, and laser-based methods, as shown in Fig. 9.1. The majority of studies presented in the literature utilize either the inkjet- or extrusion-based printing method. The inkjet-based method uses a commercially available inkjet printer for printing cell-laden hydrogel biomaterials, whereas the extrusion-based method uses a micro-scale nozzle driven by a pneumatic pressure controller or syringe pump. These two printing methods are often discussed with the associated terminology of bio-ink

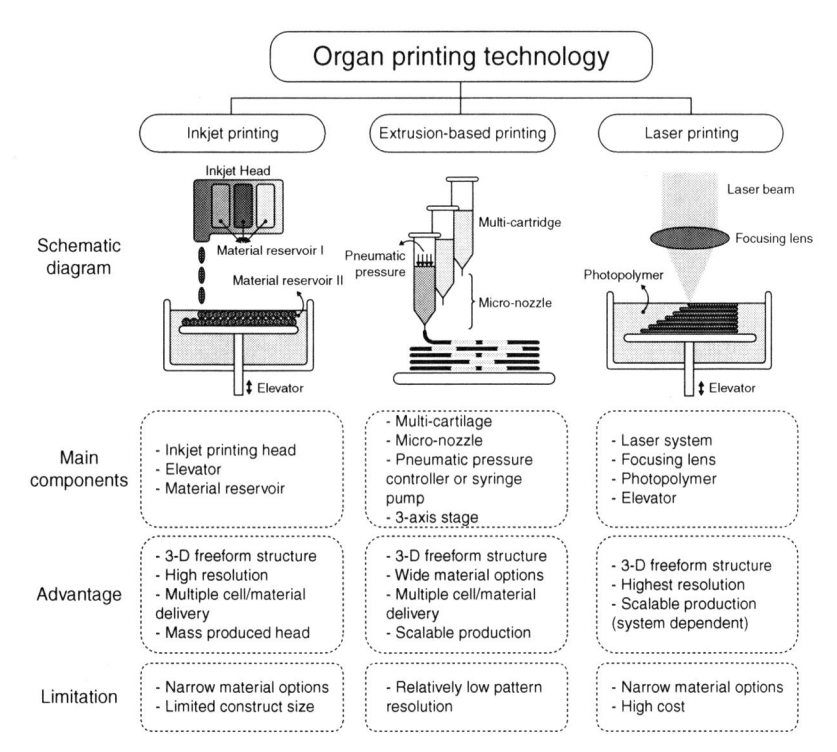

	Inkjet printing	Extrusion-based printing	Laser printing
Schematic diagram	Inkjet Head, Material reservoir I, Material reservoir II, Elevator	Multi-cartridge, Pneumatic pressure, Micro-nozzle	Laser beam, Focusing lens, Photopolymer, Elevator
Main components	- Inkjet printing head - Elevator - Material reservoir	- Multi-cartilage - Micro-nozzle - Pneumatic pressure controller or syringe pump - 3-axis stage	- Laser system - Focusing lens - Photopolymer - Elevator
Advantage	- 3-D freeform structure - High resolution - Multiple cell/material delivery - Mass produced head	- 3-D freeform structure - Wide material options - Multiple cell/material delivery - Scalable production	- 3-D freeform structure - Highest resolution - Scalable production (system dependent)
Limitation	- Narrow material options - Limited construct size	- Relatively low pattern resolution	- Narrow material options - High cost

9.1 Approaches to 3-D printing are varied, applying techniques such as inkjet printing, material extrusion, and laser-based polymerization and/or material transfer. As depicted above, each approach has significant advantages and associated limitations.

and bio-paper to describe the biological nature of both deposition materials and intended substrate. Laser-based printing methods use polymerization by irradiation of photon energy on the photopolymer to construct 3-D architectures. Other methods for 3-D printing are under intense development but are not addressed here for organ printing.

Organ printing can also be combined with other technologies developed for both medical and industrial applications. The use of 3-D computed tomography (CT) and other imaging techniques has realized the capability to render 3-D visualization of internal organs and tissue structures. Data can then be passed on to advanced computer-aided design and manufacturing (CAD/CAM) software to produce computer code specific to the anatomy of an individual patient. This level of personalized medicine is on the horizon of organ printing technology in which cells from a patient are assembled according to their specific anatomy, matching the patient genetically and physically. Therefore, organ printing can be defined as the 3-D spatial organization of cells, biomaterials, and biologics through a printing mechanism which results in a structure that anatomically and physiologically meets the need of the patient. It is important to remember that, at the end of the day, our goal is always to achieve what is truly best for the patient. These are great tools as we continue to move forward in the age of tissue engineering and regenerative medicine. It is our responsibility to deliver these treatments to those that need them in a safe and effective fashion.

9.2 Three-dimensional printing methods for organ printing

Many deposition techniques are being adapted for 3-D printing of tissues and organs. This sections looks to examine various printing methods and their limitations. Each method has strengths that can be leveraged, but further research and engineering can be done to overcome the limiting factors.

9.2.1 Inkjet-based method

Inkjet-based printing, or drop-on-demand dispensing, utilizes various technologies to eject droplets of small volumes. To develop this technique, researchers started with a modified commercial inkjet printer and applied it to cell printing and tissue engineering. They modified the ink reservoir to load cell- or drug-laden hydrogel biomaterial, termed bio-ink, and designed a stage system to achieve 3-D fabrication (Boland *et al.*, 2006; Nakamura *et al.*, 2005). Therefore, a system typically comprises the modified inkjet printer, elevator, and material reservoir (Fig. 9.1). This approach can easily achieve ejection of small volume droplets onto a substrate. The reported resolution using the inkjet printing method is about 20–100 μm (Melchels *et al.*, 2012).

3-D freeform structures can be fabricated using the layer-by-layer method which stacks printed 2-D patterns. Most often this technology uses mass-produced inkjet heads so that it is cheaper than the other types of printing systems. Further, multiple types of cells and materials can be delivered to construct a single structure. At least two approaches have been reported in the literature to achieve solidification in a desired 3-D shape: the bio-paper (layer substrate material) induces solidification of the droplet of bio-ink (Nakamura et al., 2010a) or the bio-ink initiates solidification of the bio-paper material (Xu et al., 2009). The former can result in solid spheres that become building blocks within the construct. The latter approach can form shapes with spherical pores. Nakamura et al. (2010a) used a cell and alginate solution mixture as bio-ink and calcium chloride ($CaCl_2$) solution as bio-paper. A 3-D structure was fabricated by printing the cell-mixed alginate solution on the $CaCl_2$ solution. On the other hand, Xu et al. (2009) had introduced a different method with the same materials. Calcium chloride solution, bio-ink, was printed on the cell and alginate solution bio-paper mixture. The approach in which the droplets are solidified by the bio-paper is most applicable to organ printing, as it allows the use of many different materials and cell types to be spatially organized within the construct. This printing method can be relatively fast, but it is more difficult to prevent droplet drying as the droplets become smaller and the ejection speed increases (Nakamura et al., 2010a). Many laboratories have hijacked commercial inkjet printers and customized them to print their biomaterials (Binder et al., 2011; Roth et al., 2004; Xu et al., 2005). Several groups are actively working to develop custom technologies specifically to optimize the production rates and cell viability of jetting biomaterials and cells (Guillemot et al., 2010; Nishiyama et al., 2009).

Although inkjet-based printing has many advantages as a method for organ printing, such as 3-D freeform fabrication, high resolution, heads for multiple materials, and low cost, it also has several limitations. Only biomaterials with low viscosity can be reasonably used as bio-inks. The producible construct size is also limited because it takes extended print times to fabricate anatomically and clinically relevant sizes with the small droplet size. Overcoming these difficulties is the current challenge to achieve clinical application of inkjet-based printing technology for organ printing.

9.2.2 Extrusion-based method

Extrusion-based printing technology uses a micro-nozzle and precise pressure controller or syringe pump for micro-scale patterning (Fig. 9.1). The cell-laden hydrogel biomaterials, the bio-ink, in the cartridge can be precisely dispensed by controlling either actuating pressure or piston of the syringe pump. 2-D patterns can be fabricated by precise guidance of the micro-nozzle through specified paths and then stacked in a layer-by-layer process to form a 3-D architecture. This technology can also construct a hybrid structure using a multiple-cartridge system capable of dispensing different bio-inks (Fig. 9.1). When this technology is compared with the other printing technology, it has a wider selection of biomaterials and the producible construct size is scalable. On the other hand, it has comparatively low resolution. The device used in this process is commonly referred to as a bioplotting system after the pioneering EnvisionTEC Inc. product, which was named the '3-D bioplotter.' The resulting construct is a composite of filaments stacked and/or fused together (Fedorovich et al., 2008). Biologically active and structural protein molecules can also be incorporated depending on the impact they have on the bio-ink properties.

Scaffold-free dispensing is another extrusion-based method which also utilizes the idea of bio-paper and bio-ink. Generally the bio-ink carries the cells or cell aggregates being used (Norotte et al., 2009). The bio-paper forms the cell environment and incorporates structural components to form a hydrogel that mimics aspects of an extracellular matrix (ECM). The bio-ink or bio-paper can also incorporate cytokines or other biomolecules. The resulting construct is formed by guided self-assembly driven by the innate tendencies of the biological components (Mironov et al., 2006).

9.2.3 Laser-based printing method

Several 3-D fabrication methods such as stereolithography and two-photon polymerization have also been introduced for 3-D printing (Fig. 9.1). Stereolithography makes it possible to construct micro- and meso-scale 3-D structures, whereas two-photon polymerization has sub-micron resolution in 3-D space. Some photosensitive polymers and initiators are non-toxic and allow for building of a 3-D architecture with embedded cells or cell aggregates. Fabricating designs which incorporate multiple biomaterials is a technical challenge requiring highly customized equipment. Two-photon polymerization utilizes non-linear, optically induced crosslinking of monomer units and a photoinitiator to fabricate 3-D shapes (Claeyssens et al., 2009). A laser can be focused to a small spot and traced through a volume of material. The laser induces crosslinking only within this focal region due to its high light intensity. Thus, highly precise 3-D shapes having sub-micron resolution can be made within the material volume (Ovsianikov et al., 2010). Stereolithography uses light, mostly ultraviolet (UV), in order to polymerize a layer of precursor material and photoinitiator in a specific two-dimensional pattern (Kang et al., 2012b). Precursor material is either added to the fabrication process (Dhariwala et al., 2004) or the construct being fabricated is lowered into the precursor in order to expose a new layer of material to the UV light source. The 3-D tissue construct is built layer by layer in this manner. One of the difficulties to be considered in photopolymerization-based fabrication technology is the biological properties of the photoinitiators used. Low initiator concentration (necessary to reduce cytotoxicity) may result in slow fabrication speeds or undesirable material properties.

9.2.4 Integrated organ printing system

The methods described thus far are usually used for the fabrication of cell-laden hydrogel or cell aggregates. The use of these approaches to construct a high strength, 3-D freeform structure of a clinically applicable size and shape is not easy owing to several challenges: the cell-laden hydrogel biomaterials used in the printing process have poor mechanical properties and structural stability, the feature sizes are very small requiring extended printing times, and/or the required crosslinking agents pose a threat to cell metabolism. In this regard, we have recently introduced hybrid systems for organ printing that can concurrently print high strength synthetic polymers and cell-laden hydrogels (Kang *et al.*, 2012a). The synthetic polymers can provide physical support of the 3-D architecture, and the printed cell-laden hydrogels can provide the biological components for tissue regeneration. Our group has taken the extrusion-based method used in fused deposition modeling (FDM) to design a system for 3-D printing referred to as 'integrated organ printing'. The goal of integrated organ printing is to concurrently deliver a supportive 3-D template (mostly biodegradable thermoplastic polymers) and 3-D patterned deposition of cell-laden hydrogel biomaterials in a precise manner. This approach is born out of the challenges of engineering 3-D tissue constructs.

Biomaterial challenges addressed by integrated organ printing relate to structural maintenance. The polymeric biomaterial used for the 3-D template is extruded in a molten state. After solidifying, the mechanical properties of the polymer will govern the mechanical stability of the tissue constructs. The scaffold design has as much to do with the mechanical properties of the construct as the material itself. In common with designing a civil structure, the design of the biomaterial scaffold should relate to the desired function and stability of the tissue constructs. The 3-D printing approach of FDM is well equipped to meet these design needs. Another aspect of structural maintenance is related to the cell-laden hydrogel biomaterials within the construct. Micro-extrusion allows for the deposition of 3-D lines and dots that are already in a gel state. Various techniques have been employed for enhancing the stability of these 3-D hydrogel biomaterials, and these techniques can also be employed in integrated organ printing. One approach is the use of thermosensitive hydrogels, like gelatin, to maintain structure until a crosslinkable biomaterial can be cured for longer periods of gel stability at physiological conditions (i.e., temperature, pH). Extrusion has an advantage over jetting due to its ability to dispense higher viscosity hydrogel biomaterials than most jetting valves.

Biological challenges addressed by the integrated organ printing approach include the ability to provide physiologically relevant cells to locations within the construct in a controlled process, as well as the ability to provide biomimetic patterns and mechanical environments. Based on the biological phenomena revealed in studies of development and regeneration, it is evident that cell-to-cell junctions and communications are vital for appropriate tissue formation (Artavanis-Tsakonas *et al.*, 1999). Integrated organ printing technology has the ability to place biomaterials laden with various cell types in specific arrangements for establishing heterogeneous cell-to-cell relationships. Studies have also shown that the mechanical environment impacts cell growth and development in terms of both the composition and mechanical properties of the ECM (Daley *et al.*, 2008; Rehfeldt *et al.*, 2007; Romer *et al.*, 2006). Extrusion-based technology can extrude gels and fluids with a variety of compositions and a wide range of viscoelasticity. The mechanical properties of the cell-laden hydrogel biomaterials combined with other structural components like a polymer scaffolding structure enable integrated organ printing to provide tailored mechanical environments to cells.

9.3 From medical imaging to organ printing

Human organs have arbitrary 3-D shapes composed of multiple cell types and ECM. Therefore, CAD/CAM processes are important technologies needed for the clinical application of organ printing because the processes provide an automated way to imitate the 3-D shape of a target organ or tissue (Sun *et al.*, 2005). Figure 9.2 shows the simplified process flow from medical imaging to 3-D printing. The process starts by scanning the patient to obtain 3-D volumetric information of a target tissue or organ using medical imaging modalities such as CT and magnetic resonance imaging (MRI). These imaging tools acquire information from cross-sectional slices of the body and the data is stored in the digital imaging and communications in medicine (DICOM) format, a standard format for digital imaging in medicine. This information can be transformed to 3-D CAD model data by a reverse

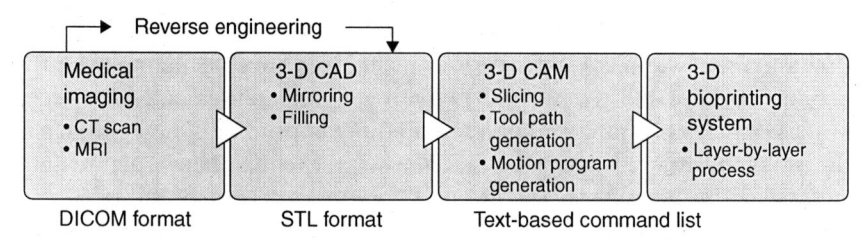

Medical imaging • CT scan • MRI	3-D CAD • Mirroring • Filling	3-D CAM • Slicing • Tool path generation • Motion program generation	3-D bioprinting system • Layer-by-layer process
DICOM format	STL format	Text-based command list	

Reverse engineering →

9.2 Computer-aided design and manufacturing (CAD/CAM) process for automated printing of 3-D shape imitating target tissue or organ.

engineering process. This process starts with interpolation of points within and between image slices to improve resolution and generate volumetric data point from the measured data. Then, a CAD model can be created by extraction of localized volumetric data from the specific tissue or organ to generate a surface model. In this step, sophisticated reconstruction of a CAD model should be performed for organ printing technology because an organ or tissue has a complex 3-D structure and consists of multiple cellular components. Fortunately, there are many commercialized software packages that can perform this reverse engineering process – Mimics (Materialise, Leuven, Belgium), Geomagic Studio® (Geomagic, Morrisville, NC, USA), Simpleware (Simpleware Ltd., Exeter, UK), and Analyze (AnalyzeDirect, Inc., Overland Park, KS, USA). After that, a motion program, which is an instructional computer code for the printing device to follow a designed path, can be generated with a CAM system. This CAM process is divided into three steps: slicing, tool path-, and motion program-generation. Slicing is to obtain information of sliced 2-D shapes of an object for layer-by-layer process. Tool path generation is for creating a path for the tool to follow in order to fill the cross-sectional space of each layer. The printed product should have the proper inner architecture constructed with multiple cellular materials for efficient tissue regeneration. Therefore, a well-organized strategy for tool path generation is required to have highly efficient organ or tissue regeneration and is an important process for automated printing. Aspects such as boundaries, interfaces, and porosity need to be carefully considered. Finally, in the case of extrusion-based printing systems, a motion program can be generated by combining the tool paths and the other fabrication conditions such as scanning speed, material dispensing rates, and layer thickness.

9.4 Applications in tissue engineering and regenerative medicine

Many studies highlight the potential of organ printing systems with the capability to spatially control the placement of cells, biomaterials, and biological factors for the generation of tissues or organs. Many types of cells and growth factors have been successfully printed and remained active and viable. However, none to date has brought all of the necessary components together in a single construct to form functional tissue systems. Table 9.1 summarizes tissue engineering applications of organ printing technology.

Blood vessels provide the pathway for blood to move from the heart to all organs. A number of studies have been performed to test the feasibility of inducing blood vessel regeneration with organ printing technology. Nakamura et al. (2005) introduced a cell printing system using an electrostatically driven inkjet system. They showed that viable bovine vascular endothelial cells (ECs) can be patterned using this system. Cui and Boland

Table 9.1 Tissue engineering applications of printing technologies

Target tissue	Printing methods	Cell types	Bioactive molecules	Biomaterials	References
Blood vessel	Electrostatically driven inkjet printing	Endothelial cells	–	–	Nakamura et al. (2005)
	Thermal inkjet printing	Endothelial cells	–	Fibrin (fibrinogen and thrombin)	Cui and Boland (2009)
	Extrusion-based printing	Endothelial cells	–	Alginate, CaCl$_2$	Khalil and Sun (2009)
	Extrusion-based printing	Cardiac and endothelial cells (multicellular spheroids)	–	Collagen	Jakab et al. (2008)
	Extrusion-based printing	Smooth muscle cells and fibroblasts (multicellular spheroids)	–	Agarose	Norotte et al. (2009)
	Extrusion-based printing	Smooth muscle cells, endothelial cells, fibroblasts (multicellular spheroids)	–	Agarose	Marga et al. (2012)
Bone	Inkjet printing	Muscle-derived stem cells	BMP-2	Fibrin (fibrinogen and thrombin)	Phillippi et al. (2008)
	Extrusion-based printing	Bone marrow stromal cells and endothelial progenitor cells	–	Agarose, alginate, methylcellulose, or Lutrol F127	Fedorovich et al. (2008)
Bone-cartilage	Extrusion-based printing	Endogenous stem cells	Transforming growth factor (TGF)-β3	Hydroxyapatite powder, poly(ε-caprolactone)	Lee et al. (2010)
Heart	Thermal inkjet printing	Feline adult or H1 cardiomyocytes	–	Alginate, CaCl$_2$	Xu et al. (2009)
Muscle	Inkjet printing	Mouse C2C12 myoblasts, C3H10T1/2 mesenchymal fibroblasts	FGF-2, BMP-2	Polystyrene, fibrin	Ker et al. (2011)

Nerve	Thermal inkjet printing	Rat embryonic motor neurons	–	Soy agar, collagen	Xu et al. (2005)
	Thermal inkjet printing	Embryonic hippocampal and cortical neurons	–	Collagen	Xu et al. (2006)
	Inkjet printing	Neural stem cells	CNTF	Polyacrylamide-based hydrogel	Ilkhanizadeh et al. (2007)
	Jetting system using pneumatic pressure	Murine neural stem cells	VEGF	Fibrin (fibrinogen and thrombin), collagen	Lee et al. (2010)
	Extrusion-based method	Bone marrow stem cells, Schwann cells (multicellular spheroids)	–	Agarose	Marga et al. (2012)
Skin	Jetting system using pneumatic pressure	Fibroblasts and keratinocytes	–	Collagen	Lee et al. (2009)

(2009) fabricated microvasculature on a 10 µm scale using human microvascular ECs and fibrin hydrogel. To do this, they used thermal inkjet printing technology. After 21 days in culture, a confluent lining of ECs could be seen in the vascular channel. Khallil et al. (2009) developed an extrusion-based multi-nozzle deposition system and showed that ECs could be printed using this system. In fact, 83% of the printed ECs remained viable after the printing process. Jakab et al. (2008) introduced a novel method for constructing a scaffold-free vascular tubular graft with a predesigned vascular structure using the self-organizing capacity of the cells and tissues. Prefabricated multicellular spheroids were printed on a collagen substrate using an extrusion-based printing machine. They showed that structures printed with cardiac cells and human umbilical vein ECs could generate primitive vasculature. Norotte et al. (2009) showed the most advanced results to date in terms of blood vessel regeneration via organ printing technology (Fig. 9.3a). They introduced a fabrication system for scaffold-free vascular tubular constructs with 300–500 µm inner diameters by co-printing smooth muscle cells (SMCs) and fibroblasts. In particular they were able to print tubes of multiple layers and complex branching geometry. Very recently, Marga et al. (2012) introduced a vascular construct printed with human aortic SMCs, human aortic ECs, and human dermal fibroblasts. The mechanical properties of these printed vascular constructs were reinforced by new ECM produced by the printed cells after 21 days *in vitro*. Interestingly, Xu et al. (2009) introduced a fabrication method that could create a 3-D heart shape with primary adult feline- or H1 cardiomyocyte-laden alginate using inkjet printing technology. Then, excitation-contraction functionality of printed cell-alginate construct caused by electrical stimulation was demonstrated.

Organ printing technology has also been applied to musculoskeletal tissue engineering. Phillippi et al. (2008) fabricated a 2-D pattern with bone morphogenic protein 2 (BMP-2) on a fibrin substrate with inkjet printing technology. The printed construct sustained BMP-2 expression for 6 days and the osteogenic differentiation of plated muscle-derived stem cells was observed only on the BMP-2 loaded printed substrate. This result indicates that it may be possible to spatially control multi-lineage differentiation of stem cells using printing technology. Fedorovich et al. (2008) demonstrated patterning of bone marrow-derived stromal cells mixed with agarose, alginate, methylcellulose, or Lutrol F127 (Poloxamer 407) using a 3D-Bioplotter (EnvisionTEC GmbH, Gladbeck, Germany), an extrusion-based multi-nozzle printing system. The results showed that the printing process did not have any significant effect on cell viability and the printed cells could undergo osteogenic differentiation. Ker et al. (2011) published a study in which aligned myocytes, tenocytes, and osteoblasts could be induced to differentiate by printing growth factors in specific locations with inkjet printing

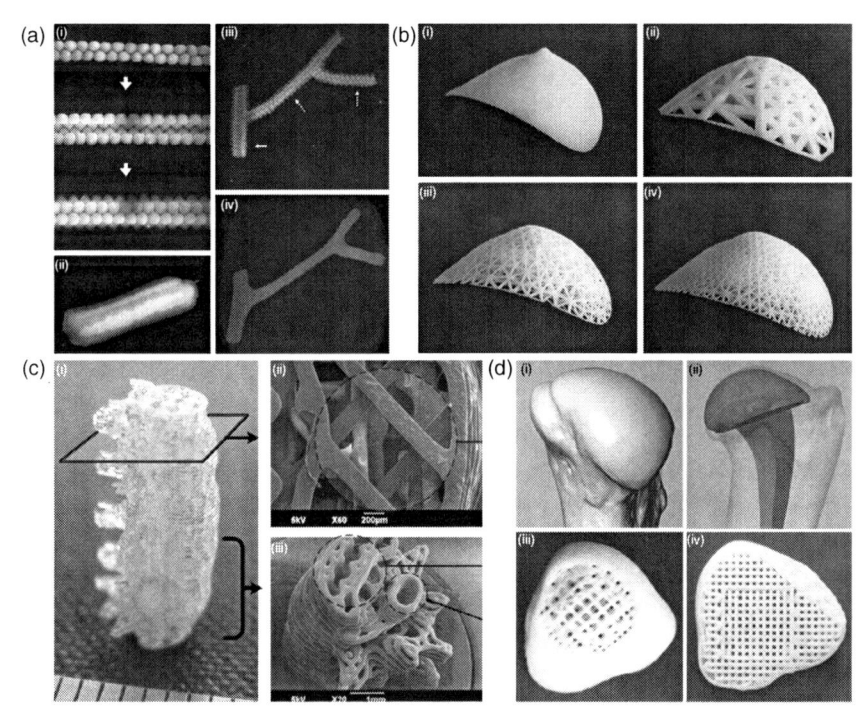

9.3 (a) Printed multicellular spheroids assembled into patterns to stimulate fusion and tubular structures. (i) Human skin fibroblast spheroids organized into 3-D structure. (ii–iv) Patterns of CHO cell spheroids printed as tubular structures and cultured to allow fusing, including branched structures to simulate vasculature. (*Source*: Reproduced with permission from Norotte *et al.*, *Biomaterials*, **30**, pp. 5910–5917 © Elsevier (Norotte *et al.* 2009).) (b) Physical prototypes of one solid and three porous models for a personalized tissue engineering scaffold, manufactured by FDM. The porous models demonstrate control of 3-D printing over shape, porosity, and structural design. (*Source*: Reproduced with permission from Melchels *et al.*, *Biofabrication*, **3**(3), p. 034114 © IOP Publishing Ltd (Melchels *et al.* 2011).) (c) Porous spine structure fabricated using micro stereolithography. (i) Anatomically relevant sizes and shapes can be demonstrated (ii, iii) while achieving spatial control over material patterns. (*Source*: Reproduced with permission from Kang *et al.*, *Biofabrication*, **4**(1), p. 015005 © IOP Publishing Ltd (Kang *et al.*, 2012).) (d) Regeneration of the articular surface of the rabbit synovial joint using (i) anatomically matched scaffold and infused growth factor (ii) utilized 3-D fabrication techniques. Guided diffusion was focused toward (iii) synovial cavity and (iv) the medullary cavity. (*Source*: Reproduced with permission from Lee *et al.*, *Lancet*, **376**(9739), pp. 7–13 © Elsevier (Lee *et al.*, 2010).)

technology. The printed constructs were composed of tendon-promoting fibroblast growth factor-2 (FGF-2) and bone-promoting BMP-2 on the aligned fibers produced by the spinneret-based tunable engineered parameters (STEP) technique. When myocytes, tenocytes, and osteoblasts were grown on these printed fibers, multiple differentiations were observed in the locations containing the printed specific growth factors.

Melchels *et al.* (2011) focused on more clinically applicable tissue constructs using advanced CAD/CAM techniques. They investigated the feasibility of employing laser scanning with CAD/CAM techniques to aid in breast reconstruction (Fig. 9.3b). A patient was imaged with laser scanning, an economical and widely available method for creating an accurate digital representation of the breasts and surrounding tissues. The model obtained was used to fabricate a customized mold that was employed as an intra-operative aid for the surgeon performing autologous tissue reconstruction of the breast removed due to cancer. Furthermore, a novel generic algorithm for creating porosity within a solid model was developed, using a finite element model as intermediate. Lee *et al.* (2009) applied printing technology to fabricate multi-layered skin-like structures using dermal fibroblasts and keratinocytes for skin tissue regeneration. The results indicated that a well-organized architecture mimicking human skin tissue was formed, and this structure was composed of dermal and epidermal layers. They also showed that the printing method could be used to fabricate freeform skin in the shape of a wound by printing a pre-designed pattern on a non-planar poly(dimethylsiloxane) (PDMS) mold.

Neural tissue exists in brain, spinal cord, and peripheral nerves. It functions to collect, transmit, and analyze information from both inside and outside the human body and controls the functions of other organs. Many researchers have recently presented preliminary results for neural tissue regeneration using organ printing technology which can alter outcomes for patients with traumatic injuries. Xu *et al.* (2005) showed a 2-D patterning containing rat embryonic motor neurons and Chinese hamster ovary (CHO) cells in a hydrogel substrate using a thermal inkjet printer. It was demonstrated that the viability of printed CHO cells and motor neuron cells could be retained with the pre-designed circular shapes. These shapes maintained the cellular properties and functionalities of the printed cells as confirmed by evaluating neuronal phenotype and electrophysiological characteristics. Ilkhanizadeh *et al.* (2007) studied differentiation of neural stem cells (NSCs) into astrocytes and SMCs using printing technology. The ciliary neurotrophic factor (CNTF) and fetal bovine serum (FBS) were incorporated in a polyacrylamide-based hydrogel substrate using an inkjet printer. This indicated that localized differentiation of NSCs occurred on the patterned area in a controlled manner. Lee *et al.* (2010) introduced a time-released growth factor delivery system that was fabricated using a jetting system based on pneumatic pressure. The printed patterns composed of NSCs and vascular endothelial growth factor (VEGF) with fibrin/collagen hydrogel. The results indicated that cell differentiation and migration could be controlled with this patterning strategy. Marga *et al.* (2012) introduced a scaffold-free, extrusion-based fabrication method that could be used to create biological nerve grafts for repair of peripheral nerve injury. The nerve grafts were printed with bone marrow-derived stem cells, Schwann cells, and agarose. *In vivo* nerve regeneration was observed in rats after three weeks of implantation. Kang *et al.* (2012c) printed a complex human spine composed of three parts: vertebrae, intervertebral disks, and spinal cord (Fig. 9.3c). The spine CAD model was divided into three sub-models in standard tessellation language (STL) format for the tissue type using 3ds Max® software by Autodesk®. The generated CAD models, along with the corresponding unit cells and incremental angles, were input into the developed unit, cell-based CAM system. Further studies are needed to combine multiple cell types within a 3-D structure.

While initial successes in developing simple-shaped tissue constructs for tissue engineering have been reported, there is an increasing demand for methods that will allow the formation of more complex, composite tissue constructs for clinical use (Mikos *et al.* 2006). Lee *et al.* (2010) fabricated an anatomically correct complex scaffold composed of a composite of poly(ε-caprolactone) (PCL) and hydroxyapatite using layer-by-layer printing (Fig. 9.3d). The result showed the regeneration of the entire articular surface of a synovial joint in a rabbit model. The ability to engineer a complex tissue construct will be more critical for the functional reconstruction of damaged tissues.

9.5 Future trends

The complexity of the natural biological structures found within the body is difficult to replicate. Moreover, current 3-D printing techniques have numerous limitations, and huge opportunities exist for research and development that could enhance 3-D printing technologies further. Though there is much work to be done to advance the field toward successful clinical translations, it will be exciting to see the interesting research that will be performed in the coming years as organ printing matures from its infancy.

Recently, there have been many attempts to improve the resolution (layer thickness) of 3-D printing. Inkjet-based method for biomaterial patterning has been shown to achieve high resolutions in the order of approximately 50 µm, but the technology lacks the ability to achieve clinically applicable organ-sized constructs with sufficient mechanical stability. On the other hand, the micro-filament extrusion-based method has been used to make filaments down to approximately 100–300 µm in layer thickness, but this is met with considerable trade-offs. At higher resolutions, the flow resistance and shear stress within the biomaterial increase dramatically due to the smaller nozzle diameter, resulting in cell death within the construct. Typically, the result is a drastic escalation in printing process time in order to maintain cell viability for the amplified number of high resolution layers required to complete a comparable structure. Related to resolution limitations, pattern intricacies become challenging, as complex tissues with many coordinating cell types need to be arranged in close proximity. In order to attain the juxtaposition of many cell types, high resolutions may be required. Other techniques may be required to create increasingly complex cell designs while reducing feature sizes required from the printing technology. In some ways, this could be analogous to fabrication in the semiconductor industry, in that higher resolutions of patterns aid in functionality, but new biomaterials are needed to perform the tasks at such high resolutions. Availability of biomaterials that can suffice as cell delivery bio-inks but which also provide mechanical support, cell-specific cues, and negligible cytotoxicity is limited. Advances in the field of suitable cell-compatible hydrogels for 3-D printing are necessary for the long-term success of organ printing.

As seen recurrently in tissue engineering and regenerative medicine, the prominent limitation to creating tissues or organs in the laboratory is vascularization of the new organ. Organ printing may have a unique ability among the varied tissue engineering technologies to overcome this limitation and is the subject of intense study. Several groups have made progress towards printing vascular structures (Cui and Boland, 2009; Mironov *et al.*, 2009; Nakamura *et al.*, 2010b). To date, integrating vascular structures into tissue- or organ-like constructs has not been achieved. Potential exists to fabricate cellularized constructs with high porosity, biologically active molecules, and organized patterns of multiple cell types to encourage vasculogenesis. Other areas of future research include tissues that are simple in nature or have few cell types. Having fewer variables will allow researchers to prove the concepts of biological interactions and achieve successful tissue fabrication on the way towards full organ regeneration.

9.6 Conclusion

Organ printing technology can construct 3-D freeform shapes with multiple cell types and biomaterials, resulting in sophisticated architectures that have the potential to replace human tissues or organs. Organ printing technology holds great promise in tissue engineering and regenerative medicine. Our efforts to deliver clinically applicable bioengineered tissues or organs will continually advance until the technology is able to improve the lives of patients.

9.7 References

Artavanis-Tsakonas, S. Rand, M.D. and Lake, R.J. (1999). Notch signaling: cell fate control and signal integration in development. *Science*, **284**(5415), pp. 770–776.

Atala, A. (2009). Engineering organs. *Curr. Opin. Biotech.*, **20**(5), pp. 575–592.

Binder, K.W., Allen, A.J., Yoo, J.J. and Atala, A. (2011). Drop-on-demand inkjet bioprinting: a primer. *Gene Ther. Regul.*, **6**(1), pp. 33–49.

Boland, T., Xu, T., Damon, B. and Cui X. (2006). Application of inkjet printing to tissue engineering. *Biotechnol. J.*, **1**, pp. 910–917.

Claeyssens, F., Hasan, E.A., Gaidukeviciute, A., Achilleos, D.S., Ranella, A., Reinhardt, C., Ovsianikov, A., Shizhou, X., Fotakis, C., Vamvakaki, M., Chichkov, B.N. and Farsari, M. (2009). Three-dimensional biodegradable structures fabricated by two-photon polymerization. *Langmuir*, **25**(5), pp. 3219–3223.

Cui, X. and Boland, T. (2009). Human microvasculature fabrication using thermal inkjet printing technology. *Biomaterials*, **30**(31), pp. 6221–6227.

Daley, W.P., Peters, S.B. and Larsen, M. (2008). Extracellular matrix dynamics in development and regenerative medicine. *J. Cell Sci.*, **121**(3), pp. 255–264.

Dhariwala, B., Hunt, E. and Boland, T. (2004). Rapid prototyping of tissue-engineering constructs, using photopolymerizable hydrogels and stereolithography. *Tissue Eng.*, **10**, pp. 1316–1322.

Fedorovich, N.E. *et al.* (2008). Three-dimensional fiber deposition of cell-laden, viable, patterned constructs for bone tissue printing. *Tissue Eng. Part A*, **14**, pp. 127–133.

Fedorovich, N.E. *et al.* (2010). 3-D-fiber deposition for tissue engineering and organ printing applications. In B. R. Ringeisen, B. J. Spargo and P. K. Wu, eds. *Cell and Organ Printing*. Dordrecht: Springer The Netherlands, pp. 225–239.

Guillemot, F., Guillotin, B., Catros, S., Souquet, A., Mezel, C., Keriquel, V., Hallo, L., Fricain, J.C. and Amedee, J. (2010). High-throughput laser printing of cells and biomaterials for tissue engineering. *Acta Biomater.*, **6**(7), pp. 2494–2500.

Ilkhanizadeh, S., Teixeira, A. and Hermanson, O. (2007). Inkjet printing of macromolecules on hydrogels to steer neural stem cell differentiation. *Biomaterials*, **28**, pp. 3936–3943.

Jakab, K., Norotte, C., Damon, B., Marga, F., Neagu, A., Besch-Williford, C., Kachurin, A., Church, K.H., Park, H., Mironov, V., Markwald, R., Vunjak-Novakovic, G. and Forgacs, G. (2008). Tissue engineering by self-assembly of cells printed into topologically defined structures. *Tissue Eng. Part A*, **14**(3), pp. 413–421.

Kang, H., Lee, S.J., Atala, A. and Yoo, J.J. (2012a). Integrated organ and tissue printing methods, system, and apparatus, US Patent US 2012/0089238 A1.

Kang, H., Park, J. and Cho, D. (2012b). A pixel based solidification model for projection based stereolithography technology. *Sensor Actuat A-Phys*, **178**, pp. 223–229.

Kang, H., Park, J., Kang, T., Seol, Y. and Cho, D. (2012c). Unit cell-based computer-aided manufacturing system for tissue engineering. *Biofabrication*, **4**(1), p. 015005.

Ker, E., Nain, A.S, Weiss, L.E., Wang, J., Suhan, J., Amon, C.H. and Campbell, P.G. (2011). Bioprinting of growth factors onto aligned sub-micron fibrous scaffolds for simultaneous control of cell differentiation and alignment. *Biomaterials*, **32**, pp. 8097–8107.

Khalil, S. and Sun, W. (2009). Bioprinting endothelial cells with alginate for 3D tissue constructs. *J. Biomech. Eng.*, **131**(11), p. 111002.

Lee, C.H. B., Polio, S., Lee, W., Dai, G., Menon, L., Carroll, R.S. and Yoo, S.S. (2010a). Regeneration of the articular surface of the rabbit synovial joint by cell homing: a proof of concept study. *Lancet*, **376**(9739), pp. 7–13.

Lee, W. *et al.* (2009). Multi-layered culture of human skin fibroblasts and keratinocytes through three-dimensional freeform fabrication. *Biomaterials*, **30**, pp. 1587–1595.

Lee, Y. B., Polio, S., Lee, W., Dai, G., Menon, L., Carroll, R.S. and Yoo, S.S. (2010b). Bio-printing of collagen and VEGF-releasing fibrin gel scaffolds for neural stem cell culture. *Exp. Neurol.*, **223**, pp. 645–652.

Marga, F., Jakab, K., Khatiwala, C., Shepherd, B., Dorfman, S., Hubbard, B., Colbert, S. and Forgacs, G. (2012). Toward engineering functional organ modules by additive manufacturing. *Biofabrication*, **4**, p. 022001.

Melchels, F., Wiggenhauser, P.S., Warne, D., Barry, M., Ong, F.R., Chong, W.S., Hutmacher, D.W. and Schantz, J.T. (2011). CAD/CAM-assisted breast reconstruction. *Biofabrication*, **3**(3), p. 034114.

Melchels, F., Domingos, M., Klein, T.J., Malda, J., Bartolo, P.J. and Hutmacher, D.W. (2012). Additive manufacturing of tissues and organs. *Prog. Polym. Sci.*, **37**(8), pp. 1079–1104.

Mikos, A.G., Herring, S.W., Ochareon, P., Elisseeff, J., Lu, H.H., Kandel, R., Schoen, F.J., Toner, M., Mooney, D., Atala, A., Van Dyke, M.E., Kaplan, D. and Vunjak-Novakovic, G. (2006). Engineering complex tissues. *Tissue Eng.*, **12**(12), pp. 3307–3339.

Mironov, V., Boland, T., Trusk, T., Forgacs, G. and Markwald, R.R. (2003). Organ printing: computer-aided jet-based 3-D tissue engineering. *Trends Biotechnol.*, **21**(4), pp. 157–161.

Mironov, V., Visconti, R.P., Kasyanov, V., Forgacs, G., Drake, C.J. and Markwald, R.R. (2009). Organ printing: tissue spheroids as building blocks. *Biomaterials*, **30**(12), pp. 2164–2174.

Mironov, V., Reis, N. and Derby, B. (2006). Review: bioprinting: a beginning. *Tissue Eng.*, **12**(4), pp. 631–634.

Nakamura, M., Kobayashi, A., Takagi, F., Watanabe, A., Hiruma, Y., Ohuchi, K., Iwasaki, Y., Horie, M., Morita, I. and Takatani, S. (2005). Biocompatible inkjet printing technique for designed seeding of individual living cells. *Tissue Eng.*, **11**, pp. 1658–1666.

Nakamura, M., Iwanaga, S., Henmi, C., Arai, K. and Nishiyama, Y., (2010a). Biomatrices and biomaterials for future developments of bioprinting and biofabrication. *Biofabrication*, **2**, p. 014110.

Nakamura, M. (2010b). Reconstruction of biological three-dimensional tissues: bioprinting and biofabrication using inkjet technology. In B. R. Ringeisen, B. J. Spargo and P. K. Wu, eds. *Cell and Organ Printing*. Dordrecht: Springer The Netherlands, pp. 23–33.

Nerem, R.M. (2006). Tissue engineering: the hope, the hype, and the future. *Tissue Eng.*, **12**(5), pp. 1143–1150.

Nishiyama, Y., Nakamura, M., Henmi, C., Yamaguchi, K., Mochizuki, S., Nakagawa, H. and Takiura, K. (2009). Development of a three-dimensional bioprinter: construction of cell supporting structures using hydrogel and state-of-the-art inkjet technology. *J. Biomech. Eng.*, **131**(3), pp. 035001–035006.

Norotte, C., Marga, F., Niklason, L. and Forgacs, G. (2009). Scaffold-free vascular tissue engineering using bioprinting. *Biomaterials*, **30**, pp. 5910–5917.

Ovsianikov, A., Gruene, M., Pflaum, M., Koch, L., Maiorana, F., Wilhelmi, M., Haverich, A. and Chichkov, B. (2010). Laser printing of cells into 3-D scaffolds. *Biofabrication*, **2**, p. 014104.

Phillippi, J., Miller, E., Weiss, L., Huard, J., Waggoner, A. and Campbell, P. (2008). Microenvironments engineered by inkjet bioprinting spatially direct adult stem cells toward muscle- and bone-like subpopulations. *Stem Cells*, **26**, pp. 127–134.

Rehfeldt, F., Engler, A.J., Eckhardt, A., Ahmed, F. and Discher, D.E. (2007). Cell responses to the mechanochemical microenvironment-Implications for regenerative medicine and drug delivery. *Adv. Drug Deliv. Rev.*, **59**(13), pp. 1329–1339.

Romer, L.H., Birukov, K.G. and Garcia, J.G.N. (2006). Focal adhesions paradigm for a signaling nexus. *Circ. Res.*, **98**(5), pp. 606–616.

Roth, E., Xu, T., Das, M., Gregory, C., Hickman, J.J. and Boland, T. (2004). Inkjet printing for high-throughput cell patterning. *Biomaterials*, **25**(17), pp. 3707–3715.

Sun, W., Starly, B., Nam, J. and Darling, A. (2005). Bio-CAD modeling and its applications in computer-aided tissue engineering. *CAD*, **37**, pp. 1097–1114.

Xu, T., Baicu, C., Aho, M., Zile, M. and Boland, T. (2009). Fabrication and characterization of bio-engineered cardiac pseudo tissues. *Biofabrication*, **1**, p. 035001.

Xu, T., Gregory, C.A., Molnar, P., Cui, X., Jalota, S., Bhaduri, S.B. and Boland, T. (2006). Viability and electrophysiology of neural cell structures generated by the inkjet printing method. *Biomaterials*, **27**, pp. 3580–3588.

Xu, T., Jin, J., Gregory, C., Hickman, J. and Boland, T. (2005). Inkjet printing of viable mammalian cells. *Biomaterials*, **26**, pp. 93–99.

10
Rapid prototyping technology for bone regeneration

J. KUNDU, F. PATI, J.-H. SHIM and D.-W. CHO, Pohang
University of Science and Technology, Korea

DOI: 10.1533\9780857097217.254

Abstract: Bone is a dynamic, highly vascularized tissue with a unique capacity to heal and remodel without scarring. Tissue engineering (TE) offers a promising new approach for bone repair, and the development of scaffolds used to restore damaged bone tissue is increasingly used in bone TE. Various methods for manufacturing scaffolds to replace bone tissue have been developed. However, scaffolds developed using conventional fabrication techniques have severe disadvantages in precision and reproducibility. To overcome these, tissue engineers have devised strategies, such as solid free-form fabrication techniques, to fabricate porous, fully interconnected scaffolds for bone TE applications. In addition, cell-printing technology has been used to manufacture cell-laden bone regeneration constructs. This chapter reviews RP-based three-dimensional bone-graft analogues fabricated using state-of-the-art computer-assisted tissue-fabrication strategies for bone TE.

Key words: rapid prototyping, 3D scaffold, cell printing, bone tissue regeneration.

10.1 Introduction

Tissue engineering (TE) is the application of biological, chemical, and engineering principles to the repair of living tissues using biomaterials, cells, and growth factors, alone or in combination (Langer and Vacanti, 1993). Bone engineering or bone growth has been practiced for centuries to repair fractures by implanting orthopedic devices and bone substitute materials such as bone grafts (Mistry and Mikos, 2005). But although a sizeable market exists for bone-graft substitutes, the availability of bone autografts is limited, and the use of allografts is associated with the potential for disease transmission and immune reactions. Considering the limitations associated with bone-graft materials, engineers and clinicians are working together to design bone-graft substitutes (Laurencin *et al.*, 1999).

Bone has a three-dimensional (3D) configuration, but cells do not grow in a 3D fashion *in vitro*; thus, a 3D structure, or scaffold, mimicking the bone structure, must be used so that new tissue grows in a 3D environment.

Material and fabrication technologies are critical in the design of scaffolds for bone TE. New methodologies such a rapid prototyping (RP) have become available for use in the TE field with advances in computer and processing technology (Hutmacher *et al.*, 2004). These methodologies are computerized fabrication techniques that can produce highly complex 3D physical objects using data generated by computer-assisted design (CAD) systems (Lee *et al.*, 2010). RP techniques use the underlying concept of layered manufacturing, whereby 3D objects are fabricated layer by layer by processing a solid sheet, liquid, or powder material stock (Lee *et al.*, 2010). The most used techniques within this field are 3D printing (Lee *et al.*, 2011a), fused deposition modeling (FDM) (Hutmacher *et al.*, 2004), 3D plotting (Mironov *et al.*, 2007), and indirect RP approaches (Peltola *et al.*, 2008). 3D printing, which employs inkjet and dispensing technology, was the first RP technique to be proposed for biomedical and TE purposes (Lee *et al.*, 2010, 2011a; Hutmacher *et al.*, 2004). During fabrication, a printer head is used to print a liquid binder onto thin layers of powder, following the object's profile as generated by a CAD file. The subsequent stacking and printing of material layers to the top of the previously printed layer re-creates the full structure of the desired object, in this case scaffolds for bone TE (Buttler *et al.*, 1996; Hutmacher, 2001).

This chapter reviews 3D printing-based bone-graft analogues developed using state-of-the-art computer-assisted tissue-fabrication strategies for bone regeneration. A brief background of bone biology is provided, followed by an exhaustive description of all relevant components of bone TE, from materials to scaffolds and from cells to state-of-the-art computer-assisted RP strategies that lead to regenerated bone.

10.2 Bone: properties, structure, and modeling

Bones are the rigid organs that comprise the endoskeleton of the human skeleton and protect the other organs of the human body. In the following section, we shall highlight the anatomy, function and pathology of bone, followed by composition, structure and mechanical properties, and finally we conclude the section with insights into bone remodeling. This background information is critical to determine the strategies in engineering of bone tissue using rapid prototyping-based technologies.

10.2.1 Anatomy, function, and pathology

The adult human skeleton has 213 bones, which are constantly undergoing modeling to help adapt to changing biomechanical forces, as well as remodeling to remove old, microdamaged bone and replace it with new, mechanically stronger bone to help preserve bone strength (Nijweide *et al.*,

1996). Bone tissue is distributed as trabecular, cancellous, or spongy bone (approximately 20% of the total skeleton), and cortical or compact bone (approximately 80% of the total skeleton) (Buttler *et al.*, 1996). Cortical bone is the more dense form of bone tissue, and is usually found on the surface of bone. It is organized in cylindrically shaped elements called osteons composed of concentric lamellae (Buttler *et al.*, 1996). The four general categories of cortical bone are long bone (femur and tibia), short bone (wrist and ankle), flat bone (skull vault), and irregular bone (Nijweide *et al.*, 1996). Trabecular bone is quite porous (50%–90%), and is organized in trabecules oriented according to the direction of the physiological load. The modulus and ultimate compressive strength of trabecular bone are approximately 20 times less than that of cortical bone (Nijweide *et al.*, 1996).

Bone serves a multitude of functions in the body, including locomotor support, providing a sequestered environment for calcium homeostasis, and as an organ and tissue (Service, 2000). The cellular component is made of osteoblasts, osteoclasts, and osteocytes, which are trapped in the extracellular matrix (ECM). The ECM, which is responsible for the mechanical strength of bone tissue, is formed of organic and mineral phases. The importance of bone regeneration is significant during diseases in which bone does not function properly, such as osteogenesis imperfecta, osteoarthritis, osteomyelitis, and osteoporosis (Howard *et al.*, 2008).

10.2.2 Composition and structure

Bone is composed of a cellular component and the ECM (Howard *et al.*, 2008; Nijweide *et al.*, 1996; Service, 2000) which contains 85%–90% collagenous proteins. The bone matrix is mostly composed of type I collagen, with trace amounts of types III and V collagens at certain stages of bone formation that may help determine collagen fibril diameter (Service, 2000). Noncollagenous proteins comprise 10%–15% of total bone protein. Approximately 25% of the noncollagenous protein is exogenously derived, including serum albumin and glycoprotein, which bind to hydroxyapatite because of their acidic properties (Buttler *et al.*, 1996). The remaining exogenously derived noncollagenous proteins are composed of growth factors and a large variety of other trace molecules that may affect bone cell activity. The noncollagenous proteins are divided broadly into several categories, including proteoglycans, glycosylated proteins, glycosylated proteins with potential cell-attachment activities, and γ-carboxylated (Gla) proteins (Nijweide *et al.*, 1996).

The main glycosylated protein present in bone is alkaline phosphatase (ALP). ALP is bound to osteoblast cell surfaces in bone via a phosphoinositol linkage, and is also found free within the mineralized matrix (Sommerfeldt and Rubin, 2001). ALP plays an important role in bone mineralization. The most prevalent noncollagenous protein in bone is osteonectin, accounting for approximately 2% of total protein in developing bone. Osteonectin is thought to affect osteoblast growth and/or proliferation and matrix mineralization (Qian et al., 2008).

The mineral content of bone is mostly hydroxyapatite [$Ca_{10}(PO_4)_6(OH)_2$], with small amounts of carbonate, magnesium, and acid phosphate, but missing hydroxyl groups that are normally present (Service, 2000). Matrix maturation is associated with the expression of ALP and several noncollagenous proteins, including osteocalcin, osteopontin, and bone sialoprotein (Ducheyne and Qiu, 1999). It is thought that these calcium and phosphate-binding proteins help regulate ordered deposition of minerals by regulating the amount and size of the hydroxyapatite crystals formed (Service, 2000).

Bone minerals provide mechanical rigidity and load-bearing strength to bone, whereas the organic matrix provides elasticity and flexibility. Matrix extracellular vesicles are synthesized by chondrocytes and osteoblasts and serve as protected microenvironments in which calcium and phosphate concentrations can increase sufficiently to precipitate crystal formation (Ducheyne and Qiu, 1999). The extracellular fluid is not normally supersaturated with hydroxyapatite, so it does not spontaneously precipitate. Matrix extracellular vesicles contain a nucleation core composed of proteins and a complex of acidic phospholipids, calcium, and inorganic phosphate that is sufficient to precipitate hydroxyapatite crystals. It is not yet certain how matrix extracellular vesicles contribute to mineralization at specific sites at the ends of collagen fibrils, because the vesicles are not apparently directly targeted to the ends of fibrils (Gonnerman et al., 2006).

10.2.3 Mechanical properties

The mechanical properties of bone are related to its complex physical structure. Bone mass accounts for 50%–70% of bone strength (Toma et al., 1997). Bone geometry and composition are important because larger bones are stronger than smaller bones, even with equivalent bone mineral density (Toma et al., 1997). As bone diameter expands radially, the strength of bone increases by the radius of the involved bone raised to the fourth power. The amount and proportion of trabecular and cortical bone at a given skeletal site affect bone strength independently (Cowan et al., 2005). Cortical bone possesses a yield strength of 78–151 MPa in tension and 131–224 MPa in compression, when tested along its longitudinal axis. Bone has yield strength of 51–66 MPa in tension and 106–131 MPa in compression when tested along its transverse axis. The Young's modulus of cortical bone is 17–20 GPa along the longitudinal axis and 6–13 GPa along the transverse axis (Toma et al., 1997). The mechanical properties of cancellous bone vary widely, and are related to the apparent density/porosity of the trabeculae

(Toma et al., 1997). Cancellous bone is highly viscoelastic, and its mechanical properties depend on the loading rate. Midrange values for cancellous bone are a strength of 5–10 MPa and modulus of 50–100 MPa (Cowan et al., 2005).

Bone material properties are important for bone strength. Mutations in certain proteins may cause bone weakness (e.g., collagen defects cause decreased bone strength in osteogenesis imperfecta and impaired γ-carboxylation of Gla proteins) (Yoshikawa and Myoui, 2005). Bone strength can be affected by osteomalacia, fluoride therapy, or hypermineralization states. Most importantly, a synthetic scaffold must have similar mechanical properties compared to native bone to simulate the dynamic mechanical physiological environment (Goldstein et al., 1983).

10.2.4 Bone remodeling

Bone remodeling is a critical process for maintaining skeletal integrity, healing, blood calcium regulation, and accommodation of changes in bone stress profiles (Hadjidakis and Androulakis, 2006). Bone adapts and remodels in response to the stress applied. Bone remodeling is a complex process by which old bone is continuously replaced by new tissue, and occurs in three distinct phases: resorption (osteoclasts are activated through paracrine pathways to digest bone), reversal (mononuclear cells appear on the surface), and formation (osteoblasts produce and secrete ECM) (Hadjidakis and Androulakis, 2006). Bone modeling and remodeling is responsible for adjusting the architecture and mechanical properties of bone as a function of mechanical and chemical signaling (Parfitt, 1984). Hormones (i.e., parathyroid hormone, calcitrol, glucocorticoids, and sex hormones) and growth factors (i.e., insulin-like growth factor (IGF), prostaglandins, transforming growth factor-βs, and bone morphogenetic proteins (BMPs)) drive and regulate bone remodeling in complicated biochemical cascades (Parfitt, 1984). Wolff's law states that bones develop a structure most suited to resist the forces acting upon them, adapting both the internal architecture and the external conformation to the change in external loading conditions. This change follows precise mathematical laws (Hayes, 1991). Bone tissue seems to be able to detect the change in strain on local bases and then adapt accordingly. The internal architecture adapts in terms of change in density and in disposition of trabecules and osteons, and the external conformation adapts in terms of shape and dimensions (Hayes, 1991). When strain is intensified, new bone is formed; on a microscopic scale, the bone density increases, and on a macroscopic scale, the bone external dimensions increase. When strain is lowered, bone resorption takes place, and the opposite occurs (Athanasiou et al., 2000).

10.3 Engineering of bone tissue

Bone TE offers a practical option for treating traumatized or diseased skeletal tissue by deliberately manipulating cellular and biological processes. TE strategies are often categorized into three broad groups: (1) direct injection of cells into the tissue of interest, (2) implantation of cell-scaffold constructs after culturing to form a 3D tissue structure, and (3) scaffold-based delivery of signaling molecules such as growth factors, capable of stimulating cell migration, growth, and differentiation (Lee *et al.*, 2011b). Cellular implantation is the classical method of cell-based reconstruction therapy, in which individual cells or small cellular aggregates from the patient are transferred to the bone defect (Lee *et al.*, 2011b). Cells are commonly placed in a hydrogel during transfer to the desired defect site. 3D tissue is grown *in vitro* using cells within a mechanically stable scaffold for extracorporeal TE. After cell growth and differentiation, the cell-scaffold constructs are implanted into the defect site (Meyer *et al.*, 2004). In the early decades of bone reconstruction, surgeons used artificial tissue substitutes containing metals, ceramics, and polymers to maintain skeletal function (Sachlos and Czernuszka, 2003). These artificial materials significantly improved the ability of surgeons to restore form, and, to some extent, function of defective bones. However, all artificial materials have disadvantages; thus, use of biomaterials is a common treatment option in clinical practice (Nöth *et al.*, 2010). Special emphasis is being given to the use of soluble therapeutic agents by different approaches for engineering bone tissue to accelerate healing. These agents have been used for both local bone repair (i.e., introduction of agents directly into a repair site) as well as systemic bone regeneration (i.e., delivery for regeneration throughout the skeletal system) (Gittens and Uludag, 2001).

10.3.1 Cells

Possible sources for bone TE are autologous, allogeneic, and xenogenic cells, which can be categorized according to the cell differentiation stage (Lee *et al.*, 2011b). Osteoblasts are the building blocks of bone in the appendicular, craniofacial, and axial skeletons. It has long been known that the vast bone regeneration capacity is due to the presence of differentiated osteoblasts. Many studies have considered osteoblast-like cells for bone TE (Jayakumar and Di Silvio, 2010). Although preosteoblasts retain some proliferative potential, osteoblasts normally do not undergo mitosis (Hughes and Aubin, 1997). The need for additional cell sources is particularly evident for severe trauma cases, cancer treatment, and maxillofacial reconstructive surgery in which large bone defects cannot be filled solely with artificial scaffolds or autografts. Skeletal developmental diseases, such as osteogenesis imperfecta, and degenerative diseases, such as osteoporosis, are associated with poor bone quality and could also benefit from cell-based therapy (Colnot, 2011).

Osteogenic differentiation of mesenchymal, umbilical cord blood, and embryonic stem cells has been characterized, is well-practiced in many laboratories, and has been successfully used in bone TE approaches (Buttery et al., 2001; George et al., 2006; Goodwin et al., 2001; Maniatopoulos et al., 1988). The majority of bone TE approaches take advantage of bone marrow-derived cells, which are easily accessible and have been extensively described in the literature. These cells can differentiate into chondrocytes and osteoblasts in vitro, and appear to be an ideal autologous cell type (Bianco and Robey, 2001, Viateau et al., 2007). Other autologous cell types are similarly attractive, such as adipose-derived cells, which are also very accessible, and exhibit osteogenic and chondrogenic potential in vitro (Bodle et al., 2011).

Various sources of stem cells are now being employed for bone TE. Among them, some cell sources contribute normally to bone repair (bone marrow and periosteum), whereas others may or may not participate in repair (fat and muscle). Embryonic stem cells, induced-pluripotent stem cells, and cord blood cells have also been investigated, but do not usually participate in adult tissue repair (Gruenloh et al., 2011; Kahle et al., 2010; Kuznetsov et al., 2011). Skeletal stem cells have therapeutic potential as they are able to differentiate into osteoblasts and/or chondrocytes in vitro. However, little is known about the fate of these cells during bone repair. The swift integration of skeletal stem/progenitor cells into a fracture callus or a bone defect and their differentiation toward chondrogenic and osteogenic pathways in situ remains a challenge (Kuznetsov et al., 2011).

Bone marrow-derived stem cells (BMSCs) have been commonly used for both clinical and experimental purposes with varying degrees of success. Regrettably, BMSCs alone are not very effective as exogenous osteoblast progenitors, and their capacity decreases with age (Muschler et al., 2001; Stolzing et al., 2008). Although BMSCs are easy to collect compared with other adult stem cells, there is risk of donor site morbidity (Sen and Miclau, 2007). Periosteum-derived cells generally have high regenerative potential and contribute directly to cartilage and bone (Roberts et al., 2011; Zhang et al., 2005), but they are difficult to harvest (Colnot, 2011). Additionally, other bone cell lines, such as genetically altered cell lines (sarcoma cells, immortalized cells, and nontransformed clonal cell lines), have been developed and used to evaluate basic aspects of in vitro cell behavior in nonhuman settings (Handschel et al., 2006). A number of animal models have been developed for cell-based therapies, including stabilized and nonstabilized fractures, distraction osteogenesis, segmental defects, cortical and calvarial defects, implant osseointegration, bone grafting, and bone transplantation (Colnot, 2011).

10.3.2 Matrix materials

To augment the therapeutic effects of cells delivered locally, many efforts have been focused on scaffold designs to create a biocompatible environment and to provide a surface for cell adhesion and migration. An engineered scaffold is more critical for bone TE than other systems, such as cell delivery without scaffolds. Engineered scaffolds not only serve as structural conduits for bone regeneration, but also provide mechanical support for load bearing (Sundaramurthy and Mao, 2006). The classic orthopedic carriers include allogeneic bone, demineralized bone matrix, and various bone-graft substitutes made of both natural and synthetic originating polymers, hydroxyapatite, and calcium phosphate. New scaffolds are now being developed using nanotechnologies to combine nanofiber mesh with biocompatible carriers such as hydrogels (Wang et al., 2011).

Natural biomaterials

These are promising for applications in cell delivery scaffolds, as they mimic certain aspects of native ECM. Besides exhibiting optimal biocompatibility and biodegradability, they facilitate cell adherence, migration, differentiation, and ECM deposition. In contrast, there are certain limitations of natural polymers, such as their variable range of physicochemical properties, a requirement for extensive purification protocols, and potential pathogen contamination when they are harvested from animal or human sources.

Collagen is the major structural component of the ECM of different connective tissues, including tendons/ligaments, cartilage, bone, and skin. This protein regulates essential cellular events, such as proliferation, migration, and differentiation via cell–matrix interactions (Friess, 1998). Collagen contains specific adhesion domain sequences (i.e., arginine-glycine-aspartic acid (RGD)) that may function in retaining the cell phenotype and regulating important cellular events via integrin binding (Ruoslahti, 1996). Specifically, collagen type I has been extensively investigated as a scaffold material and can be processed as nanofibrous, nonwoven mesh, sponge, membrane, fleece, and hydrogel for cartilage, bone, tendon, ligament, meniscus, and vertebral disc constructs (Glowacki and Mizuno, 2008).

Gelatin is derived from collagen via a denaturing process using either alkali or acid treatment, resulting in a charged polyelectrolyte with different isoelectric point values of 5.0–9.0. These positively or negatively charged polyelectrolytes interact with oppositely charged biomolecules to form polyionic complexes (Ikada and Tabata, 1998, Young et al., 2005). This offers the possibility of binding and releasing growth factors or incorporating proteins and peptides (e.g., vitronectin, fibronectin, and RGD peptides) that can influence cell adhesion and growth in a controlled manner and have been used in TE targeting different tissues including bone (Van Vlierberghe et al., 2011).

Fibrin, a protein matrix, is derived from fibrinogen under the enzymatic action of thrombin (Wolberg, 2007). Fibrin glue (fibrinogen plus thrombin) has been extensively used as an injectable biomaterial due to its biomimetic and physical properties. This glue can be mixed with cells and is rapidly invaded, remodeled, and replaced by the transplanted or host cells (Ahmed et al., 2008).

Hyaluronic acid (HA), a nonsulfated glycosaminoglycan, is a major ECM macromolecule in many connective tissues. The physiological role of HA is associated with ECM fluid regulation and the structural integrity of the tissue (Liao et al., 2005). HA can be applied as an injectable, gel-like cell carrier or as a pre-shaped (nano)fibrous scaffold for cartilage, bone, and osteochondral regeneration (Van Vlierberghe et al., 2011).

Chondroitin sulfate (CS) is another glycosaminoglycan that is present in the ECM of many connective tissues. CS is covalently linked via a link protein to a core protein to form proteoglycans (Jackson et al., 1991; Liao et al., 2005). Similar to other glycosaminoglycans, CS modulates the binding of growth factors and cytokines as well as protease inhibition, and, thus regulates cell adhesion, migration, proliferation, and differentiation (Wang et al., 2007). CS is used alone or in combination with other natural or synthetic biomaterials to obtain a more stable design and retain the favorable characteristics of CS (Wang et al., 2007).

Chitosan, a cationic polymer, is the partially or fully deacylated form of chitin, a natural polysaccharide found in the shell of crustaceans and insects (Khor and Lim, 2003). Chitosan is extensively used in a variety of TE and drug-delivery applications owing to its renewable, biodegradable, biocompatible, nonantigenic, nontoxic, and biofunctional qualities (Khor and Lim, 2003; Martino et al., 2005). Furthermore, chitosan reveals certain structural similarities to glycosaminoglycans, and is of particular interest for use in cartilage regeneration (Martino et al., 2005). Additionally, chitosan can be processed as fibers, granules, sponge, or hydrogel due to its biochemical properties, and has been used for bone TE (Van Vlierberghe et al., 2011).

Alginate is a polysaccharide polymer harvested principally from marine brown algae. Alginates undergo reversible gelation and form a hydrogel upon interaction with divalent cations, such as Ca^{2+} in aqueous solution (George and Abraham, 2006), which offers the possibility of homogeneous encapsulation of cells and/or retention of growth factors within the hydrogel. Alginate hydrogel beads promote cell growth and deposition of newly synthesized ECM for primary chondrocytes and MSCs *in vitro* and *in vivo* (Nicodemus and Bryant, 2008).

Synthetic biomaterials

These include man-made polymers (poly(α-hydroxy esters)) and ceramics, and are used extensively in bone TE. They offer high primary stability and are more amenable to macro/microstructure formation. Among the poly(α-hydroxy esters), polyglycolic acid (PGA), polylactic acid (PLA), their copolymer poly(lactic-co-glycolic acid) (PLGA), and poly-ε-caprolactone (PCL) are widely used and have gained Food and Drug Administration (FDA) approval for human use in a variety of applications (Yu et al., 2010). Degradation of these polymers occurs through bulk erosion by hydrolysis of ester bonds. The degradation products, such as lactic and glycolic acid, are generally nontoxic and physiologically eliminated from the body via metabolic pathways to form carbon dioxide and water (Rezwan et al., 2006). The degradation rates of these polymers can be adjusted from weeks to several years by altering the initial molecular weight, crystallinity, and co-polymer ratio (Dawson et al., 2008; Rezwan et al., 2006). Furthermore, 3D scaffolds with desired microarchitecture, porosity, biomechanical properties, and gross shape can be manufactured due to the physicochemical nature of poly(α-hydroxy esters) using various techniques, including crystal leaching, porogen melting, gas-foaming, sintering, and nanofiber electrospinning (Li and Tuan, 2009; Rezwan et al., 2006).

Ceramics

Ceramics such as calcium phosphates, calcium sulfates, and bioactive glass, exhibit high compressive strength, variable degradation time from weeks (calcium sulfate) to nondegradable (crystalline hydroxyapatite), and have been used widely in bone TE (Khan et al., 2008). Calcium phosphates have been used particularly owing to their association with calcium hydroxyapatite (Hak, 2007; Sarkar et al., 2001). Calcium phosphates stimulate the formation, precipitation, and deposition of calcium phosphate from solution due to their high protein binding affinity, and result in enhanced bone–matrix interface strength (Goyenvalle et al., 2003).

10.3.3 Growth factors

Growth factors are soluble, cell-secreted, signaling polypeptides capable of initiating specific cellular responses in a biological environment, and can result in a very wide range of cell actions, including cell survival, migration, and differentiation or proliferation of a specific subset of cells (Cross and Dexter, 1991). Growth factors provide signals at local injury sites of bone allowing progenitors and inflammatory cells to migrate and trigger the healing process (Furth et al., 2007, Lee and Shin, 2007). Several attempts have been made to promote bone regeneration using growth factors and morphogen (Table 10.1) included within bioactive scaffolds to stimulate cellular adhesion, proliferation, and differentiation (Lee *et al.*, 2011b). Growth factors, added either exogenously or through genetic manipulation, stimulate angiogenesis and osteogenic differentiation. Purified BMSCs grown on various materials reveal a better response in most animal models when combined with exogenous growth factors (Gittens and Uludag, 2001; Lee *et al.*, 2011b). Enhanced bone repair in a rat critical-sized defect occurs when BMP-7 is added to human BMSCs (Burastero *et al.*, 2010). In another animal study, BMSCs were transfected with an adenovirus vector to overexpress BMP-2, leading to improved union of a mouse critical-sized defect (Lieberman *et al.*, 1999). A number of animal and clinical studies have provided experimental evidence for potential use of growth hormone (GH/IGF) system components to increase bone mass (Kasukawa *et al.*, 2004). Genetic modification is also a powerful method to drive osteogenesis via either the expression of BMPs, key transcription factors, and/or angiogenic factors (Kumar *et al.*, 2010). Although the main aim of these *ex vivo* genetic manipulations is to further enhance the regenerative potential of cells (Steinert *et al.*, 2007), they may also induce major alterations in their inherent osteogenic capacity. These approaches are still in their infancy, as new vectors are required to prevent tumorigenesis, which may occur due to insertional mutations (Illich *et al.*, 2011). New vectors are being developed and nonviral approaches, such as gene activated matrices, are also promising for bone TE (Evans, 2011).

10.4 Conventional scaffolds for bone regeneration

Relevant cell(s) are allowed to grow *in vitro* into 3D organs or tissues by seeding the cells onto porous matrices, known as scaffolds, to which the cells attach and colonize (Langer and Vacanti, 1993). The scaffold is a very important TE component, as cells lack the ability to grow in 3D orientations, and they randomly migrate to form a 2D layer of cells, thus lacking the desired anatomical shape of the tissue (Ikada, 2006). Several requirements have been identified as crucial for fabricating TE scaffolds (Hutmacher, 2001): the scaffold should (1) possess interconnecting pores of appropriate scale to favor tissue integration and vascularization, (2) be made from material with controlled biodegradability or bioresorbability so that the tissue will eventually replace the scaffold, (3) have appropriate surface chemistry to favor cellular attachment, differentiation, and proliferation, (4) possess adequate mechanical properties to match the intended site of implantation and handling, (5) should not induce any adverse response, and (6) be easily fabricated into a variety of shapes and sizes. Several techniques have been developed to process natural and synthetic materials into porous structures to fulfill these requirements. There are both top-down and bottom-up

Table 10.1 Growth factors in tissue regeneration

Growth factor	Tissues treated	Representative function
FGF-2	Blood vessel, bone, skin, nerve, spine, muscle	Migration, proliferation and survival of endothelial cells, inhibition of differentiation of embryonic stem cells
BMP-2	Bone, cartilage	Differentiation and migration of osteoblasts
BMP-7	Bone, cartilage, kidney	Differentiation and migration of osteoblasts, renal development
HGF	Bone, liver, muscle	Proliferation, migration and differentiation of mesenchymal stem cells
IGF-1	Muscle, bone, cartilage, bone liver, lung, kidney, nerve, skin	Cell proliferation and inhibition of cell apoptosis
PDGF-AB (or -BB)	Blood vessel, muscle, bone, cartilage, skin	Embryonic development, proliferation, migration, growth of endothelial cells
TGF-β	Bone, cartilage	Proliferation and differentiation of bone-forming cells, antiproliferative factor for epithelial cells

bFGF, basic fibroblast growth factor; BMP, bone morphogenetic protein; FGF, fibroblast growth factor; HGF, hepatocyte growth factor; IGF, insulin-like growth factor; PDGF, platelet-derived growth factor; TGF, transforming growth factor.

Sources: Gittens and Uludag, 2001, Lee *et al.*, 2011.

approaches for scaffold manufacturing strategies using conventional and advanced techniques (Tabata, 2009).

10.4.1 Conventional method based scaffold for bone regeneration

Conventional scaffold fabrication techniques are defined as processes by which scaffolds with a continuous, uninterrupted pore structure can be made, but which lack any long-range channeling microarchitecture (Sachlos and Czernuszka, 2003). A description of the different techniques, such as solvent-casting, particulate-leaching, gas foaming, fiber mesh/fiber bonding, phase separation, emulsion freeze drying, solution casting, freeze drying, and solid freeform (SFF) fabrication, have been reviewed elsewhere in detail (Sachlos and Czernuszka, 2003; Tabata, 2009; Velema and Kaplan, 2006). The resulting interior scaffold architectures produced by traditional methods through material processing and casting are determined by the processing technique. For example, in the case of particulate leaching, the internal architecture is determined by the embedded salts in the dissolved polymer matrix. The salt particles are subsequently leached out to leave interconnecting interior channels. Thus, by regulating the size of the salt particles, the pore diameters of the scaffolds can be controlled; however, agglomeration of salt particles can alter the eventual pore size and pore distribution during leaching (Sachlos and Czernuszka, 2003). In the case of freeze drying, the internal architecture of scaffolds is determined by the size and microstructures of the formed ice crystals, which, in turn, are governed by the freezing rate (Tabata, 2009).

Scaffolds produced by conventional techniques have an enormous number of applications in bone TE. For example, chemically crosslinked collagen-glycosaminoglycan scaffolds for bone regeneration have been developed by freeze drying, and the ability of these scaffolds to heal calvarial bone defects has also been demonstrated (Lyons *et al.*, 2010). Biomimetic collagen-hydroxyapatite scaffolds have been developed by freeze-drying followed by a biphasic calcium phosphate immersion process based on the two primary constituents of bone to facilitate the repair of load-bearing regions (Lyons *et al.*, 2010). Several researchers have successfully fabricated PLGA scaffolds for bone TE using a wide variety of approaches, such as emulsion, freeze-drying, nano- and microparticle encapsulation, double emulsion, solvent extraction, electrospinning, compression molding, and others (Porter *et al.*, 2009). Several groups have fabricated PLA-PEG scaffolds, which have been successfully loaded with recombinant human BMPs, and demonstrated accelerated healing and osteogenesis *in vivo* (Porter *et al.*, 2009).

Despite the ease of processing, conventional scaffold fabrication techniques are incapable of precisely controlling pore size, pore geometry, spatial distribution of pores, and construction of internal channels within the scaffold. Many of the conventional techniques produce scaffolds with random porous architecture, which do not necessarily produce a suitable homogeneous environment for bone formation (Hutmacher, 2001). Moreover, only thin scaffold cross-sections can be produced using this technique. While scaffolds prepared by gas foaming have only 10%–30% interconnecting pores (Mooney *et al.*, 1996), the nonwoven fiber mesh has poor mechanical integrity, despite having a well-interconnected porous structure. Most conventional scaffold fabrication techniques, excluding gas foaming and melt molding, use organic solvents such as chloroform and methylene chloride at some stage in the process. This is a significant problem due to the risks of toxicity and carcinogenicity it poses to cells. Additionally, scaffolds produced by a conventional fabrication technique are typically foam-like structures, and it is expected that seeded cells will grow in these scaffolds. However, the *in vitro* growth of tissues on these type of scaffolds occurs with cross-sections of < 500 µm from the external surface due to the diffusion constraints of nutrients and oxygen and insufficient removal of waste products (Freed and Vunjak-Novakovic, 1998). The cells cannot migrate deep into the scaffold; rather, cells colonize at the scaffold periphery, which further acts as an effective barrier to oxygen and nutrient diffusion into the core of the scaffold (Sachlos and Czernuszka, 2003). Thick cross-sections of tissue are required for bone TE, which require high oxygen and nutrient concentrations (Ikada, 2006). Mass transfer to the interior of the scaffold is further limited by mineralization of the scaffold surface with high rates of nutrient and oxygen transfer at the surface (Tabata, 2009). Thus, cells are able to survive only close to the surface. It should be noted that no cells except chondrocytes can exist more than 25–100 μm away from a blood supply (Guyton and Hall, 1996). Bone TE scaffolds should support the mass transport of oxygen and nutrients deep within, and the removal of waste products from, the core, and should have some form of artificial vascular system present within them. Osteoblasts exhibit a greater cellular response when pore sizes of 200–400 μm are used (Chang et al., 2000). This may be due to the curvature of the pores, which may provide optimum compression and tension on the cell mechanoreceptors and allow them to migrate into openings of such a size (Chang et al., 2000). Researchers have also demonstrated that control over the interior architecture is crucial to ensure scaffold vascularization and bone deposition (Chu et al., 2002).

SFF strategies are being used to produce scaffolds by which these requirements can be optimized. For example, SFF techniques can be used to produce scaffolds with customized external shapes, and predefined and reproducible internal morphologies (Seol et al., 2012).

10.4.2 Rapid prototyping based scaffold for bone regeneration

TE scaffolds should support the mass transport of oxygen and nutrients deep within and the removal of waste products from the core, and should have some form of artificial vascular system present. RP strategies are being used to produce scaffolds by which these requirements can be optimized. RP techniques can be used to produce scaffolds with customized external shapes and predefined internal morphology. Furthermore, besides controlling pore size, porosity, and pore distribution, this technique can also produce structures to increase the mass transport of oxygen and nutrients throughout the scaffold (Sachlos and Czernuszka, 2003). Additionally, RP technology can be used to create parts having compositional variation within them with local composition control, potentially creating new classes of components. Thus, local control of properties can be achieved by tailoring the material composition within a component. Using such local control, monolithic components can be created that integrate the function of multiple discrete components, saving space and weight, and enabling concepts that would otherwise be impractical. Controlling the spatial distribution of properties via composition allows for control of the entire component, which can be used to study the influence of composition on tissue regeneration and spatial delivery of growth factors (Hollister, 2005). Some designer scaffolds with controlled architecture are being developed using this technology, and are being used to specifically study the influence of geometry on tissue regeneration (Rumpler *et al.*, 2008). Moreover, RP techniques can also be used to fabricate customized scaffolds based on the prerequisite shape and structure from medical imaging data obtained by diagnostic imaging techniques such as magnetic resonance imaging (MRI) and computerized tomography (CT) (Seol *et al.*, 2012). Various RP technologies have been demonstrated to be effective methods for free-form fabrication of 3D scaffolds in the bone TE field. In particular, FDM, precise extrusion manufacturing (PEM), and the multi-head deposition system (MHDS), which extrude strands or filaments under heating conditions, can be used in a variety of FDA-approved biomaterials, including PCL, PLA, PGA, and PLGA.

Abbah and co-workers reported various scaffolds made of PCL, PCL/HA, and PCL/tricalcium phosphate (TCP) for bone TE using an FDM machine (Fig. 10.1a and 10.1d) (Hollister 2005; Abbah *et al.*, 2009). The FDM method enables easy and flexible material handling and processing for scaffold fabrication. In addition, a toxic solvent is not required to prepare the material to fabricate a 3D scaffold. However, the use of intermediate precursor filaments, which often causes filament buckling failures, is necessarily included in the FDM method. Shor *et al.* (2009) have developed the PEM and applied it to bone TE. The uniqueness of PEM is such that thermoplastic biomaterials are melted and extruded by the turning, screw-driven pressure without filament preparation (Fig. 10.1b). A variety of 3D scaffolds have been fabricated to regenerate bone using PEM with various biomaterials, such as PCL and PCL/HA (Fig. 10.1e). Additionally, Kim and Cho (2009) developed pneumatic, pressure-driven, 3D printing technology without the use of intermediate filament (Kim and Cho, 2009). The MHDS, which is specially equipped with four dispensing heads, can dispense various biomaterials into a 3D scaffold (Fig. 10.1c). Indeed, fabrication of PLGA, PCL, PCL/PLGA, PCL/PLGA/TCP, and even PCL/PLGA/collagen hybrid scaffolds has been attempted for bone regeneration using MHDS (Fig. 10.1f).

Stereolithography (SL) is another key RP technology. In the SL process, a photocurable polymer is selectively solidified via ultraviolet (UV) laser irradiation. A 2D pattern on the surface of the photopolymer is solidified, and the generated 2D patterns are sequentially stacked up to achieve the desired 3D structure (Fig. 10.2a). The resolution of the fabricated structure is dependent on the resolution of the laser source, and micro- to nano-sized

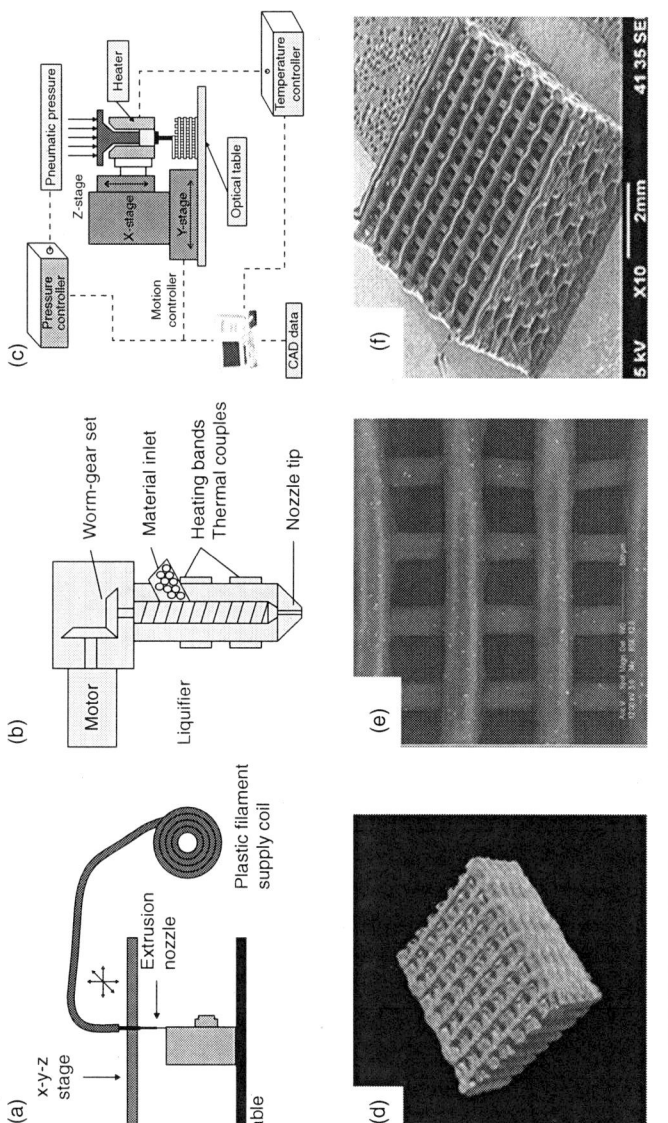

10.1 Schematic diagram of 3D printing technologies: (a) fused deposition modeling (FDM) (Hollister 2005), (b) precise extrusion manufacturing (PEM) (Shor et al., 2009), and (c) multi-head deposition system (MHDS) (Kim and Cho, 2009). Various 3D printed scaffolds: (d) PCL-TCP scaffold by FDM (Abbah et al., 2009), (e) PCL-TCP scaffold by PEM (Shor et al., 2009), and (f) PCL/PLGA scaffold by MHDS (Kim and Cho, 2009).

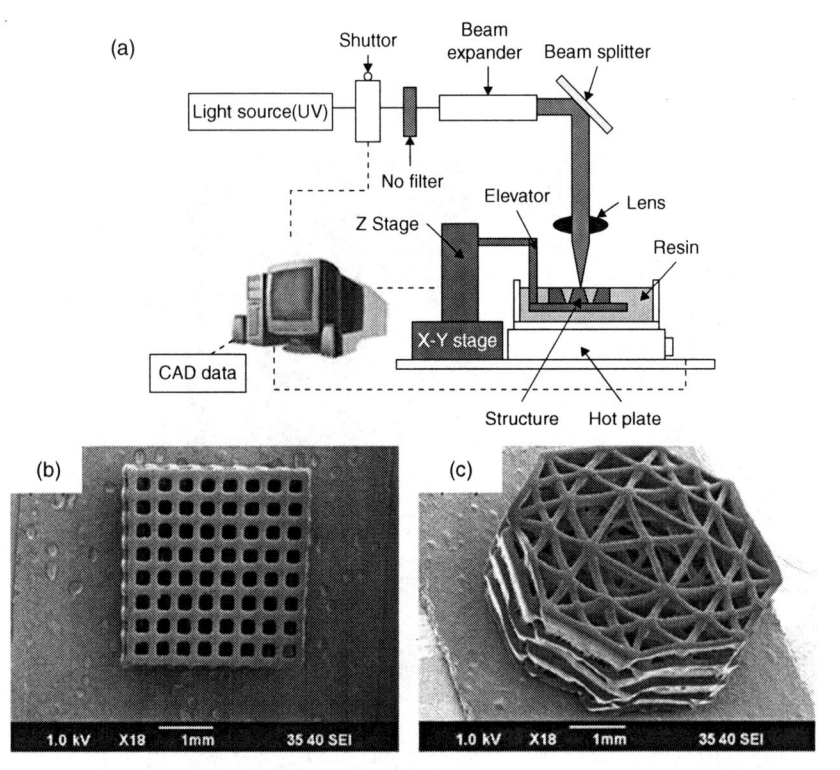

10.2 Schematic diagram of (a) SL system, (b) fabricated PPF/DEF scaffold (Lee *et al.*, 2007), (c) fabricated TMC/TMP scaffold (Lee *et al.*, 2008).

structures can be fabricated using various laser systems. However, only photocurable materials may be used, including poly(propylene fumarate)/diethyl fumarate (PPF/DEF) (Lee *et al.*, 2007; 2011a) (Fig. 10.2b) and trimethylene carbonate/trimethylolpropane (TMC/TMP) (Fig. 10.2c) (Lee *et al.*, 2008), which have not yet been approved by the FDA for biomedical applications.

Selective laser sintering (SLS) is also widely used as an RP technology for TE. The process of SLS for 3D scaffold fabrication involves powders of polymers, ceramics, and metals, which are heat-fusible, and are selectively sintered by the scanning of a CO_2 laser with extremely high energy. The surface of the powder bed is sintered to create a 2D cross-section pattern of a 3D structure. Then, the sintered object moves down, and the new powder is spread with a roller for the next layer (Fig. 10.3a). The layer-by-layer process is repeated until the desired 3D structure is formed. The SLS method achieves a controlled, totally interconnected, highly porous microstructure using PCL and PLGA (Williams *et al.*, 2005) (Fig. 10.3b). Moreover, inorganic ceramics,

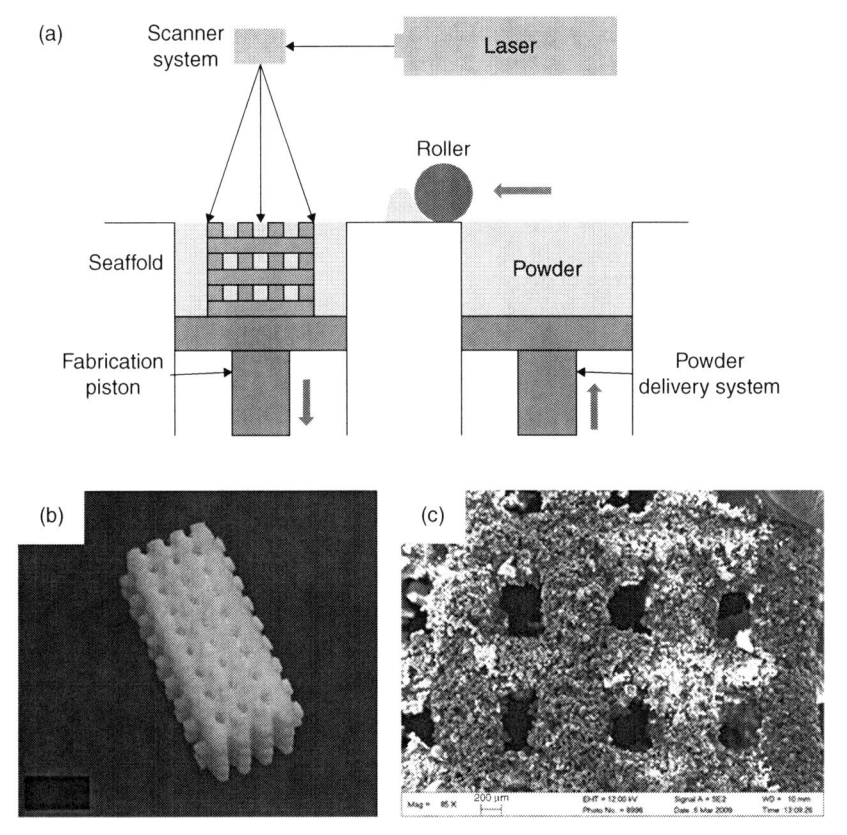

10.3 Schematic diagram of (a) SLS system (Hollister 2005). Scaffolds fabricated by SLS: (b) PCL (Williams *et al.*, 2005) and (c) carbonated hydroxyapatite (CHAp)/poly(L-lactic acid) (PLLA) (Duan *et al.*, 2010).

which promote structural and functional properties for bone regeneration, can be used in SLS technology (Duan *et al.*, 2010) (Fig. 10.3c).

10.5 Cell printing technology for bone regeneration

Although well-defined 3D scaffolds consisting of relevant biomaterials have been produced by various RP technologies, scaffolds with a hydrophobic surface cannot provide an ideally favorable environment for cells and growth factors. Moreover, selectively placing different cells and growth factors according to anatomical arrangement is practically impossible using a cell seeding-based approach. In common with other tissues or organs, bone is also composed of complex components, including ECM, cells, and growth factors. Therefore, these three components should be harmoniously integrated using a novel method to create a clinically ideal TE bone graft.

From this perspective, 3D cell printing technology, which can dispense the biologically relevant biomaterials and living cells in an anatomical shape according to medical imaging data, is a promising solution to integrate these three components (Fedorovich et al., 2011a). Various 3D cell printing technologies, including inkjet, dispensing, and laser-based printing technologies, have been attempted for bone-tissue regeneration.

10.5.1 Inkjet based cell printing

Cell printing using inkjet printing has become an effective method to deliver multiple cell types to target specific regions. Inkjet printing technology has been a focus due to its advantages of high-throughput capability, low cost, repeatability, and ease of use. Inkjet printing technology is a practical and efficient method to dispense biological and/or material elements, including living cells. Xu *et al.* (2005) have showed that Chinese hamster ovary (CHO) and embryonic motoneuron cells can be printed on substrate made from soy agar and collagen gel using a modified Hewlett Packard printer, and the printed cells are 92% viable. In addition, for the bone TE application, Philippi and co-workers reported that inkjet bioprinting technology can be used to create spatial patterns of BMP-2 for studying muscle-derived stem cells (MDSCs) possessing multiple lineages, such as bone and muscle-like cells (Boland *et al.*, 2006; Phillippi *et al.*, 2008) (Fig. 10.4). BMP-2 bio-ink (10 µg/mL) was loaded into a sterilized print head and subsequently printed on a fibrin surface to create the spatial patterns. MDSCs seeded on the BMP-2 pattern exhibited ALP activity indicating osteogenic differentiation, but MDSCs seeded outside the BMP-2 pattern remained undifferentiated (Fig. 10.5).

10.5.2 Laser assisted cell printing

Catros *et al.* (2011) have demonstrated laser-assisted bioprinting (LAB), which includes a pulsed laser source, a target, and a receiving substrate for patterning and assembling the biomaterials and cells (Fig. 10.6a). The target, which is transparent to the laser radiation wavelength, is coated with biological materials (bioink) to print. A laser-absorbing interlayer with a high heat transfer coefficient is necessary to transfer heat to the bioink. Thus, the laser-absorbing interlayer is generally located between the target and the bioink. Inorganic nano-hydroxyapatite (nHA) and living human osteoprogenitors (HOPs) were deposited on the substrate for bone TE using LAB. The target was composed of a quartz slide, an absorbing interlayer of titanium (60 nm), and a bioink layer (20–30 µm). An on-demand nHA and HOP pattern was achieved by LAB. In addition, the physicochemical properties of nHA were

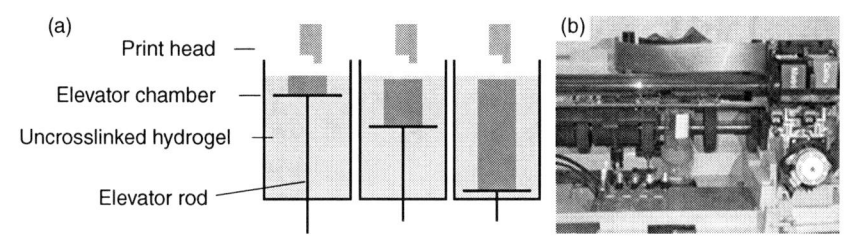

10.4 (a) Schematic view and (b) photograph of the moving platform inside the chamber under a modified HP Deskjet. (Boland *et al.*, 2006).

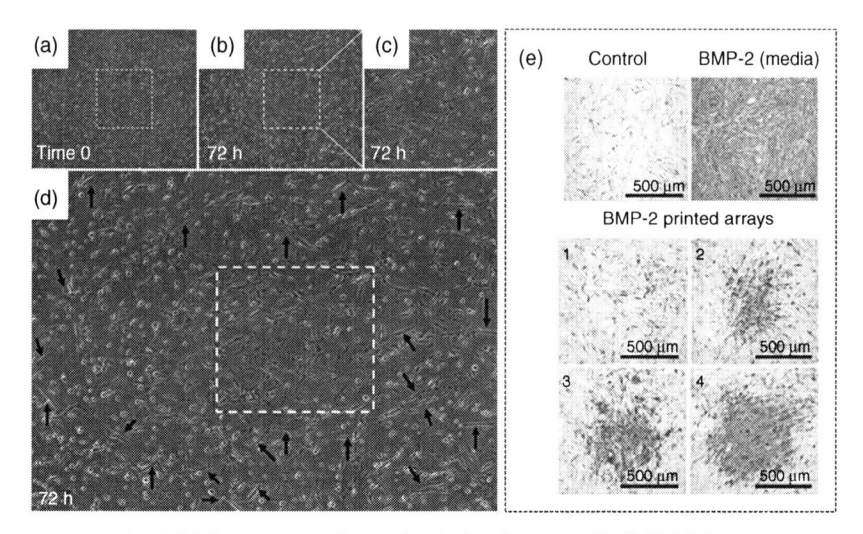

10.5 (a) Microscopy of muscle-derived stem cells (MDSCs) on the printed BMP-2 and (b) image of the MDSCs after 72 h; dashed box indicates BMP-2 printed area. (c) Magnified box of (b), and (d) enlargement of (b). Arrows indicate myotubes. (e) ALP stained MDSC on the BMP-2 pattern (Philippi *et al.*, 2008).

not changed, and the HOPs retained their cellular activities such as adhesion, spreading, proliferation, and differentiation after the printing process (Fig. 10.6b to 10.6g).

10.5.3 Extrusion based cell printing

Although inkjet and LAB technologies have shown potential in bone TE using spatial patterning of cells, growth factors, and inorganic materials, these technologies remain limited to 2D thin-patterns or structures, as it is difficult to achieve clinically relevant, 3D-sized constructs using these technologies (Fedorovich *et al.*, 2011c). In contrast, a dispensing-based bioprinting system

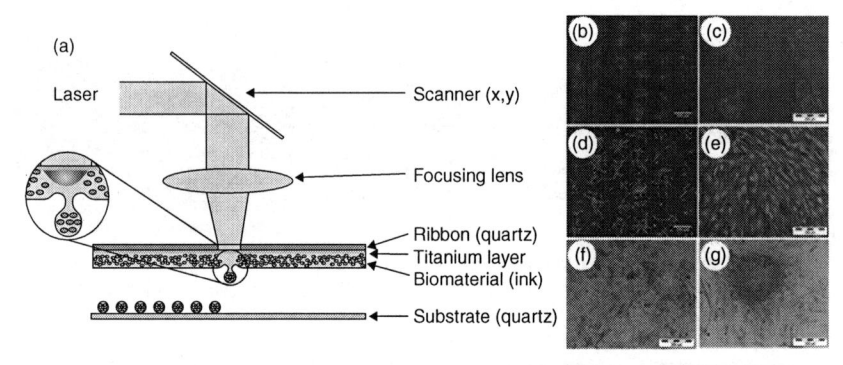

10.6 (a) Schematic diagram of laser-assisted bioprinting (LAB). (b) Micro-pattern of HOP cell and (c) its magnified image by Hoechst dye at day 0. (d) Live/dead cell assay of HOPs on day 2 and (e) on day 15. (f) Alkaline phosphatase staining of printed HOPs and (g) osteocalcin immune-straining result of the HOPs on day 15 (Catros *et al.*, 2011).

enables rapid and easy dispensing of a clinically relevant-sized construct. In particular, Fedorovich *et al.* (2008) demonstrated a pneumatic dispensing system-based bioplotter (EnvisionTEC GmbH, Gladbeck, Germany) for 3D fiber deposition of cells and hydrogels. They printed various hydrogels, such as alginate, agarose, Matrigel, and Lutrol F127, in which BMSCs obtained from iliac bone marrow aspirates of goat were encapsulated for TE bone grafts. The BMSCs were not damaged during the printing process. Moreover, the BMSCs retained osteogenic differentiation ability 2 weeks after printing.

Advanced 3D organ printing for bone-tissue regeneration has been attempted by creating heterogeneous constructs consisting of two distinct cell types. Alblas *et al.* (2009) have achieved the cell-laden hydrogel constructs containing endothelial progenitor cells (EPCs) and multipotent stromal cells (MSCs) to promote neovascularization during the bone regeneration process. Neovascularization plays an important role in the survival of large functional bone grafts by ensuring supply of oxygen and nutrients. They demonstrated that separately printed EPCs and MSCs retain their functionality after *in vivo* implantation and form blood vessels and bone, respectively. Interestingly, the newly formed vascular tubules were filled with erythrocytes, which may have been perfused from the native vessel. This heterogeneous construct comprising distinct parts with different cell types is possible only in RP-based 3D printing technology.

In addition, Fedorovich *et al.* (2011b) reported that fully interconnected pores, which are readily adjustable by RP technology, are critical for retaining viability and functionality of MSCs encapsulated in a hydrogel for bone-tissue regeneration. A sufficient supply of oxygen and nutrients can be guaranteed through the interconnected pores. In contrast, hypoxia of MSCs was observed when the MSCs were encapsulated in the hydrogel construct. This hypoxic phenomenon can be explained by the limited diffusion depth causing cell necrosis in the center of the hydrogel (Ling et al., 2007). Therefore, the interconnected pores are particularly crucial for printing of clinically relevant-sized bone grafts in which the size of the hydrogel is larger than the limited diffusion depth.

A cell-laden construct fabricated using 3D organ-printing technology should have adequate mechanical properties similar to those of native bone tissue. However, a printed construct comprising only a soft hydrogel is likely to be insufficient to meet the appropriate mechanical requirements. To overcome this limitation, a variety of photo-crosslinkable hydrogels have been introduced by many researchers to improve mechanical stability of the hydrogel structure (Rouillard et al., 2011; Slaughter et al., 2009). However, the possible cytotoxicity caused by ultraviolet exposure, chemical photoinitiators, and organic solvents needs to be carefully investigated (Rouillard et al., 2011). Hybrid 3D viable constructs that combine the uses of thermoplastic biomaterials and cell-laden hydrogels have been suggested to increase mechanical stability (Schuurman et al., 2011; Shim et al., 2011). They have illustrated the 3D printing process using layer-by-layer deposition of two different biomaterials to create a mechanically relevant construct. First, thermoplastic biomaterial such as PCL was melted and subsequently extruded by pneumatic pressure (Fig. 10.7). Then, the cell-laden alginate solution was dispensed between the PCL lines. The dispensing of PCL and alginate solution was repeated layer by layer to produce a reinforced 3D cell-laden construct. In this way, mechanical stability was considerably enhanced and a high aspect ratio of the 3D cell-laden construct was achieved.

10.6 Future trends

Bone growth and remodeling involves a surplus of growth factors, recruitment of MSCs, and the action of three different mature cell types, as well as other factors yet to be unveiled. This is probably one of the most challenging aspects of developing a bioartificial tissue engineered bone. New RP methodologies have overcome some of the limitations of conventional scaffold fabrication strategies, and are very promising for the future of TE applications. Improved 3D printing methodologies should be studied to develop scaffolds with improved cell-material responses without influencing porosity and interconnectivity. In the future, 3D printed cell-laden bone constructs, in which potential bone regeneration cell sources are encapsulated in the hydrogel, will be implanted into *in vivo* defects with complex shapes. Subsequently, in-depth *in vivo* studies on the fate of implanted

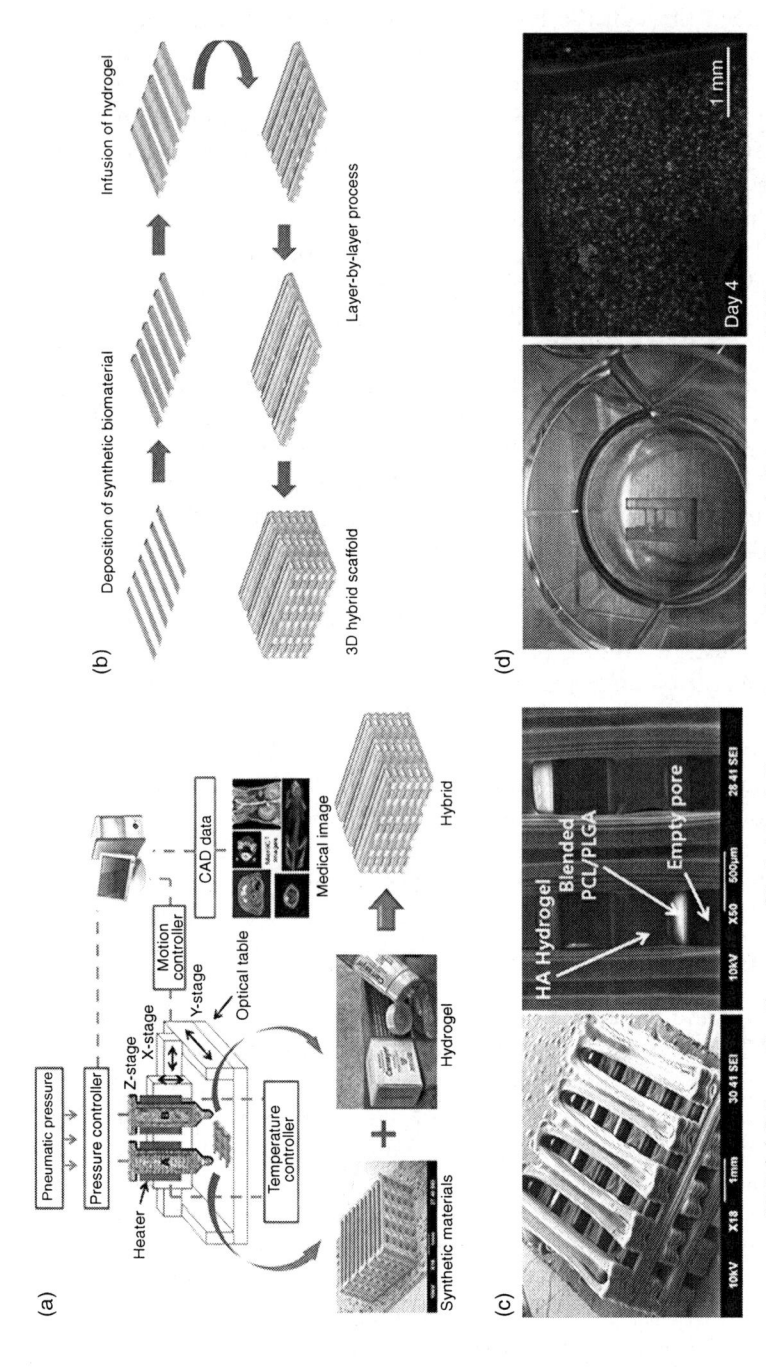

10.7 (a) Schematic diagram of hybrid printing system and (b) the printing process. (c) Scanning electron microscopy image of the hybrid construct fabricated by the printing process. (d) Live/dead cell assay result of printed MC3T3-E1 cells in the PCL framework (Shim et al., 2011).

cells should be performed. In addition, the potential of various hydrogels should be investigated to find the ideal hydrogel mimicking the native ECM structure of bone.

10.7 Conclusion

Although great progress has been made in the field of bone regenerative medicine during the past few years, current therapies, such as bone grafts, still have many limitations. Moreover, although material science technology has resulted in clear improvements in the field of bone substitution medicine, no adequate bone substitute has been developed and, hence, large bone defects/injuries still represent a major challenge for orthopedic and reconstructive surgeons. TE has been emerging as a valid approach to current therapies for bone regeneration/substitution. Many researchers have shown improvements in tissue regeneration using scaffolds fabricated by SFF compared to results using scaffolds fabricated by conventional methods. Future developments in SFF-based 3D cell printing technologies will play a key role in meeting future challenges.

10.8 Acknowledgement

This work was supported by the National Research Foundation of Korea (NRF) grant funded by the Korea government (MEST) (No. 2012–0001235).

10.9 References

Abbah, S. A., Lam, C. X. L., Hutmacher, D. W., Goh, J. C. H. and Wong, H.-K. (2009). Biological performance of a polycaprolactone-based scaffold used as fusion cage device in a large animal model of spinal reconstructive surgery. *Biomaterials*, **30**, 5086–5093.

Ahmed, T. A., Dare, E. V. and Hincke, M. (2008). Fibrin: a versatile scaffold for tissue engineering applications. *Tissue Eng Part B Rev*, **14**, 199–215.

Athanasiou, K. A., Zhu, C., Lanctot, D. R., Agrawal, C. M. and Wang, X. (2000). Fundamentals of biomechanics in tissue engineering of bone. *Tissue Eng*, **6**, 361–381.

Bianco, P. and Robey, P. G. (2001). Stem cells in tissue engineering. *Nature*, **414**, 118–121.

Bodle, J. C., Hanson, A. D. and Loboa, E. G. (2011). Adipose-derived stem cells in functional bone tissue engineering: lessons from bone mechanobiology. *Tissue Eng Part B*, **17**, 195–211.

Boland, T., Xu, T., Damon, B. and Cui, X. (2006). Application of inkjet printing to tissue engineering. *Biotechnol J*, **9**, 910–917.

Burastero, G., Scarfi, S., Ferraris, C., Fresia, C., Sessarego, N., Fruscione, F., Monetti, F., Scarfo, F., Schupbach, P., Podesta, M., Grappiolo, G. and Zocchi, E. (2010).

The association of human mesenchymal stem cells with BMP-7 improves bone regeneration of critical-size segmental bone defects in athymic rats. *Bone*, **47**, 117–126.

Buttery, L. D., Bourne, S., Xynos, J. D., Wood, H., Hughes, F. J., Hughes, S. P., Episkopou, V. and Polak, J. M. (2001). Differentiation of osteoblasts and in vitro bone formation from murine embryonic stem cells. *Tissue Eng*, **7**, 89–99.

Buttler, W. T., Ridall, A. L. and Mckee, M. D. (1996). *Principles of Bone Biology*, San Diego, CA, Academic.

Catros, S., Fricain, J. C., Guillotin, B., Pippenger, B., Bareille, R., Remy, M., Lebraud, E., Desbat, B., Amédéé, J. and Guillemot, F. (2011). Laser-assisted bioprinting for creating on-demand patterns of human osteoprogenitor cells and nano-hydroxyapatite. *Biofabrication*, **3**, 025001.

Chang, B. S., Lee, C.-K., Hong, K.-S., Youn, H.-J., Ryu, H.-S., Chung, S.-S. and Park, K.-W. (2000). Osteoconduction at porous hydroxyapatite with various pore configurations. *Biomaterials*, **21**, 1291–1298.

Chu, T. M. G., Orton, D. G., Hollister, S. J., Feinberg, S. E. and Halloran, J. W. (2002). Mechanical and in vivo performance of hydroxyapatite implants with controlled architectures. *Biomaterials*, **23**, 1283–1293.

Colnot, C. (2011). Cell sources for bone tissue engineering: insights from basic science. *Tissue Eng Part B*, **17**, 449–457.

Cowan, C. M., Soo, C., Ting, K. and Wu, B. (2005). Evolving concepts in bone tissue engineering. *Current Topics in Developmental Biology*. Waltham, Massachusetts, Academic Press.

Cross, M. and Dexter, T. M. (1991). Growth factors in development, transformation, and tumorigenesis. *Cell*, **64**, 271–280.

Dawson, E., Mapili, G., Erickson, K., Taqvi, S. and Roy, K. (2008). Biomaterials for stem cell differentiation. *Adv Drug Deliv Rev*, **60**, 215–228.

Duan, B., Wang, M., Zhou, W. Y., Cheung, W. L., Li, Z. Y. and Lu, W. W. (2010). Three-dimensional nanocomposite scaffolds fabricated via selective laser sintering for bone tissue engineering. *Acta Biomater*, **6**, 4495–4505.

Ducheyne, P. and Qiu, Q. (1999). Bioactive ceramics: the effect of surface reactivity on bone formation and bone cell function. *Biomaterials*, **20**, 2287–2303.

Evans, C. (2011). Gene therapy for the regeneration of bone. *Injury*, **42**, 599–604.

Fedorovich, N. E., Alblas, J., Hennink, W. E., Öner, F. C. and Dhert, W. J. A. (2011a). Organ printing: the future of bone regeneration? *Trends Biotechnol*, **29**, 601–606.

Fedorovich, N. E., DE Wijn, J. R., Verbout, A. J., Alblas, J. and Dhert, W. J. A. (2008). Three-dimensional fiber deposition of cell-laden, viable, patterned constructs for bone tissue printing. *Tissue Eng Part A*, **14**, 127–133.

Fedorovich, N. E., Kuipers, E., Gawlitta, D., Dhert, W. J. and Alblas, J. (2011b). Scaffold porosity and oxygenation of printed hydrogel constructs affect functionality of embedded osteogenic progenitors. *Tissue Eng Part A*, **17**, 2473–2486.

Fedorovich, N. E., Schuurman, W., Wijnberg, H. M., Prins, H. J., Van Weeren, P. R., Malda, J., Alblas, J. and Dhert, W. J. (2011c). Biofabrication of osteochondral tissue equivalents by printing topologically defined, cell-laden hydrogel scaffolds. *Tissue Eng Part C: Methods*, **18**, 33–44.

Freed, L. E. and Vunjak-Novakovic, G. (1998). Culture of organized cell communities. *Adv Drug Deliv Rev*, **33**, 15–30.

Friess, W. (1998). Collagen-biomaterial for drug delivery. *Eur J Pharm Biopharm*, **45**, 113–136.
Furth, M. E., Atala, A. and Van Dyke, M. E. (2007). Smart biomaterials design for tissue engineering and regenerative medicine. *Biomaterials*, **28**, 5068–5073.
George, J., Kuboki, Y. and Miyata, T. (2006). Differentiation of mesenchymal stem cells into osteoblasts on honeycomb collagen scaffolds. *Biotechnol Bioeng*, **95**, 404–411.
George, M. and Abraham, T. E. (2006). Polyionic hydrocolloids for the intestinal delivery of protein drugs: alginate and chitosan – a review. *J Control Release*, **114**, 1–14.
Gittens, S. A. and Uludag, H. (2001). Growth factor delivery for bone tissue engineering. *J Drug Target*, **9**, 407–429.
Glowacki, J. and Mizuno, S. (2008). Collagen scaffolds for tissue engineering. *Biopolymers*, **89**, 338–344.
Goldstein, S. A., Wilson, D. L., Sonstegard, D. A. and Matthews, L. S. (1983). The mechanical properties of human tibial trabecular bone as a function of metaphyseal location. *J Biomech Eng*, **16**, 965–969.
Gonnerman, K. N., Brown, L. S. and Chu, T. M. (2006). Effects of growth factors on cell migration and alkaline phosphatase release. *Biomed Sci Instrum*, **42**, 60–65.
Goodwin, H. S., Bicknese, A. R., Chien, S.-N., Bogucki, B. D., Oliver, D. A., Quinn, C. O. and Wall, D. A. (2001). Multilineage differentiation activity by cells isolated from umbilical cord blood: expression of bone, fat, and neural markers. *Biol Blood Marrow Transplant*, **7**, 581–588.
Goyenvalle, E., Guyen, N. J.-M., Aguado, E., Passuti, N. and Daculsi, G. (2003). Bilayered calcium phosphate coating to promote osseointegration of a femoral stem prosthesis. *J Mater Sci Mater Med*, **14**, 219–227.
Gruenloh, W., Kambal, A., Sondergaard, C., Mcgee, J., Nacey, C., Kalomoiris, S., Pepper, K., Olson, S., Fierro, F. and Nolta, J. A. (2011). Characterization and in vivo testing of mesenchymal stem cells derived from human embryonic stem cells. *Tissue Eng Part A*, **17**, 1517–1525.
Guyton, A. C. and Hall, J. E. (1996). *Textbook of Medical Physiology*, Philadelphia, WB Saunders.
Hadjidakis, D. J. and Androulakis, I. I. (2006). Bone remodeling. *Ann N Y Acad Sci*, **1092**, 385–396.
Hak, D. J. (2007). The use of osteoconductive bone graft substitutes in orthopaedic trauma. *J Am Acad Orthop Surg*, **15**, 525–536.
Handschel, J., Wiesmann, H. P., Depprich, R., K Bler, N. R. and Meyer, U. (2006). Cell-based bone reconstruction therapies-cell sources. *Int J Oral Maxillofac Implants*, **21**, 890–898.
Hayes, W. C. (1991). *Basic Orthopedic Biomechanics*. New York, Raven Press.
Hollister, S. J. (2005). Porous scaffold design for tissue engineering. *Nat Mater*, **4**, 518–524.
Howard, D., Buttery, L. D., Shakesheff, K. M. and Roberts, S. J. (2008). Tissue engineering: strategies, stem cells and scaffolds. *J Anat*, **213**, 66–72.
Hughes, F. J. and Aubin, J. E. (1997). Culture of cells of the osteoblast lineage methods in bone biology. In: *Methods in Bone Biology*, Arnett, T. R. and Henderson, B. (eds.). US, Springer.

Hutmacher, D. W. (2001). Scaffold design and fabrication technologies for engineering tissues state of the art and future perspectives. *J Biomater Sci, Polym Ed*, **12**, 107–124.

Hutmacher, D. W., Sittinger, M. and Risbud, M. V. (2004). Scaffold-based tissue engineering: rationale for computer-aided design and solid free-form fabrication systems. *Trends Biotechnol*, **22**, 354–362.

Ikada, Y. (2006). Review: challenges in tissue engineering. *J R Soc Interface*, **3**, 589–601.

Ikada, Y. and Tabata, Y. (1998). Protein release from gelatin matrices. *Adv Drug Deliv Rev*, **31**, 287–301.

Illich, D. J., Demir, N., Stojkovic, M., Scheer, M., Rothamel, D., Neugebauer, J., Hescheler, J. and Zoller, J. E. (2011). Induced pluripotent stem (iPS) cells and lineage reprogramming: prospects for bone regeneration. *Stem Cells*, **29**, 555–563.

Jackson, R. L., Busch, S. J. and Cardin, A. D. (1991). Glycosaminoglycans: molecular properties, protein interactions, and role in physiological processes. *Physiol Rev*, **71**, 481–539.

Jayakumar, P. and Di Silvio, L. (2010). Osteoblasts in bone tissue engineering. *Proc Inst Mech Eng H*, **224**, 1415–1440.

Kahle, M., Wiesmann, H. P., Berr, K., Depprich, R. A., Kubler, N. R., Naujoks, C., Cohnen, M., Ommerborn, M. A., Meyer, U. and Handschel, J. (2010). Embryonic stem cells induce ectopic bone formation in rats. *Biomed Mater Eng*, **20**, 371–380.

Kasukawa, Y., Miyakoshi, N. and Mohan, S. (2004). The anabolic effects of GH/IGF system on bone. *Curr Pharm Des*, **10**, 2577–2592.

Khan, Y., Yaszemski, M. J., Mikos, A. G. and Laurencin, C. T. (2008). Tissue engineering of bone: material and matrix considerations. *J Bone Joint Surg Am*, **90**, 36–42.

Khor, E. and Lim, L. Y. (2003). Implantable applications of chitin and chitosan. *Biomaterials*, **24**, 2339–2349.

Kim, J. Y. and Cho, D. W. (2009). Blended PCL/PLGA scaffold fabrication using multi-head deposition system. *Microelectron Eng*, **86**, 1447–1450.

Kumar, S., Wan, C., Ramaswamy, G., Clemens, T. L. and Ponnazhagan, S. (2010). Mesenchymal stem cells expressing osteogenic and angiogenic factors synergistically enhance bone formation in a mouse model of segmental bone defect. *Mol Ther*, **18**, 1026–1034.

Kuznetsov, S. A., Cherman, N. and Robey, P. G. (2011). In vivo bone formation by progeny of human embryonic stem cells. *Stem Cells Dev*, **20**, 269–287.

Langer, R. and Vacanti, J. P. (1993). Tissue engineering. *Science*, **260**, 920–926.

Laurencin, C. T., Ambrosio, A. A., Borden, M. D. and Cooper, J. A. (1999). *Tissue Engineering: Orthopedic Applications*, Palo Alto, CA, Annual reviews.

Lee, J. W., Lan, P. X., Kim, B., Lim, G. and Cho, D. W. (2007). 3D scaffold fabrication with PPF/DEF using micro-stereolithography. *Microelectron Eng*, **84**, 1702–1705.

Lee, J.-W., Kim, J.-Y. and D-W, CHO. (2010). Solid free-form fabrication technology and its application to bone tissue engineering. *Inter J Stem Cells*, **3**, 85–93.

Lee, J. W., Kang, K. S., Lee, S. H., Kim, J.-Y., Lee, B.-K. and Cho, D.-W. (2011a). Bone regeneration using a microstereolithography-produced customized poly(propylene fumarate)/diethyl fumarate photopolymer 3D scaffold incorporating BMP-2 loaded PLGA microspheres. *Biomaterials*, **32**, 744–752.

Lee, K., Silva, E. A. and Mooney, D. J. (2011b). Growth factor delivery-based tissue engineering: general approaches and a review of recent developments. *J R Soc Interface*, **8**, 153–170.

Lee, S. H. and Shin, H. (2007). Matrices and scaffolds for delivery of bioactive molecules in bone and cartilage tissue engineering. *Adv Drug Deliv Rev*, **59**, 339–359.

Lee, S. J., Kang, H. W., Park, J. K., Rhie, J. W., Hahn, S. W. and Cho, D. W. (2008). Application of microstereolithography in the development of three-dimensional cartilage regeneration scaffolds. *Biomed Microdevices*, **10**, 233–241.

Li, W. J. and Tuan, R. S. (2009). Fabrication and application of nanofibrous scaffolds in tissue engineering. *Curr Protoc Cell Biol*, **25**, 2.

Liao, Y. H., Jones, S. A., Forbes, B., Martin, G. P. and Brown, M. B. (2005). Hyaluronan: pharmaceutical characterization and drug delivery. *Drug Deliv*, **12**, 327–342.

Lieberman, J. R., Daluiski, A., Stevenson, S., Wu, L., Mcallister, P., Lee, Y. P., Kabo, J. M., Finerman, G. A., Berk, A. J. and Witte, O. N. (1999). The effect of regional gene therapy with bone morphogenetic protein-2-producing bone marrow cells on the repair of segmental femoral defects in rats. *J Bone Joint Surg Am*, **81**, 905–917.

Ling, Y., Rubin, J., Deng, Y., Huang, C., Demirci, U., Karp, J. M. and Khademhosseini, A. (2007). A cell-laden microfluidic hydrogel. *Lab Chip*, **7**, 756–762.

Lyons, F. G., Al-Munajjed, A. A., Kieran, S. M., Toner, M. E., Murphy, C. M., Duffy, G. P. and O'Brien, F. J. (2010). The healing of bony defects by cell-free collagen-based scaffolds compared to stem cell-seeded tissue engineered constructs. *Biomaterials*, **31**, 9232–9243.

Maniatopoulos, C., Sodek, J. and Melcher, A. H. (1988). Bone formation in vitro by stromal cells obtained from bone marrow of young adult rats. *Cell Tissue Res*, **254**, 317–330.

Martino, A. D., Sittinger, M. and Risbud, M. V. (2005). Chitosan: a versatile biopolymer for orthopaedic tissue-engineering. *Biomaterials*, **26**, 5983–5990.

Meyer, U., Joos, U. and Wiesmann, H. P. (2004). Biological and biophysical principles in extracorporal bone tissue engineering. Part I. *Int J Oral Maxillofac Surg*, **33**, 325–332.

Mironov, V., Prestwich, G. and Forgacs, G. (2007). Bioprinting living structures. *J Mater Chem*, **17**, 2054–2060.

Mistry, A. S. and Mikos, A. G. (2005). Tissue engineering strategies for bone regeneration. *Adv Biochem Eng Biotechnol*, **94**, 1–22.

Mooney, D. J., Baldwin, D. F., Suh, N. P., Vacanti, J. P. and Langer, R. (1996). Novel approach to fabricate porous sponges of poly(D,L-lactic co-glycolic acid) without the use of organic solvents. *Biomaterials*, **17**, 1417–1422.

Muschler, G. F., Nitto, H., Boehm, C. A. and Easley, K. A. (2001). Age- and gender-related changes in the cellularity of human bone marrow and the prevalence of osteoblastic progenitors. *J Orthop Res*, **19**, 117–125.

Nöth, U., Rackwitz, L., Steinert, A. F. and Tuan, R. S. (2010). Cell delivery therapeutics for musculoskeletal regeneration. *Adv Drug Deliv Rev*, **62**, 765–783.

Nicodemus, G. D. and Bryant, S. J. (2008). Cell encapsulation in biodegradable hydrogels for tissue engineering applications. *Tissue Eng Part B Rev*, **14**, 149–165.

Nijweide, P. J., Burger, E. H., Nulend, J. K. and Van Der Plas, A. (1996). *Principles of Bone Biology*, San Diego, CA, Academic.

Parfitt, A. M. (1984). The cellular basis of bone remodeling: the quantum concept reexamined in light of recent advances in the cell biology of bone. *Calcif Tissue Int*, **36**, S37–S45.

Peltola, S. M., Melchels, F. P., Grijpma, D. W. and Kellom Ki, M. (2008). A review of rapid prototyping techniques for tissue engineering purposes. *Ann Med*, **40**, 268–280.

Phillippi, J. A., Miller, E., Weiss, L., Huard, J., Waggoner, A. and Campbell, P. (2008). Microenvironments engineered by inkjet bioprinting spatially direct adult stem cells toward muscle- and bone-like subpopulations. *Stem Cells*, **26**, 127–134.

Porter, J. R., Ruckh, T. T. and Popat, K. C. (2009). Bone tissue engineering: a review in bone biomimetics and drug delivery strategies. *Biotechnology Progress*, **25**, 1539–1560.

Qian, J., Kang, Y., Zhang, W. and Li, Z. (2008). Fabrication, chemical composition change and phase evolution of biomorphic hydroxyapatite. *J Mater Sci: Mater Medicine*, **19**, 3373–3383.

Rezwan, K., Chen, Q. Z., Blaker, J. J. and Boccaccini, A. R. (2006). Biodegradable and bioactive porous polymer/inorganic composite scaffolds for bone tissue engineering. *Biomaterials*, **27**, 3413–3431.

Roberts, S. J., Geris, L., Kerckhofs, G., Desmet, E., Schrooten, J. and Luyten, F. P. (2011). The combined bone forming capacity of human periosteal derived cells and calcium phosphates. *Biomaterials*, **32**, 4393–4405.

Rouillard, A. D., Berglund, C. M., Lee, J. Y., Polacheck, W. J., Tsui, Y., Bonassar, L. J. and Kirby, B. J. (2011). Methods for photocrosslinking alginate hydrogel scaffolds with high cell viability. *Tissue Eng Part C Methods*, **17**, 173–179.

Rumpler, M., Woesz, A., Dunlop, J. W. C., Dongen, J. T. V. and Fratzl, P. (2008). The effect of geometry on three-dimensional tissue growth. *J R Soc Interface*, **5**, 1173–1180.

Ruoslahti, E. (1996). RGD and other recognition sequences for integrins. *Annu Rev Cell Dev Biol*, **12**, 697–715.

Sachlos, E. and Czernuszka, J. T. (2003). Making tissue engineering scaffolds work. Review on the application of solid freeform fabrication technology to the production of tissue engineering scaffolds. *Eur Cell Mater*, **5**, 29–40.

Sarkar, M. R., Wachter, N., Patka, P. and Kinzl, L. (2001). First histological observations on the incorporation of a novel calcium phosphate bone substitute material in human cancellous bone. *J Biomed Mater Res*, **58**, 329–334.

Schuurman, W., Khristov, V., Pot, M., Van Weeren, P., Dhert, W. and Malda, J. (2011). Bioprinting of hybrid tissue constructs with tailorable mechanical properties. *Biofabrication*, **3**, 021001.

Sen, M. K. and Miclau, T. (2007). Autologous iliac crest bone graft: should it still be the gold standard for treating nonunions? *Injury*, **38**, S75–S80.

Seol, Y.-J., Kang, T.-Y. and Cho, D.-W. (2012). Solid freeform fabrication technology applied to tissue engineering with various biomaterials. *Soft Matter*, **8**, 1730–1735.
Service, R. F. (2000). Tissue engineers build new bone. *Science*, **289**, 1498–1500.
Shim, J. H., Kim, J. Y., Park, M., Park, J. and Cho, D. W. (2011). Development of a hybrid scaffold with synthetic biomaterials and hydrogel using solid freeform fabrication technology. *Biofabrication*, **3**, 034102.
Shor, L. Güçeri, S., Wen, X., Gandhi, M. and Sun, W. (2009). Fabrication of three-dimensional polycaprolactone/hydroxyapatite tissue scaffolds and osteoblast-scaffold interactions in vitro. *Biomaterials*, **28**, 5291–5297.
Slaughter, B. V., Khurshid, S. S., Fisher, O. Z., Khademhosseini, A. and Peppas, N. A. (2009). Hydrogels in regenerative medicine. *Adv Mater*, **21**, 3307–3329.
Sommerfeldt, D. W. and Rubin, C. T. (2001). Biology of bone and how it orchestrates the form and function of the skeleton. *Eur Spine J*, **10**, S86–S95.
Steinert, A. F., Palmer, G. D., Capito, R., Hofstaetter, J. G., Pilapil, C., Ghivizzani, S. C., Spector, M. and Evans, C. H. (2007). Genetically enhanced engineering of meniscus tissue using ex vivo delivery of transforming growth factor-beta 1 complementary deoxyribonucleic acid. *Tissue Eng*, **13**, 2227–2237.
Stolzing, A., Jones, E., Mcgonagle, D. and Scutt, A. (2008). Age related changes in human bone marrow-derived mesenchymal stem cells: consequences for cell therapies. *Mech Ageing Dev*, **129**, 163173.
Sundaramurthy, S. and Mao, J. J. (2006). Modulation of endochondral development of the distal femoral condyle by mechanical loading. *J Orthop Res*, **24**, 229–241.
Tabata, Y. (2009). Biomaterial technology for tissue engineering applications. *J R Soc Interface*, **6**, S311–S324.
Toma, C. D., Ashkar, S., Gray, M. L., Schaffer, J. L. and Gerstenfeld, L. C. (1997). Signal transduction of mechanical stimuli is dependent on microfilament integrity: identification of osteopontin as a mechanically induced gene in osteoblasts. *J Bone Miner Res*, **12**, 1626–1636.
Van Vlierberghe, S., Dubruel, P. and Schacht, E. (2011). Biopolymer-based hydrogels as scaffolds for tissue engineering applications: a review. *Biomacromolecules*, **12**, 1387–1408.
Velema, J. and Kaplan, D. (2006). Biopolymer-based biomaterials as scaffolds for tissue engineering. *Adv Biochem Engin/Biotechnol*, **102**, 187–238.
Viateau, V., Guillemin, G., Bousson, V., Oudina, K., Hannouche, D., Sedel, L., Logeart-Avramoglou, D. and Petite, H. (2007). Long-bone critical-size defects treated with tissue engineered grafts: a study on sheep. *J Orthop Res*, **25**, 741–749.
Wang, D.-A., Varghese, S., Sharma, B., Strehin, I., Fermanian, S., Gorham, J., Fairbrother, D. H., Cascio, B. and Elisseeff, J. H. (2007). Multifunctional chondroitin sulphate for cartilage tissue-biomaterial integration. *Nat Mater*, **6**, 385–392.
Wang, G., Zheng, L., Zhao, H., Miao, J., Sun, C., Ren, N., Wang, J., Liu, H. and Tao, X. (2011). In vitro assessment of the differentiation potential of bone marrow-derived mesenchymal stem cells on genipin-chitosan conjugation scaffold with surface hydroxyapatite nanostructure for bone tissue engineering. *Tissue Eng Part A*, **17**, 1341–1349.
Williams J. M., Adewunmi A., Schek R. M., Flanagan C. L., Krebsbach P. H., Feinberg S. E. Hollister S. J. and Das S. (2005). Bone tissue engineering using polycapro-

lactone scaffolds fabricated via selective laser sintering. *Biomaterials*, 26, 4817–4827.

Wolberg, A. S. (2007). Thrombin generation and fibrin clot structure. *Blood Rev*, **21**, 131–142.

Xu, T., Jin, J., Gregory, C., Hickman, J. J. and Boland, T. (2005). Inkjet printing of viable mammalian cells. *Biomaterials*, **26**, 93–99.

Yoshikawa, H. and Myoui, A. (2005). Bone tissue engineering with porous hydroxyapatite ceramics. *J Artif Organs*, **8**, 131–136.

Young, S., Wong, M., Tabata, Y. and Mikos, A. G. (2005). Gelatin as a delivery vehicle for the controlled release of bioactive molecules. *J Control Release*, **109**, 256–274.

Yu, N. Y. C., Schindeler, A., Little, D. G. and Ruys, A. J. (2010). Biodegradable poly(α-hydroxy acid) polymer scaffolds for bone tissue engineering. *J Biomed Mater Res Part B: Appl Biomater*, **93B**, 285–295.

Zhang, X., Naik, A., Xie, C., Reynolds, D., Palmer, J., Lin, A., Awad, H., Guldberg, R., Schwarz, E. and O'keefe, R. (2005). Periosteal stem cells are essential for bone revitalization and repair. *J Musculoskelet Neuronal Interact*, **5**, 360–362.

11
Additive manufacturing of a prosthetic limb

S. SUMMIT, Bespoke Products/3D Systems, USA

DOI: 10.1533/9780857097217.285

Abstract: The chapter begins with a review of the shortcomings found in the modern prosthetic leg. It then reviews how the extensive new advances in digital data acquisition, parametric modeling, and additive fabrication stand to improve the quality of care for amputees. The chapter also reviews a range of techniques and benefits offered by the flexibility found in these digital methods.

Key words: prosthetics, additive fabrication, additive manufacturing, 3D scanning, 3D printing, biomimicry, parametrics, parametric modeling, finite element analysis.

11.1 Introduction

Everyone alive is, has been, or will be disabled. When we come to accept this, we realize the profound need for the highest attention to improving the quality of living for those with specific needs. Yet despite this, many of the challenges faced by individuals are met with entirely medical or mechanically-driven devices, neglecting the more nuanced and complex nature of the human spirit. While a utilitarian product may solve an immediate challenge, the solution remains incomplete unless it explores the totality of the task.

It is assumed, for example, that a prosthetic leg must address the utilitarian and mechanical constraints, though this typically leaves it devoid of aesthetic content, individual form, or expression. We accept this because of the arc of modern engineering tradition, where a break from stoic utility suggests a certain frivolity belonging to the world of the arts, far from the realm of mechanical functionality. This stands to reason, since the most pressing need of an amputee is to return the basic, biomechanical functionality to their life. But to stop the exploration at this stage neglects the human desire for body symmetry, for aesthetic expression, and for a degree of normalcy despite the change in appearance that an amputation represents. Amputees frequently hide their prosthetic limbs beneath layers of clothing, or shroud them with bubble wrap and tape in an effort to approximate body symmetry. This speaks of the profound need to exceed the base mechanical specification and explore the human spirit that is not being met with the current offerings in prosthetic devices.

Fortunately, many new tools have matured to a stage where they allow a reevaluation of some traditional methods, and invite an exploration of an improved quality of living for all amputees. As is often the case, it is now the design process and the approach which must catch up with the recent advances in technology in order to take full advantage of the opportunities now presented.

11.1.1 The prosthetics that surround us

Prosthetic devices, when we use the term in its broadest sense, augment our base, human feature set with additional functionality. Those designed by Ducati and Porsche, for example, provide us mobility while igniting our passions. Eyewear designed by Prada improves vision, enhances our facial features, and communicates our sense of style. A prosthetic device, in this sense of the word, simply offers us that functionality we lack. But devices from the consumer realm celebrate the product, flatter the user, and turn the solution into art intended to excite us. We accept this and expect it from the products we purchase.

A prosthetic limb is not considered – nor is it treated – as a consumer product, and therefore remains exempt from the level of attention that would be demanded from other consumer products. Most appear as an amalgam of unrelated parts created by competitive manufacturers, bolted together and adjusted to fit. The grace of the human form ends with such a leg, and the resulting user often appears to be an unlikely hybrid of organic and mechanical elements. One option for them includes disguising the limb beneath shaped foam in order to give the leg a degree of human characteristics, suggesting that the leg is a facsimile of the original. However, this response implies that the best way to address a prosthetic limb is through disguise and a quality camouflage, as if to deny the circumstance. For this reason, many amputees prefer to leave the mechanical hardware 'naked', as rudimentary and oddly juxtaposed as this may be.

11.1.2 Today's tools

The modern prosthetic was shaped as much by the manufacturing tools of the twentieth century as by the design approach that evolved to accommodate those methods. It is modular, made from many mass-produced components bolted together. Mass-manufacturing favors this approach, since identical parts can be designed *en masse* to be adjusted to individual needs, to some degree, endowing the product with a degree of flexibility. The tradeoff comes in the form of a resulting product which can be impersonal, unnecessarily heavy, and unnecessarily harmful to the environment.

The newest tools available to designers, however, stand to rewrite our fundamental approach to product design, offering solutions from methods unthinkable before their existence. Furthermore, the tools have become increasingly easy to use and inexpensive, lowering the barrier to development and improving the accessibility. Three that may be integrated specifically for amputees include 3D scanning, parametric modeling, and additive manufacture (AM). When combined, they open the doors to innovative new approaches to product creation and greater efficiencies.

11.2 The aim in designing a prosthetic limb

The 'monocoque' prosthetic leg concept was begun in 2007, as a way to apply the great design flexibility offered by the additive process to the unique and user-specific needs of transfemoral amputees. Combining 3D scanning, complex parametric 3D computer modeling, and 3D printing, while drawing from the unique attributes of each technology, would enable it. The end goal would be to create a leg from a process and parametric model, and not from an overseas tooling house. This leg would be designed in 'the cloud' in order to diminish the need for powerful local computing. The process would allow any unskilled person anywhere in the world to initiate the process, thereby availing the technology to the many people in need by removing the medical practitioner as a gating factor. Finally, the leg would be created additively, taking advantage of its vast flexibility and the potential drop in fabrication costs.

11.2.1 The process

In order to connect the new functionality and methodology with those in need, the necessary digital tools must be easily accessed throughout the world, with minimal barriers to access. Every effort in the design of the leg and the process to create it aimed to meet this degree of ubiquitous access. Only recently, however, has the cost and ease of use reached a state where this goal may be achieved. The following are some of the areas where very recent technological advances have paved the way for the creation of a monocoque prosthetic leg easily and at a low cost.

11.2.2 Simplified 3D scanning

The process begins with a 3D scan made from both legs of the user. This guarantees that any part that results will inevitably be as unique as a fingerprint;

in fact, the process is then incapable of creating identical parts. The scan captures both the 'sound side' (the biological) leg and the residual limb.

Traditionally, this would be a daunting task, as the type of scanner required was expensive, cumbersome, and suitable only for professional use. In the fall of 2011, however, Autodesk® Inc. introduced a long-awaited application capable of transforming multiple 2D images taken with a common digital camera into an accurate 3D data set. In short, the scan process merely requires that 30–50 digital pictures be taken circumferentially about the person, in increments of ten degrees or so. The application, known as 123D® Catch, requires minimal user skills, and trivial requirements of both the subject being scanned and the scanning environment. Most remarkably, the application is free, since all computation is performed on remote servers hosted by Autodesk. Suddenly, in the span of one year, the ability to generate accurate 3D data at no cost became a reality.

Once the sound side limb has been scanned, the digital surface data may then be mirrored and superimposed over the residual limb data. This now creates the reference geometry that will be used through the process, guaranteeing the creation of a leg that accurately re-creates the morphology of the lost limb, and provide the proportions and morphology needed to re-create body symmetry with the prosthetic limb.

11.2.3 Parametrics

Parametric modeling performs the 'heavy lifting' in terms of geometry creation. In short, parametrics describes a process where minimal user input may drive global changes to a master template and, ultimately, to the end product. In the case of a prosthetic leg, for example, by entering Cartesian coordinates at four variable locations in the template, an entirely user-specific leg may be generated digitally.

These measurements are taken from landmarks of the body, using commonly accepted 'bony prominences', which can be seen or found by palpating. These are entered into a standard spreadsheet, which then updates a parametric 3D computer aided design (CAD) model with the necessary dimensions to individualize the template. While the template may have been created for a generic 170 cm overall body height, for example, the newly entered data will adjust the various pivot points to approximate the dimensions of the original limb. Additionally, body weight and activity level may be entered into the parametric model to further influence the structure. Each resulting leg is therefore entirely unique, and closely tailored to the specific body and needs of the individual user. With the updated template and the mirrored morphology from the sound side limb, enough data exists to generate entirely user-specific data that will drive the creation of the new limb.

11.2.4 Cloud-based computing

Traditionally, the 3D computation needed to create and analyze geometry of this complexity required a powerful computer running elaborate combinations of software applications, requiring maintenance and frequent upgrades. This translated to a high cost per site and significant technical difficulties, further limiting the overall scalability of the process. But with the shift toward cloud-based computing, the geometry creation and analysis can now be performed on far more powerful machines than an individual would likely possess, running software that is automatically updated. And so internet access replaced individual computation power as the sole gateway to the creation of complex geometry. It is likely, in fact, that low-cost, touch-interface tablet computers will soon be able to provide the front-end to what had until recently been a prohibitively complex and expensive process.

11.2.5 Selective laser sintering (SLS) and preferred materials

The 3D CAD model for the limb is generated remotely in 'the cloud', in order to offload the need for powerful computing at any individual site. Once the data is created, it can be exported into the common stereolithographic tessellation language (STL) format and queued to be fabricated using selective laser sintering (SLS).

Once translated into the tessellated format, the model is ready for fabrication. Until recently, however, this task was not likely to be undertaken by anyone but an engineer with a relationship with the proper vendors. Worse, the material choices, until around 2007, were limited to less-than-durable thermoplastics, and unusable for anything beyond temporary prototypes.

The advent of SLS, however, changed the landscape, since it enabled the creation of large, durable, final parts of great complexity. Polyamide – also known as nylon – can be both strong and flexible, making it versatile enough to meet many engineering needs. It is also among the most common materials used in the SLS process, allowing fabrication to be bid among multiple vendors about the world. This change, therefore, added a new degree of flexibility to the AM method.

SLS machines now exist throughout the world. These must be in use as often as possible in order to amortize their purchase and maintenance costs. As a result, services have arisen to connect the needs of the designers with the available bandwidth of the devices. This has become an informal process, where a file may be uploaded to a site, immediately quoted, and placed in queue for the next available SLS machine worldwide. The prosthetic limb can then be fabricated at one of many locations and shipped directly to the user, potentially within days of the scan.

11.3 A biomimetic approach to design

Because additive fabrication is capable of creating individual parts of great complexity, it opens the door to the application of natural principles in order to enhance the mechanical characteristics of a product. As a result, we may now enjoy engineering benefits traditionally found only in nature. A skeletal structure, for example, typifies such an idealized natural solution in nature, since it provides structure at the appropriate amount for the intended need, with the minimum volume of material invested. The trabecular (complex lattice structure found within bones) increases in density at the proximal and distal ends of each bone where the stress concentration is greatest, leaving the middle of the bone more airy to reduce weight and mass. Several techniques allow this degree of efficiency to be created in a part created additively. (see Fig. 11.1)

11.3.1 Spiral internal lattice

Some of my first load-bearing leg concepts comprised tibial components, which were mostly hollow, yet the contoured walls were connected with internal counter-rotating double-helical structures at a 60° pitch. A three-dimensional triangular lattice structure results from the intersecting spirals, creating a lattice of three-dimensional triangles with edge angles of roughly 60° throughout, enhancing the overall strength of the part with minimal material usage.

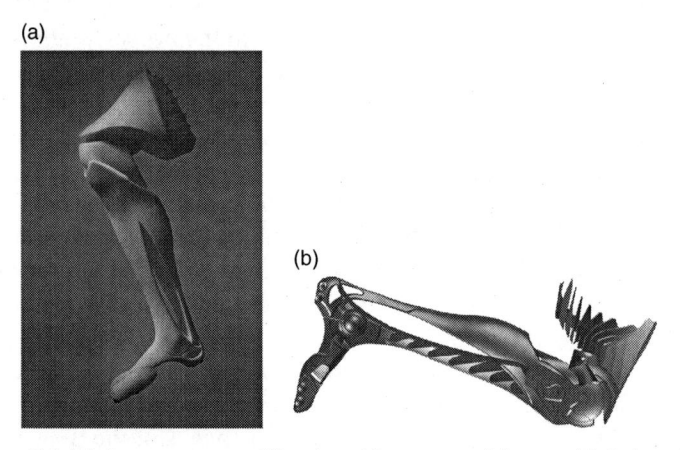

11.1 (a) A monocoque, 3D printed leg, comprising multiple interlocking parts, printed in the assembled state. (b) A cross-section of this monocoque leg, depicting the internal load-bearing structure, the multiple knee linkages, and the ball-and-socket foot assemble.

The spirals constructed in this way also allow a void running down the middle, where an axis would be, allowing the trapped 'cake' (excess powder condensed in the process) to escape the individual triangular chambers and vent via special egress ports at the ankle.

11.3.2 Voxel-based lattices

Software now exists that allows the user to create a single 'voxel' or 3D cell, which can propagate three-dimensionally so that the resulting mesh populates a volume. The advantage is that the designer has the flexibility to design that voxel to meet either isotropic or anisotropic needs, depending on the mechanical demands of the part. For example, a lattice that can expect more compressive pressure than rotational will look rather different from one that is designed to resist or promote torque. A creative designer can fine-tune the voxel – or the blending of voxels throughout the structure – so that the resulting lattice accurately addresses the demands of a highly specialized environment.

11.3.3 Finite element analysis (FEA)-driven lattice

Nature designs structures that react to the demands placed upon them. Increasing impact – taking up running or boxing, for example – will cause the bones to increase their structural characteristics, adjusting to the new normal accordingly. Although a digitally fabricated part cannot currently react to its environment once it has been created, it can, through computer simulations, anticipate structural demands and self-design accordingly before the fabrication process begins. In other words, it can accurately analyze the demands it may expect, and design its own structural needs to meet or exceed them. Through finite element analysis (FEA) we can see the effects of structural loading on a designed part while it exists only in its virtual CAD stage. Through a recursive simulation process, however, the part can be designed to 'learn' where it fails, strengthen itself in such a way to minimize that failure condition, and test again. This generative cycle may be repeated until it is suitably strong for its demands.

11.4 Integrating functionality

Every mechanically complex product has been assembled from a plurality of smaller parts or subassemblies. Gears, linkages, motor mounts – all such parts are held in place with machine screws, creating a collage of components, often created by distant and unrelated manufacturers. A typical mountain bike, for example, requires 11 adjustments in order to accommodate even the most basic variation between individual users. A prosthetic leg, similarly, may include a piston created by one company, an outer housing from another, and pylons from another. Furthermore, each of these assemblies may still be created from yet more parts from yet more places held together by mechanical fasteners.

Unlike traditional manufacturing, complex assemblies created additively may be fabricated in their final, assembled state. Because the process creates all parts simultaneously – layer by layer – additional complexity comes at no additional cost or assembly steps. All mechanisms may often be fabricated as-is, *in situ*, suspended during the fabrication process by unused powder, from which they are excavated upon completion. The additively created prosthetic leg, for example, includes a knee comprising many interlocking mechanical components, all interlinked. Despite this, it requires no part lines dividing the leg into its constituent groups, since no subassemblies exist within it. Furthermore, the multiplicity of parts comes at no additional cost or weight.

This integration of functionality offers benefits on many levels. First, it stands to reduce part weight and mass. By eliminating all extraneous bosses, screws, housings, or anything that would typically create an independent component, the overall part complexity and part count may be reduced. Additionally, product strength may be increased, since part failures often occur at the junctions where components are assembled. Finally, the assembly process may be reduced or simplified. When parts are created as completed, final assemblies, there may be no additional steps required.

But a subtler dividend results from this approach. An otherwise complex assembly of parts may instead be created as related, interconnected components bounded and defined by common surfaces, appearing more integrated and fluid. The part is not divided into unrelated elements; instead, each part flows fluidly into those adjacent, and relates both proportionally and contextually to the whole. The resulting product appears to have been born of nature, rather than of the assembly line. It bears more design resemblance to the body itself, transversed by tendons, bound by muscles and tissues, enveloped in skin. When creating a part that lives in the context of the body, such a visual language naturally feels so much more in keeping with the environment that it is intended to complement. (see Fig. 11.2)

11.5 A 'greener' approach to design

A modern prosthetic leg, like most complex products that we live with, bears a high 'carbon footprint'. This fact is exacerbated by the short lifespan of a typical leg (1–4 years), and the exotic materials required in its construction. Its components include cast titanium (melting point: 1660°C), stainless steel screws, thermoset carbon fiber, aluminum, etc. The final assembly is

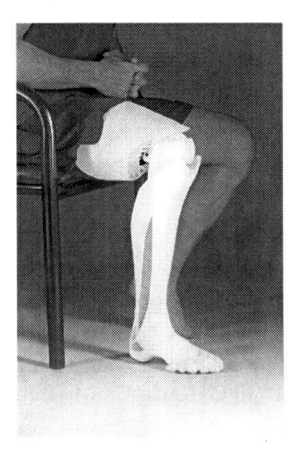

11.2 Concept prototype created in 2009. This proved the possibility of 3D printing a monocoque prosthetic limb, based on the contralateral scan of the patient.

not recyclable until disassembled, and even then, only very few parts are actually likely to be recycled or reused. Considering also that each part must also be shipped, stored, and assembled, the resulting carbon footprint can be rather high.

Creating a prosthetic leg additively, on the other hand, simplifies the process, thereby reducing the total carbon footprint in a variety of ways. Since all functionality can be created by a single machine in a single printing operation, one 150 W laser running for 28 h can create what would traditionally require many materials coming from many machines. Furthermore, since AM favors the monocoque-style assembly, the fabrication run may create the load-bearing tibial component, the flexible gastrocnemius (calf muscle), and the durable bearing race may all be fabricated simultaneously, in the same contiguous part. These are differentiated only by the design, which endows regions of the part with specific mechanical attributes. Finally, the entire leg may be recycled at the curbside, along with other plastics.

11.6 Tactile dividends of additively manufactured parts

Many of the assistive devices – chairs, eyewear, and toilet seats – add pressure to locations on the body, whether that force is low (as in the pads on the ear and nosepieces of eyewear) or higher (prosthetic limb sockets, motorcycle seats). Pressure applied to a specific location over an extended period at 7 psi, however, may lead to bruising, circulation loss, and tissue breakdown. By creating a conformal surface, however, that pressure may be distributed over a broader surface of the body, thereby reducing the specific point-loading risks.

Furthermore, pressurized surfaces against the body cause additional problems and discomfort if they trap heat or moisture. Since the skin surface aids the thermal regulation of the body, it becomes vital that skin is well ventilated wherever possible.

Unwanted humidity and heat buildup can be solved with generous perforations at those contact parts. This perforation, however, must abide by rules driven by human physiology. Holes larger than 5 mm in diameter risk causing 'window edema', a symptom where tissue is extruded into the void in a pillow-like shape, thereby diminishing circulation at the outer surface. Further, these holes must contact the skin such that the sidewall of each hole remains nearly normal to the skin's surface, since an acute angle may lead to increased abrasion or discomfort.

Using additive fabrication, a densely ventilated, conformal surface can be created easily. Better still, perforation adds no additional time or cost to the product, since the process simply creates more 'nothing', in a manner of speaking. In fact, while adding perforation to aid the comfort of the user, the part is actually made less expensive, faster to fabricate, and lighter in overall weight.

11.7 Vast design flexibility

Henry Ford once offered that users could 'have any color [of Model T] that they like, as long as it's black'. By contrast, the modern BMW Mini may be co-created by the user with a total of over 10 million permutations. This trend points to increased user involvement in the design process, and greater emotional attachment to the resulting product. Prosthetic legs, though far more personal and intimate by the nature of their usage than almost any other product, remain generally free of any form of personalization. This is because the market is small, the part variability is great, and the cost of custom craftsmanship would typically render this price-prohibitive.

Additive manufacturing, however, may shift the calculus. A complex graphical pattern or design may now exist simply as a template to be introduced into a CAD model and integrated into the overall design. Not only can artistic or creative details be infused into any part on a per-user basis, the user may even specify or create the detailing, as he prefers. A user may, for example, find black and white artwork for a tattoo, which may be debossed into a surface for adornment. Or it may be used to punch through a wall, creating lacy filigree. We have had clients work with artists, send images of patterns, or even photograph manhole covers that appeal, and we can then infuse that into the overall style of the limb. (Fig. 11.3)

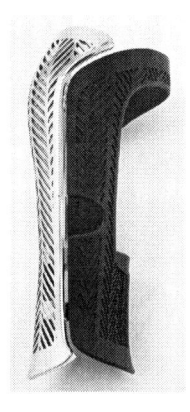

11.3 A filigree pattern, while difficult to create using traditional manufacturing technologies, is not only easily fabricated with 3D printing, but it helps to reduce weight and cost, while increasing strength.

11.8 Conclusion

The manufactured parts that we live with have been shaped as much by their intended function as by the constraints of the manufacturing process that created them. As a result, we accept parts that are overly impersonal, complex, and environmentally harmful. When the demands of the part must address the specific or unique needs of an individual, however, these shortcomings greatly compromise the efficacy of the solution. The new tools of the digital fabrication process offer an entirely new approach to meeting these needs. By rewriting the process and factoring in the new design flexibility, the end user may now be involved in the first part of the process, instead of the last. A bespoke part may now be created without the high cost that would have traditionally accompanied a custom, 'guild-style' production model. The additive process enables the kind of flexibility in the design and creation of final parts that invites improvements in quality of life, strength to weight ratio, and environmental areas that would not be possible within the limitations of traditional manufacturing.

Many current research efforts in the prosthetics area are expected to mature in the next 5–10 years. These include prosthetics that integrate directly with the skeletal system, increasingly intelligent active joints, and smaller, lighter components. But while these advances in particular apply mainly to the engineering and bioengineering aspects of the device, similar efforts are being made to address the social stigma and off-putting appearance associated with prosthetic limbs. Artists and designers are being drawn to this area as fertile ground for an opportunity for expression. Designer Alexander McQueen, for example, designed highly sculptural legs for amputee model Aimee Mullins, showing the range of possibilities available when the arts drive the process.

Advances in AM stand to further improve the quality of living and the accessibility of high quality components for lower-income amputees or those living beyond the reach of high quality care. As the process of creating a prosthetic limb shifts from an entirely physical craft to a largely digital process, the high quality care can be expected to proliferate globally, offering quality products to those who would traditionally find them beyond reach. At the time of writing, additive fabrication machines capable of creating a prosthetic limb suitable for extended use remain prohibitively expensive for many of those in need around the world, in the order of $800 000 USD. At the same time, we are seeing a surge in the quality and material options found in lower-end machines, and a nascent AM industry forming in India and China. As machine costs drop, we can expect a day when the cost of printing an entire leg will become competitive with the currently low cost of hand fabrication in the poorer nations. At this point, we can expect to see a great improvement in the quality of devices available to people, since the cloud-based CAD template will determine the efficacy, instead of the limitations of hand fabrication and mass production.

Index

ABL9000, 211–12
ablation, 79
acoustic energy driven system, 59
additive manufacturing, 2
   prosthetic limb, 285–96
      aim in designing a prosthetic limb, 287–90
      biomimetic approach to design, 290–1
      'greener' approach to design, 292–3
      integrating functionality, 291–2
      prosthetic device, 286
      tactile dividends of additively manufactured parts, 293–4
      today's tools, 286–7
      vast design flexibility, 294–5
aerosol jet process, 29–30
agarose hydrogels, 211
alginate, 184–5, 262
alginate hydrogel, 135
alkaline phosphatase (ALP), 256
Alpha Prototypes Inc., 91
amine-reactive silanes, 83–4
Analyze, 243
AutoCAD, 3, 211–12
Autodesk, 211–12, 249, 288

beam quality, 161
benchtop 3D printer, 7
bio-blueprint, 203–8
bio-electrospraying, 52–3
bio-ink, 130
bio-nanocomposites, 22–3
bio-papers, 124, 134
bioabsorbable, 180

BioAssembly Tool (BAT), 123
BioCell Printing, 123
bioceramics, 36, 110–11
   scaffolds, 49, 50–1
biocompatibility, 179
biocompatible hydrogel blends, 182
biocompatible inks, 48–9
biocompatible metals, 37
biocompatible scaffolds, 100
biodegradable, 180
biodegradable polymers, 37
BioExtruder, 44
biofabrication, 2, 221, 232–3
biological arrangements, 207–8
biological laser printing (BioLP), 52–3, 118–19
biomarkers, 90
biomaterials, 178–82
   compatibility, 89
biomaterials scaffold
   extrusion freeforming, 35–57
      scaffold fabrication, 40–57
      scaffold materials and macro-microstructure design, 35–40
biomimetic model, 202–3
biomolecular cues, 102
bioplotter, 123–6
   fusion of multicellular spheroids, 125
   schematic for tubular structure fabrication and multicellular spheroids, 126
bioplotting system, 240
biopolymers, 9
bioprinted structures, 135–7
bioprinters, 10–11

297

Index

bioprinting
  constructing microvascular systems for organs, 201–17
    bio-blueprint for microvasculature printing, 203–8
    biomimetic model for microvasculature printing, 202–3
    future trends, 217
    microvasculature post-printing stage, 215–17
    microvasculature printing strategies, 208–15
  resolution, 132–5
biosensors fabrication
  rapid prototyping techniques, 75–91
    biomaterials compatibility, 89–90
    functionalisation, 82–89
    future trends, 90–1
    RP of microfluidic systems, 77–82
    schematic outlining common bioelement recognition strategies and transformation methods, 76
blood-typing, 80–1
bone implants fabrication, 21
bone marrow cells, 213
bone marrow-derived stem cells (BMSCs), 260
bone morphogenic protein 2 (BMP-2), 246
bone regeneration, 254–77
  bone properties, structure and modelling, 255–8
    anatomy, function and pathology, 255–6
    bone remodelling, 258
    composition and structure, 256–7
    mechanical properties, 257–8
  cell printing technology, 271–5, 276
    extrusion based cell printing, 273–5
    inkjet based cell printing, 272
    laser-assisted bioprinting schematic diagram, 274
    laser assisted cell printing, 272–3
    moving platform inside the chamber under a modified HP Deskjet, 273
    muscle-derived stem cells on printed BMP-2, 273
    schematic diagram of hybrid printing system and the printing process, 276
  conventional scaffolds, 264–71
    conventional method based scaffold, 265–7
    3D printing technologies schematic diagram, 269
    rapid prototyping based scaffold, 267–71
    schematic diagram of SL system, fabricated PPF/DEF scaffold and fabricated TMC/TMP scaffold, 270
    SLS system schematic diagram, 271
  engineering of bone tissue, 259–64, 265
    cells, 259–60
    growth factors, 263–4
    growth factors in tissue regeneration, 265
    matrix materials, 261–3
  future trends, 275–7
bone tissue engineering, 259–64
  cells, 259–60
  growth factors, 263–4
  matrix materials, 261–3
    ceramics, 263
    natural biomaterials, 261–2
    synthetic biomaterials, 263
Boolean operation algebra, 206
bovine serum albumin (BSA), 84
BurrPlug, 120

calcium alginate hydrogels, 211
calcium chloride solution, 239
calcium phosphate (CaP), 120–1
  nanofillers, 23
cancellous bone, 257–8
Canon i905D, 84
CAPTAL R Sintering grade HA, 60
CAPTAL S Sintering grade HA, 60
carbon nanotubes (CNT), 20, 82
cell encapsulation, 111–12
cell-level resolution, 167, 170
cell patterning, 170

cell printing, 164–5
cell seeded scaffold, 193
cell sheets, 136
cell sources, 213–15
   engineered vasculature, 213–15
      vasculogenic potential of hESC-derived endothelial cells, 214
ceramics, 263
chemically induced binding, 113
chitosan, 262
chondroitin sulfate, 262
cloud-based computing, 289
collagen, 183, 210, 261
complex tissues
   rapid prototyping and laser assisted bioprinting (LAB), 156–71
      applications, 169–71
      high resolution and high throughput needs and limits, 165–9
      LAB parameters for cell printing, 164–5
      rationale in tissue engineering, 158–60
      terms of reference, 160–4
computational topology design (CTD), 39–40
computed tomography (CT), 37, 192, 205
computer-aided design (CAD), 160
   model, 3, 77
computer-assisted manufacturing (CAM), 160
computerised tomography (CT) scan, 3–4
conjugate gradient (CG) algorithm, 207
constructing microvascular systems
   bioprinting for organs, 201–17
      bio-blueprint for microvasculature printing, 203–8
      biomimetic model for microvasculature printing, 202–3
      future trends, 217
      microvasculature post-printing stage, 215–17
      microvasculature printing strategies, 208–15

continuous inkjet (CIJ), 25
conventional scaffold fabrication techniques, 265
cortical bone, 256, 257

3D Bioplotter, 46, 240, 246
3D-bioplotting, 33–4, 45–6, 189–91
123D Catch, 288
3D fibre deposition, 33–4, 43, 121
3D microvascular system, 206
2D microvascularised tissue constructs generate, 208–9
   images of patterned PGS capillary network and endotheliased capillary network, 209
3D microvascularised tissue constructs generate, 209–12
   schematics of omnidirectional printing within a hydrogel reservoir, 212
   SEM images of printed fibrin fibre, 211
3D organ printing technologies
   applications in tissue engineering and regenerative medicine, 243–9
      printed multicellular spheroids assembled into patterns, 247
      tissue engineering applications of printing technologies, 244–5
   approaches to 3D printing, 237
   from medical imaging to organ printing, 242–3
      CAD/CAM process for automated printing of 3D shape imitating target tissue or organ, 242
   future trends, 249–50
   three-dimensional printing methods, 238–42
      extrusion-based method, 239–40
      inkjet-based method, 238–9
      integrated organ printing system, 241–2
      laser-based printing method, 240
   tissue engineering applications, 236–50
3D printing, 2, 80, 255
   freeform fabrication of nanobiomaterials, 16–63

Index

3D printing (*cont.*)
  droplet-based SFF techniques, 25–30
  dry powder printing, 57, 59–62
  extrusion freeforming of biomaterials s scaffold, 35–57
  laser-based solid freeform fabrication (SFF) techniques, 18–25
  nozzle-based SFF techniques, 30–5
3D scaffolds
  biofabrication, 232–3
  feasibility for organs, 221–34
  material types, 230–2
  organ fabrication overview, 222–5
    fabricated tissue voxel as building block for complex organs or tissue test system, 224
    potential of biofabrication to produce organ products, 222
  organ repair/replacement paradigm, 234
  physical properties, 225–7
    scaffold fabrication effects on non-scaffold components, 228–30
    sources of interaction that has potential detrimental effects on cellular behaviours, 230
    two broad categories of biofabrication process, 229
    temporal expectations on the scaffold, 227–8
    scaffold strength and mass loss to match the needs and capability of the evolving tissue, 227
3D tissues, 124
data acquisition, 205–6
  computed tomography (CT) and micro-CT, 205
  magnetic resonance imaging (MRI), 205
  optical microscopy, 206
  single photon emission computed tomography (SPECT), 206
degradable hydrogels, 103
desktop robot based rapid prototyping (DRBRP) systems, 43
digital imaging and communications in medicine (DICOM), 242

digital light processing (DLP), 19
digital micro-mirror device (DMD), 19
Dimatrix Materials Printer, 29
Dimatrix model DMP-2822 inkjet printer, 86
dip-coating, 83
direct bioprinting, 181–2
direct ink writing, 33, 208
direct laser ablation, 208
direct-write assembly, 33–4, 47–8
direct writing, 2
DNA ultrafiltration system, 78–9
drop on demand (DOP), 25
droplet-based solid freeform fabrication (SFF) techniques, 25–30
  aerosol jet process, 29–30
  schematic illustration, 29
  inkjet printing process, 25–9
droplet ejection mechanism, 161–4
droplet formation, 28
dry powder printing, 57, 59–62
  different nanobiomaterials which have been tested, 61
  experimental arrangement of ultrasonic vibration controlled micro-feeding system, 60
  fine features printed, 60
  high speed camera image of Merck HA nanopowder flow and uneven packing of β-TCP, 62
  mean dose mass of Merck R HA, CAPTAL S HA, CAPTAL R HA and β-TCP, 63
  SEM images of different nanobiomaterial powders, 62
Dulbecco's Modified Eagles Medium (DMEM), 191

electrochemiluminescence, 79
electrospinning, 192–3
electrostatic discharge, 208
engineered stem cell niche, 169–70
Envirion TEC Inc, 240
enzyme aggregates, 88
EOSINT, 8
Epilog Laser, 79
ethylenediamine tetraacetic acid (EDTA), 136

extracellular matrix (ECM), 158–9, 256
extrusion based cell printing, 273–5
extrusion-based printing technology, 239–40
extrusion-based solid freeform fabrication (SFF) systems, 33–5
extrusion-based techniques, 119–29
  bioplotter, 123–6
  fused deposition modelling (FDM), 119–23
  three-dimensional printing (3DP), 126–9
    high flowability of plasma treated powders and homogeneous bed, 127
extrusion pressure, 52

Fab@Home, 123, 211
fibrin, 135, 183–4, 210, 262
fibrin glue, 262
filamentary materials, 42
formation, 258
freeform fabrication
  nanobiomaterials using 3D printing, 16–63
    droplet-based SFF techniques, 25–30
    dry powder printing, 57, 59–62
    extrusion freeforming of biomaterials scaffold, 35–57
    laser-based solid freeform fabrication (SFF) techniques, 18–25
    nozzle-based SFF techniques, 30–5
full melting, 113
fumaric acid-based hydrogels, 187–8
functionalisation, 82–89
  4000 droplets of glucose oxidase solution inkjet printed, 86
  biotinylated BSA printed on microcantilevers images, 88
  schematic of microring resonator biosensor, 85
  sensors response curves for ten drop-cast pH sensors and inkjet printed pH sensors, 87
fused deposition modelling (FDM), 5, 21–2, 119–23, 189–91, 241

galvanometric mirrors, 161
Gaussian beam spatial distribution, 114
gelatine, 261
Geomagic Studio, 243
geometrical cues, 101–2
good interfacial adhesion, 23
Graphtec FC5100A-75, 78
9340 GRAY Hysol Epoxi-Patch Structural Adhesive, 59

hard tissue regeneration, 117
Healon, 185–6
heating, 183
heterogeneous injectable implant, 225
Hewlett-Packard graphics language (HP/GL), 3–4
high cell density printing, 167
human umbilical vein endothelial cells (HUVEC), 118–19
Hyalograft C, 185–6
hyaluronan, *see* hyaluronic acid (HA)
hyaluronic acid (HA), 185–6, 262
hydrogel-based printing, 102–3
hydrogels, 124, 210
  scaffold, 122
  structure, 103
hydroxyapatite (HA), 110–11

in vitro transplantation, 170
in vivo printing, 171
in vivo transplantation, 170
infrared (IR), 108
initial graphics specification (IGES), 3–4
ink viscosity, 133
inkjet-based cell printing, 272
inkjet-based printing technology, 237, 238–9
inkjet droplet, 88
inkjet printing, 25–9, 80–1, 84, 129–37, 184, 189–91
  bioprinted structures, 135–7
    heart shape structures and bubble structures and clear and dyed gels and cell aggregate fusion, 136
  bioprinting resolution, 132–5
  fundamentals, 129–32

inkjet printing (*cont.*)
　　possible mechanism for inkjet-induced gene transfection and satellite drops formation, 131
　　SEM images of sintered TiO_2 micro-pillars and 3D ceramic micropattern, 26
　　typical piezoelectric DOD printing system for nanobioparticle printing, 27
integrated organ printing system, 241–2
ion-sensitive field effect transistors (ISFET), 84

jet-based printing, 208

lab on a chip systems, 77
Labview Software, 49
laminated object manufacturing (LOM), 7
laser assisted bioprinting (LAB), 118–19, 272–3
　　applications, 169–71
　　　　addressing the perfusion issue in tissue engineering, 170
　　　　engineered stem cell niche and tissue chips, 169–70
　　　　engineering transplants, 170
　　　　in vivo printing, 171
　　　　multicolour printing at micro-scale, 169
　　high resolution and high throughput needs and limits, 165–9
　　　　high cell density printing with cell-level resolution, 167
　　　　spatio-temporal limit of cell printing, 166
　　　　spatio-temporal limit of jet formation, 166
　　　　three-dimensional printing layer-by-layer approach, 167–9
　　　　viscosity of bioink and laser energy influence on printing resolution, 166–7
　　LAB parameters for cell printing, 164–5
　　　　bioink composition considerations, 164
　　　　viscosity of bioink and laser energy influence on cell viability, 164–5
　　rapid prototyping of complex tissues, 156–71
　　　　high-throughput LAB and optomechanical set-up and high resolution positioning system, 159
　　　　mechanism for laser-induced droplet ejection, 158
　　　　rationale in tissue engineering, 158–60
　　　　terms of reference, 160–4
　　　　typical LIFT experimental set-up is generally composed of three elements, 157
laser assisted cell printing, 272–3
laser-assisted techniques, 108–19
　　other technologies, 118–19
　　　　using an excimer laser-based device for adding features on polymer scaffolds, 119
　　photopolymerisation, 108–13
　　selective laser sintering (SLS), 113–18
laser-based printing method, 240
laser-based solid freeform fabrication (SFF) techniques, 18–25
　　classification of different SFF systems suitable for processing of nanobiomaterials, 18
　　selective laser sintering (SLS) process, 21–5
　　　　SEM images of PLLA/CHAp nanocomposites scaffold, 24
　　stereolithography (SL) process, 19–21
laser-based workstation, 161
laser engineered net shaping (LENS), 8
laser induced forward transfer (LIFT), 118–19
laser machining, 79
laser repetition rate, 161
laser writing, 52–3
LaserCUSING, 8
layer-by-layer fabrication techniques, 2
layer-by-layer inkjet printing, 80
layer-by-layer rapid prototyping device, 114

layered manufacturing, 2
Linear Mold Inc., 91
liquid-based patterning technique, 84
liquid-based systems, 6
liquid phase sintering partial melting (LPS), 113
live structure printing, 103–7
  main types of rapid prototyping systems, 106–7
  organisation of microvascular networks from building blocks, 105
  tissue fusion and cell aggregates indicates liquidity properties, 104
local cell environment, 160
low-temperature deposition manufacturing (LDM), 33–4, 121

magnetic resonance imaging (MRI), 5, 37, 205
manufacturing technology, 180–1
Matrigel, 164, 233
matrix-assisted pulsed laser evaporation direct write (MAPLE DW), 118–19, 169–70
maturation, 215
mechanical arrangements, 207–8
mechanical cues, 101–2
mechanical properties, 180
MED610, 5
Med-Link, 207
MedCAD, 207
medical imaging data, 204
micro-computed tomography (CT), 205
microelectromechanical systems (MEMS), 77
MicroFab inkjet, 88
microfluidic systems, 77–82
  microfluidic network produced by laser micromachining, 81
  PDMS flow cell for plasmonic biosensor and PDMS cell lysis chamber cast, 82
  simulations of sample concentration along microfluidic channel, 78
micromoulding technique, 77
micropatterning, 118–19
micropipette spotters, 89

microvasculature modelling, 206–7
microvasculature post-printing stage, 215–17
  implantation, 215–17
    three-dimensional scaffolds directly after implantation onto striated muscle tissue, 216
  maturation, 215
microvasculature printing
  bio-blueprint, 203–8
    blue-print of heart depicting cross section and corresponding side view, 204
    3D microvascular system, 206
    data acquisition, 205–6
    mechanical and biological arrangements, 207–8
    modelling, 206–7
  biomimetic model, 202–3
    geometry of vascular network, 203
  strategies, 208–15
    cell sources for engineered vasculature, 213–15
    generating 2D microvascularised tissue constructs, 208–9
    generating 3D microvascularised tissue constructs, 209–12
Mimics, 207, 243
modelling simulation algorithm, 206–7
'monocoque' prosthetic leg concept, 287
multi-head bioprinter, 193
Multi-jet Modelling System (MMS), 7
multicolour printing, 169
multinozzle deposition manufacturing (MDM), 45, 123
multinozzle low-temperature deposition and manufacturing (M-LDM), 45
multiphase jet solidification (MJS), 33–4
multiple head deposition system (MHDS), 123
multiplexing, 85–6
multiwalled carbon nanotubes (MWCNT), 20

Nano-Plotter NP2.1, 89
nanobioceramic ink, 28–9

nanobiomaterials
  freeform fabrication and 3D printing, 16–63
  droplet-based SFF techniques, 25–30
  dry powder printing, 57, 59–62
  extrusion freeforming of biomaterials scaffold, 35–57
  laser-based solid freeform fabrication (SFF) techniques, 18–25
  nozzle-based SFF techniques, 30–5
nanocomposite deposition system (NCDS), 30–3
nanofilter distribution, 23
natural biomaterials, 261–2
natural polymers, 36–7
non-contact patterning technique, 84
non-uniform rational B-splines (NURBS), 206
nozzle-based solid freeform fabrication (SFF) techniques, 30–5
  extrusion-based SFF systems, 33–5
    different techniques, 33
    SEM morphologies of scaffolds produced by robocasting process, 35
  nanocomposite deposition system (NCDS), 30–3
    hardware and components deposition head and fabrication process sequence, 31–2
nozzle fouling, 133
nozzle size, 61–2
nScrypt, 123
nucleic acid detection, 79
nylon, 289

Objet Geometries, 6
Ohnesorge number, 26–7
oligopoly(ethylene glycol fumarate) (OPF), 187–8
Olivetti I-Jet, 84
optical microscopy, 206
OPTOMEC, 29–30
organ bioprinting, 52–3
organ fabrication, 222–5
organ printing, 125, 238
organs

bioprinting for constructing microvascular systems, 201–17
  bio-blueprint for microvasculature printing, 203–8
  biomimetic model for microvasculature printing, 202–3
  future trends, 217
  microvasculature post-printing stage, 215–17
  microvasculature printing strategies, 208–15
osteoblasts, 259
osteonectin, 257
osteons, 256

paper-based biochemical tests, 80–1
paper-based immunoassay tests, 80–1
particulate leaching method, 118–19
photopatterning technique, 191
photopolymerisation, 5–6, 108–13
  stereolithography (SLA), 108–12
    photoinitiator type and concentration and energy dosage effect and cytotoxicity, 109
    SLA fabricated scaffolds, 111
  two photon polymerisation (2PP), 112–13
piezoelectric printers, 25–6, 130–1
piston assisted microsyringe (PAM2), 44
plastic sheet lamination (PSL), 7
Pluronic, 132
Pluronic F-127, 120, 182, 186
Poloxamer, 132
Poloxamer 407, see Pluronic F-127
polyamide, 289
polydimethylsiloxane (PDMS), 77–8
poly(ethylene)-based hydrogels, 186–7
Polyjet, 6
polymerisation, 5–6
poly(propylene fumarate-co-ethylene glycol) (PPF-co-EG), 187–8
poly(propylene fumarate) (PPF), 187–8
poly(vinyl alcohol), 188–91
  basic structural formulae of some commonly used synthetic hydrogel-forming polymers, 189
  crosslinking and RP processing-based characteristics of hydrogel-forming polymers, 190

pore size, 37, 39
porosity, 37, 39
post-processing, 4
powder-based system, 126
powder variables, 114–15
pre-processing, 4
precise extrusion manufacturing (PEM), 33–4
precision extrusion deposition (PED), 33–4, 121
pressure-assisted microsyringe (PAM), 33–4, 128–9
printing bioactive factors, 169–70
printing resolution
 viscosity of bioink and laser energy influence, 166–7
  rapid prototyping of soft free form of fibrin, 168
Pro/Engineer, 3
prosthetic devices, 286
prosthetic limb
 additive manufacturing, 285–96
  the prosthetics that surround us, 286
  today's tools, 286–7
  aim in designing a prosthetic limb, 287–90
  cloud-based computing, 289
  parametrics, 288
  process, 287
  selective laser sintering and preferred materials, 289–90
  simplified 3D scanning, 287–8
  biomimetic approach to design, 290–1
  3D printed monocoque leg, 290
  finite element analysis-driven lattice, 291
  spiral internal lattice, 290–1
  voxel-based lattices, 291
 'greener' approach to design, 292–3
 integrating functionality, 291–2, 293
  concept prototype created in 2009, 293
 tactile dividends of additively manufactured parts, 293–4
 vast design flexibility, 294–5
  filigree pattern, 295
ProtoCAM, 91

pulse duration, 161
pulse-to-pulse stability, 161
Pyrex, 208–9

rapid freeze prototyping technique, 123
rapid prototyping
 biomaterials, 1–11
  basic process, 3–5
  conventional RP systems and classification, 5–8
  definition of RP systems, 2–3
  future trends, 9–11
 biosensors fabrication, 75–91
  biomaterials compatibility, 89
  functionalisation, 82–89
  future trends, 90–1
  RP of microfluidic systems, 77–82
 complex tissues with laser assisted bioprinting (LAB), 156–71
  applications, 169–71
  high resolution and high throughput needs and limits, 165–9
  LAB parameters for cell printing, 164–5
  rationale in tissue engineering, 158–60
  terms of reference, 160–4
 feasibility of 3D scaffolds for organs, 221–34
  biofabrication, 232–3
  material types, 230–2
  organ fabrication overview, 222–5
  physical properties of scaffolds, 225–7
  scaffold fabrication effects on non-scaffold components, 228–30
  temporal expectations on the scaffold, 227–8
 scaffolding hydrogels and tissue engineering, 176–95
  applications, 191–4
  biomaterials, 178–82
  review of commonly used hydrogel-forming scaffolding biomaterials, 182–91
 rapid prototyping robot dispensing (RPBOD), 46–7

rapid prototyping technologies
  bone regeneration, 254–77
    bone properties, structure and modelling, 255–8
    cell printing technology, 271–5
    conventional scaffolds, 264–71
    engineering of bone tissue, 259–64
    future trends, 275–7
  tissue regeneration, 97–143
    extrusion-based techniques, 119–29
    inkjet printing, 129–37
    laser-assisted techniques, 108–19
    overview of RP systems, 99
rapid tooling, 2
Rayleigh-Plesset equation, 162
repetition rate, 161
resin, 109
resorption, 258
reversal, 258
Reynolds number, 26–7
robocasting, 33–4, 47

scaffold architecture, 180
scaffold-based tissue engineering, 52–3
scaffold fabrication, 40–57, 100–2
  extrusion freeforming with material melting, 40, 42–4
  extrusion freeforming without material melting, 44–57
    comparison of key extrusion-based SFF techniques, 54–6
    examples used in scaffold-based and scaffold-free tissue engineering, 58
    paste extrusion freeforming experimental set-up and schematic of extrusion axis, 50
    plane view of HA scaffold and fracture of sintered filament near weld area, 51
    schematic of different nozzle design, 53
    schematic illustration of different extrusion-based systems, 41–2
scaffold-free dispensing, 240
scaffold geometry, 180
scaffolding hydrogels
  applications, 191–4

    PCL scaffold with chondrocytes suspended in alginate, 194
  rapid prototyping based tissue engineering, 176–95
    biomaterials, 178–82
    cells noting the major components of the process, 177
    review of commonly used hydrogel-forming scaffolding biomaterials, 182–91
scaffolding materials
  background, 178–81
    examples of tissue engineering scaffolds fabricated by RP methods, 178
  macro-microstructure design, 35–40
    properties of biodegradable polymers suitable for TE scaffolds, 38
scanning electrochemical microscopy, 88–9
scanning electron microscopy (SEM), 60–1, 116–17
Scienion S3 Flexarrayer, 84–5
sciFLEXARRAYER DW, 88
selective laser melting (SLM), 7–8
selective laser sintering (SLS), 5, 21–5, 113–18, 270, 289–90
  liquid phase sintering partial melting and neck form between two adjacent particles, 115
  PCL scaffolds fabricated using SLS, 117
selective mask sintering (SMS), 8
self-assembled monolayer (SAM) systems, 83
shear-thinning behaviour, 122
silicon photonic wire biosensors, 83–4
Simpleware, 243
single photon emission computed tomography (SPECT), 206
single-wall carbon nanotubes (SWCNT), 28–9
sintering, 7–8
soft lithography, 208
Solid Creation System (SCS), 6

solid freedom fabrication, 2
solid ground curing system, 189–91
Solid Object Ultraviolet-laser Printer (SOUP), 6
solid state sintering (SSS), 113
Solid Works, 3
solvent-based extrusion freeforming technique, 33–4, 49
solvent casting, 118–19
spatial mode, 161
spatial organisation, 159–60
spin-coating, 83
spinneret-based tunable engineered parameters (STEP) technique, 248
Standard Terminology for Additive Manufacturing Technologies, 2
standard tessellation language (STL) format, 249
stereolithography apparatus (SLA), 6
stereolithography contour (SLC), 3–4
stereolithography (SL), 3–4, 19–21, 80, 108–12, 189–91, 208, 240, 268
  collapsed MWCNT at fracture surface of nanocomposite produced by SL, 21
  direct and indirect fabrication of bio-nanocomposite scaffolds, 22
  SEM images of two identical micro parts produced via micro SL from composites, 20
sterolithographic tessellation language, 289
SurgiCad, 207
synthetic biomaterials, 263

temperature controlled syringe (TCS), 44
TheriForm, 128
thermal inkjet printing, 84
thermal printers, 25–6
three-dimensional printing, 126–9, 167–9
tissue chips, 169–70
tissue engineering, 158–60, 176–95, 254
  biomaterials, 178–82
    evolution towards hydrogel-based scaffolds, 181–2
    scaffolding materials background, 178–81
  3D organ printing technologies, 236–50
    applications in tissue engineering and regenerative medicine, 243–9
    from medical imaging to organ printing, 242–3
    future trends, 249–50
    three-dimensional printing methods for organ printing, 238–42
  scaffolding hydrogels for rapid prototyping, 176–95
    applications, 191–4
    review of commonly used hydrogel-forming scaffolding biomaterials, 182–91
tissue fusion, 103–5
tissue healing, 102
tissue printing, 103
tissue regeneration
  rapid prototyping technologies, 97–143
    comparison of major RP techniques, 141–2
    extrusion-based techniques, 119–29
    hydrogel-based printing, 102–3
    inkjet printing, 129–37
    laser-assisted techniques, 108–19
    live structure printing, 103–7
    properties of common biomaterials used, 139–40
    scaffold fabrication, 100–2
tissue-resembling structure, 100–1
trabecular bone, 256
tricalcium phosphate (TCP), 110–11
two-photon polymerisation (2PP), 80, 112–13, 208, 240

Uformia, 5
ultrahigh molecular weight polyethylene (UHMWPE), 21–2
ultraviolet (UV), 5–6, 19, 108
ultraviolet (UV) lightning, 160–1
ultraviolet (UV) radiation, 183

ultraviolet (UV) range, 5–6
uniform base powder, 23

Vinnapas polymer solution, 88–9
Visual Basic program, 211–12

Weber number, 26–7
welding, 80
window edema, 294

workstation requirements, 160–1
  example of CAD/CAM software and printed pattern, 162

X-ray attenuation, 205
X-ray computed tomography (CT), 205
Xurography, 78